Advances in Water-Based Nanolubricants

Advances in Water-Based Nanolubricants

Guest Editors

Hui Wu
Pradeep Menezes

Basel • Beijing • Wuhan • Barcelona • Belgrade • Novi Sad • Cluj • Manchester

Guest Editors

Hui Wu
University of Wollongong
Wollongong
Australia

Pradeep Menezes
University of Nevada, Reno
Reno
USA

Editorial Office
MDPI AG
Grosspeteranlage 5
4052 Basel, Switzerland

This is a reprint of the Special Issue, published open access by the journal *Lubricants* (ISSN 2075-4442), freely accessible at: https://www.mdpi.com/journal/lubricants/special_issues/ Water-Based_Nanolubricants.

For citation purposes, cite each article independently as indicated on the article page online and as indicated below:

Lastname, A.A.; Lastname, B.B. Article Title. *Journal Name* **Year**, *Volume Number*, Page Range.

ISBN 978-3-7258-2799-2 (Hbk)
ISBN 978-3-7258-2800-5 (PDF)
https://doi.org/10.3390/books978-3-7258-2800-5

Contents

About the Editors

Hui Wu

Dr. Hui Wu is currently an Honorary Fellow in the School of Mechanical, Materials, Mechatronic and Biomedical Engineering at the University of Wollongong (UOW). He was appointed as a full-time Associate Research Fellow and then a Research Fellow after receiving his PhD at UOW in 2017. His expertise falls into the fields of materials, mechanical, and manufacturing engineering. His specific research interests include tribology and nanolubrication in metal forming, advanced rolling technology, composite materials, self-lubricating materials, wear-resistant materials, material characterization, phase transformation of steels, and micromanufacturing.

As a chief or key investigator, Dr. Wu has been involved in two Australian Research Council (ARC) DPs, two ARC LPs, four Baosteel-Australia Joint Research & Development Centre (BAJC) projects, three Australian Coal Association Research Program (ACARP) projects, and one China Coal Technology & Engineering Group Corp (CCTEG) project, which covered interdisciplinary research revolving around tribology, metal forming, micromanufacturing, and novel materials. His research findings obtained in these projects have been published in over 80 quality papers, which attracted ¿1,700 citations with an h-index of 23 (Google Scholar). As an early–mid career researcher, Dr. Wu's research achievements have been highly recognized by the international community. For example, one of his publications in *Tribology International* was recognised as a "Highly Cited Paper" by Web of Science in both 2020 and 2021, and another two publications won the "Best Paper Award". He is serving as a Youth/Guest/Review/Topic Editor for four international journals (including *International Journal of Extreme Manufacturing*, IF¿16) and as a member of the Editorial Board/Reviewer Board/Youth Editorial Board/Topical Advisory Panel for six international journals. He is also serving as an active referee for over 30 international journals, including some top journals in engineering such as *Materials & Design*, *Journal of Materials Research and Technology*, *Tribology International*, *Wear*, *Friction*, *Journal of Cleaner Production*, *International Journal of Mechanical Sciences*, *Journal of Alloys and Compounds*, etc.

Pradeep Menezes

Dr. Pradeep Menezes is a distinguished Associate Professor in the Mechanical Engineering Department at the University of Nevada, Reno (UNR). He earned his Ph.D. from the prestigious Indian Institute of Science, Bangalore, India, laying the foundation for an illustrious academic and research career. Dr. Menezes has an extensive research portfolio, with over 200 peer-reviewed journal publications that have collectively garnered more than 12,000 citations on Google Scholar, reflecting a remarkable h-index of 53. His scholarly contributions extend beyond journals, including over 50 book chapters, 10 authored books, and a significant patent. A dedicated academic, Dr. Menezes actively contributes to the scholarly community, serving as a reviewer for more than 100 leading journals and as a member of the Editorial Board of 10 distinguished publications. He holds key editorial positions as an Associate Editor for both the *ASME Journal of Tribology* and *STLE Tribology Transactions*. Dr. Menezes is deeply involved in academic events and professional service, participating in numerous national and international conferences as a paper reviewer, technical committee member, session chair, and review committee member. His expertise has been sought in evaluating books, book chapters, research grants, and master's and doctoral theses. As a committed member of professional organizations like the American Society of Mechanical Engineers (ASME) and the Society of Tribologists and Lubrication Engineers (STLE), Dr. Menezes continues to advance the fields of mechanical engineering and tribology through his research, teaching, and professional engagement.

Review

A Comprehensive Review of Water-Based Nanolubricants

Afshana Morshed, Hui Wu * and Zhengyi Jiang *

School of Mechanical, Materials, Mechatronic and Biomedical Engineering, University of Wollongong,
Wollongong, NSW 2522, Australia; am294@uowmail.edu.au
* Correspondence: hwu@uow.edu.au (H.W.); jiang@uow.edu.au (Z.J.)

Abstract: Applying nanomaterials and nanotechnology in lubrication has become increasingly popular and important to further reduce the friction and wear in engineering applications. To achieve green manufacturing and its sustainable development, water-based nanolubricants are emerging as promising alternatives to the traditional oil-containing lubricants that inevitably pose environmental issues when burnt and discharged. This review presents an overview of recent advances in water-based nanolubricants, starting from the preparation of the lubricants using different types of nanoadditives, followed by the techniques to evaluate and enhance their dispersion stability, and the commonly used tribo-testing methods. The lubrication mechanisms and models are discussed with special attention given to the roles of the nanoadditives. Finally, the applications of water-based nanolubricants in metal rolling are summarised, and the outlook for future research directions is proposed.

Keywords: water-based nanolubricant; nanoadditive; dispersion stability; tribology; metal forming

Citation: Morshed, A.; Wu, H.; Jiang, Z. A Comprehensive Review of Water-Based Nanolubricants. *Lubricants* **2021**, *9*, 89. https://doi.org/10.3390/lubricants9090089

Received: 31 July 2021
Accepted: 23 August 2021
Published: 10 September 2021

Publisher's Note: MDPI stays neutral with regard to jurisdictional claims in published maps and institutional affiliations.

1. Introduction

Friction and wear occur between moving materials in contact, the study of which is of fundamental importance in many applied sciences [1]. Lubricants, such as neat oils [2–4] and oil-in-water emulsions [5,6], have been extensively used to reduce the friction and wear, and satisfactory results have been obtained. To further enhance the friction-reduction and anti-wear properties of the oil-containing lubricants, great efforts have been directed towards incorporating different types of nanoadditives into the base lubricants [7–9]. These nanoadditives include metals, metal oxides, metal sulphides, non-metallic oxides, carbon materials, composites, and others such as nitrides [10,11]. The use of these oil-containing lubricants, however, unavoidably leads to adverse effects on the environment, especially when burnt and discharged, and regular maintenance of oil nozzles is always a laborious task. It is thus desirable to use eco-friendly lubricants as alternatives to the oil-containing ones without compromising on the lubrication performance in terms of the decreases in both friction and wear.

Over the past decade, water-based nanolubricants have been emerging as promising eco-friendly lubricants by dispersing nanoadditives into water [12–23], which integrates superb cooling capacity of water with excellent lubricity contributed by the nanoadditives. The use of water-based nanolubricants not only provides protection against friction and wear between the tool and the workpiece, but also improves overall quality of the product, demonstrating a great potential in engineering applications, such as metal forming [24–29]. Despite an increasing number of experimental studies on various nanomaterials as nanoadditives in water, several aspects including dispersion stability, tribological behaviour, and lubrication mechanism have not yet been fully understood. Most importantly, a review of the advances in water-based nanolubricants is still in its infancy [30,31], which brings an urgent need for comprehensive summary of the fundamental knowledge in the field of water-based nanolubrication.

In this review, the recent advances in water-based nanolubricants will be systematically summarised. To begin with, preparation of water-based nanolubricants will be introduced, including the preparation methods and the nanoadditives used in water. Based on the classified lubricants, we will then discuss the dispersion stability of the nanoadditives and the enhancement techniques. Subsequently, the tribo-testing methods used for tribological characterisation of the water-based lubricants will be compared, with a focus on the roles of nanoadditives, to propose the lubrication mechanism. We finally present the applications of the water-based nanolubricants in metal rolling, followed by proposing the outlook for future research directions.

2. Preparation of Water-Based Nanolubricants

2.1. Preparation Methods

Preparation of industrial lubricants usually requires the consideration of using various additives, such as antioxidants, corrosion inhibitors, defoamers, emulsifiers, extreme pressure (EP) agents, pour point depressants, and viscosity index improvers for different purposes in practical applications [32]. In particular, water-based nanolubricants are basically prepared by dispersing nano-scale solid particles into base water with the aid of a dispersant or surfactant, followed by mechanical agitation and ultrasonic treatment. In general, two primary techniques including one-step and two-step methods are adopted by most researchers to prepare nanolubricants. These two methods are subdivided in Figure 1.

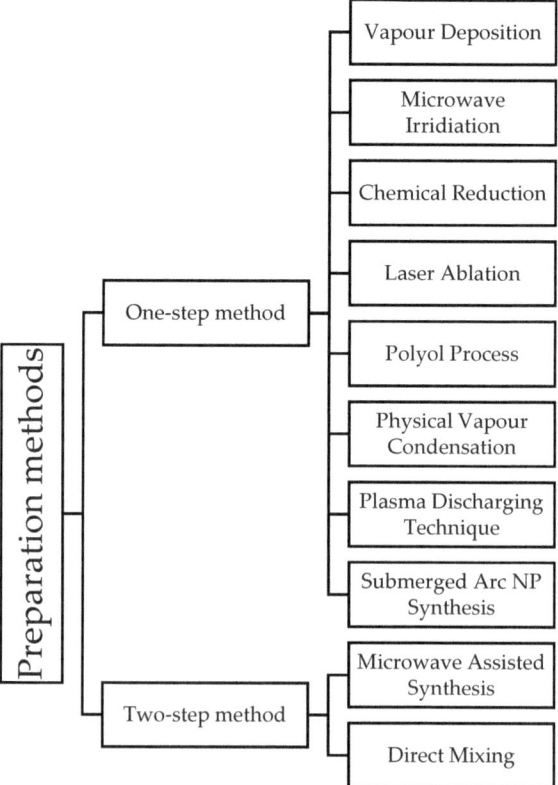

Figure 1. Subdivision of preparation methods of nanolubricants.

One-step method is a procedure that simultaneously combines the production of nanoparticles (NPs) with dispersion of NPs in base lubricant. One of the most commonly used methods is named vapour deposition, which was developed by Choi and Eastman [33]. The schematic of this method is shown in Figure 2a. First, a flowing thin film made of base liquid is formed on the vessel wall under centrifugal force of the rotating disk. Then the raw material is placed in the resistively heated crucible with heating for evaporation. The produced vapour is condensed into nano-sized particles when contacting the cold base liquid film, and nanolubricant is thus obtained. Another one-step direct evaporation method named vacuum evaporation onto a running oil substrate (VEROS) was developed by Akoh et al. [34], which aimed to produce ultrafine NPs with an average size of around 0.25 nm. Additionally, other techniques in the one-step method include microwave irradiation [35], chemical reduction [36], laser ablation [37], polyol process [38], physical vapour condensation method [39], plasma discharging technique [40], and submerged arc NP synthesis system [41]. However, the disadvantage in the one-step method is that there may be residual reactants such as impurities left in the nanolubricants due to the incomplete reaction which is difficult to avoid.

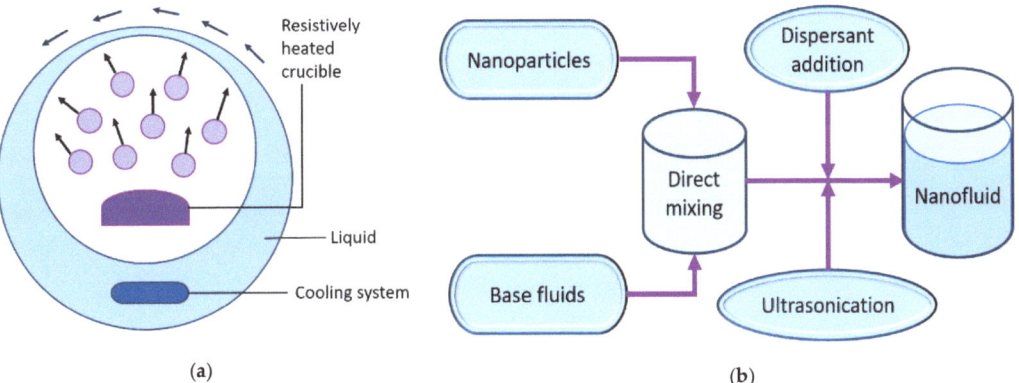

(a)　　　　　　　　　　　　　　　　　　(b)

Figure 2. Schematics of preparation methods of nanolubricants: (**a**) one-step method (vapour deposition) [33], and (**b**) two-step method (direct mixing technique) [42].

In contrast, the two-step method is more widely used in the preparation and synthesis of nanolubricants with consideration of raw materials provided by manufacturing companies due to large scalability and cost effectiveness. The schematic of this method involves two procedures (see Figure 2b). The first step is to directly mix NPs with base fluid, followed by subsequent addition of dispersant or surfactant with ultrasonication. If necessary, extra treatments such as magnetic force agitation, mechanical stirring, high-shear mixing, and homogenising at certain temperatures should be combined with the second step to enhance the dispersion stability of the final nanolubricant. Two typical representatives in the two-step methods comprise microwave assisted synthesis and direct mixing technique [42].

2.2. Nanoadditives in Water

Nanomaterials have emerged as one of the most interesting materials in the areas of chemistry, physics, and materials science, and they have been widely applied in many engineering fields. Several varieties of nanoadditives can be dispersed in water, including pure metals, metal and non-metal oxides, metal sulphides, carbon-based materials, composites, and some others such as metal salts, nitrides, and carbides.

Figure 3 reveals a summary of different types of nanoadditives used in water-based lubricants in the past decade. Among all the nanoadditives the most used one is carbon-based materials, accounting for 35%, followed by metal and non-metal oxides (18%) and composites (16%), whereas the least used ones are pure metals (3%) and carbides (2%). More details about various nanoadditives used in water-based lubricants are discussed below.

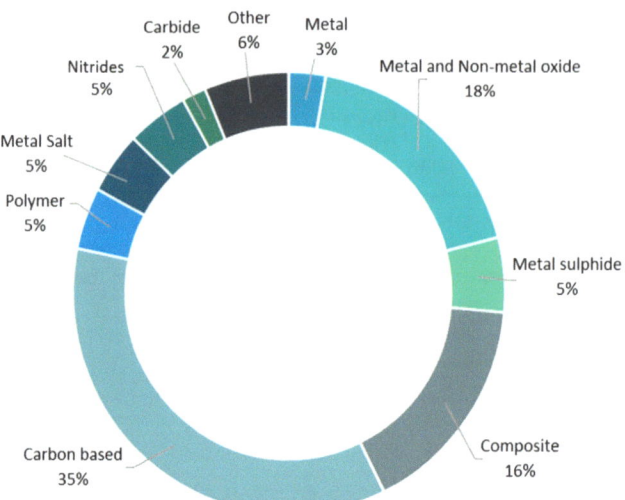

Figure 3. Statistics of nanomaterials used as nanoadditives in water in the past decade.

2.2.1. Pure Metals

Metallic NPs such as Au, Ag, and Cu have been considered to be important additives in base water due to their unique optical, electrical, and photo thermal properties in the fields of physics, chemistry, and biology [43]. These NPs, however, are only weakly compatible with base lubricant because of their high surface activity. This issue can be resolved by surface modification techniques [44]. Among all the candidates, Cu NPs have been suitably used in aqueous lubricants due to their low cost and superb tribological characteristics by acting as nano-bearings and forming metallic and/or tribo-sintered film with low shear stress on rubbed surfaces [45]. For example, Zhao et al. [46] fabricated Cu NPs through in situ surface modification, and then uniformly dispersed varying concentrations of Cu NPs (0.1–2.0 wt.%) into distilled water using prepared water-soluble bis (2-hydroxyethyl) dithiocarbamic acid (HDA) as a capping ligand. A reddish-brown uniform water-based nanolubricant with superior dispersion stability was thus synthesised. Zhang et al. [47] also prepared water-based nanolubricant containing nano-Cu surface-capped with methoxyl-polyethylene-glycol xanthate potassium. The surface-capped Cu NPs had an average diameter of 2 nm and showed no sign of apparent agglomeration in water.

2.2.2. Metal and Non-Metal Oxides

Metallic oxide NPs including Al_2O_3, TiO_2, and ZnO have been widely used as possible lubricating additives in base water. Radice et al. [48] investigated the lubrication behaviour of globular-shaped Al_2O_3 NPs (AKP50) with approximately 0.197 μm in diameter. The lubricant was prepared by diluting AKP50 with an acetate buffer solution (0.1 M $NaCH_3CO_2$ + 0.1 M CH_3CO_2H) in deionised water (DW) under 15 min ultrasound stirring prior to each tribo-test to avoid sedimentation. He et al. [16] used spherical Al_2O_3 NPs of 30, 150, and 500 nm in water to prepare water-based nanolubricants containing 0.2–8 wt.% Al_2O_3 and 10 wt.% glycerol through 400 W ultrasonic agitation for 10 min, and no evident agglomeration was observed until 3 days. It is also of vital importance to report several examples for TiO_2 NPs as water-based nanoadditives. For example, Wu et al. [13–15,25,27] prepared the lubricant by mixing 0.2–8.0 wt.% TiO_2 NPs (20 nm in diameter) into DW by mechanical stirring, stepwise adding 0.002–0.08 wt.% polyethyleneimine (PEI) and 10% glycerol under 30 min high-speed centrifuge at 2000 rpm followed by 10 min ultrasonication. The as-prepared lubricants had no significant sedimentation even after one week. In their recent studies, they prepared industrial-scale water-based nanolubricants using coarse TiO_2 NPs of 300 nm, with the aid of 0.1–0.2 wt.% of sodium dodecyl benzene sulfonate (SDBS) and 1 wt.% Snailcool by mechanical stirring only. The as-prepared lubricants could be stably dispersed for 24 h [24]. Sun et al. [49–52] also prepared 0.1–5.0 wt.% TiO_2 water-based nanolubricants by adding 20–90 nm TiO_2 particles and different additives, including sodium hexametaphosphate (SHMP) and SDBS which exhibited good stability for 7 days. Additionally, there has been some work conducted on the preparation of water-based nanolubricants using other metal oxide nanoadditives such as CeO_2 [53], CuO [54], γ-Fe_2O_3 [55], MoO_3 [56], and WO_3 [57], and all these nanoadditives can also be well dispersed in water under mechanical agitation and ultrasonic processing.

Non-metallic oxides have also been used extensively as nanoadditives in base water. Silica (SiO_2), one of the typical representatives, is a widely used ceramic material both as a precursor to the fabrication of other ceramic products and as a material on its own. Silica has good abrasion resistance, electrical insulation, and high thermal stability [43]. Ding et al. [58] followed four steps of preparation, including synthesis, modification, purification, and dispersion, to obtain ceramic water-based lubricant with 100 nm SiO_2, which showed no apparent sedimentation for 1 h. Bao et al. [59] also prepared SiO_2 water-based lubricants using 0.1–1 wt.% surface-modified spherical SiO_2 nanoparticles (30 nm in diameter) in 15% ammonia solution under 5 min stirring at room temperature, followed by 30 min stirring at 60 °C with addition of polyethylene glycol-200. The solution was finally mixed using 20 kHz ultrasonic disperser for 5 min, and finally a stably dispersed SiO_2 water-based lubricant was obtained.

To date, many experimental studies have reported the use of metal and non-metal oxides as nanoadditives, as listed in Table 1.

2.2.3. Metal Sulphides

In recent years, lubricant additives based on metal sulphides have received considerable interest in the lubricant industry. It is generally accepted that metal sulphides such as Ag_2S, CuS, and MoS_2 present outstanding lubrication performance when used both as solid lubricants and as additives in liquid lubricants. These materials offer a low-shear resistance to an external shear stress due to their layered structure with strong-interlayer and weak-interlayer bonds [60]. It was reported that Kuznetsova et al. [61] synthesised MPS-capped Ag_2S NPs (2–10 nm) by a simple one-step process as per the following reaction: $2AgNO_3 + Na_2S = Ag_2S\downarrow + 2NaNO_3$, eventually adding MPS (3-mercaptopropyl trimethoxysilane) and ethanol in water by sonication in an ultrasonic bath for 10 min to avoid sedimentation in the solution, and the solution remained stable for up to several months at room temperature. Zhao et al. [62] used HAD-CuS nanoparticles as nanoadditives in water. The nanolubricant was prepared by pouring PEG-400 and HAD in a solution of CTAB, $Cu(NO_3)_2 \cdot 3H_2O$ and distilled water under 10 min stirring, followed by 170 °C heating under 1 h vigorous stirring. The final black homogeneous solution was centrifuged, cleaned, and dried to obtain HAD-CuS NPs that could be uniformly dispersed in water for at least 2 days within the concentration of 0.1–2.0 wt.%.

In the last few years, special attention has been focused on using MoS_2 NPs in water due to their excellent chemical and thermal stability. For example, Meng et al. [63] added multilayer-MoS_2 of 100 nm in water under 10 min magnetic stirring to prepare 0.3–0.5% MoS_2 nanolubricants, and the as-prepared nanolubricants presented good stability for 16 h. To further enhance the dispersion stability of nano-MoS_2 in water, Wang et al. [64] conducted the exfoliation and modification processes on bulk MoS_2 (15 μm in thickness) to prepare functional MoS_2 nanosheets (3.5 nm in thickness) which can be stably dispersed in water for 10 days after ultrasonication for 1 h. In contrast, the unfunctional MoS_2 nanosheets would aggregate in water within a very short time.

2.2.4. Carbon-Based Materials

When compared with metal sulphides, carbon-based nanomaterials have higher chemical stability and superior mechanical properties, which provides outstanding tribological performance as well as environmentally friendly characteristics for a renewable future. In light of this, carbon-based nanomaterials, such as pure carbon nanomaterials and carbon derivatives, have become potential lubricating additives dispersed in base water.

According to different characteristics, pure carbon nanomaterials include carbon nanotubes (CNTs), carbon dots (CDs), graphene, and nanodiamonds (NDs). Peña-Parás et al. [65] dissolved functionalised CNTs of 30–50 nm in DW and applied a magnetic stirring in an ice bath for 60 h. The as-prepared water-based nanolubricants containing 0.1–2.0 wt.% CNTs remained stable for 4 months even without the aid of dispersing agents. Hu et al. [66] prepared CDs by dissolving thermally carbonized ammonium citrate in DW, followed by high-speed centrifugation, dialysis purification, and freeze-drying. The as-prepared CDs (3–4 nm in diameter) had hydrophilic oxygen-containing groups, thereby exhibiting superb dispersion stability in water for over 6 months. Liang et al. [67] used in situ exfoliated graphene as water-based nanoadditive which remained stable in water for over a month. The authors prepared the graphene enhanced lubricants by dissolving 1.5, 3, and 6 g pristine graphite powder and non-ionic surfactant (Triton X-100) into 1 mg/mL DI-water under 8 h bath sonication, 12 h magnetic stirring, 24 h sedimentation, and 1 h centrifugation. Jiao et al. [68] synthesised a bioinspired copolymer consisting of dopamine and 2-methacryloyloxyethyl phosphorylcholine for surface modification of NDs. Although the modified NDs exhibited remarkable lubricity when added to water-based lubricants, the dispersion stability of NDs in water was not investigated.

Among a variety of carbon-based nanoadditives, graphene and its derivatives stand out owing to their unprecedented structural, chemical, and physical properties. However, it is difficult to prepare stable water-based nanolubricants with graphene due to the formation of irreversible agglomerate caused by its strong π–π stacking and van der Waals interaction [69]. In contrast, graphene oxide (GO) has excellent hydrophilicity because it contains a large number of oxygen-containing functional groups, which enables it to become an ideal nanoadditive in water. Song and Li [70] prepared graphene oxide nanosheets with a diameter of 20–30 nm and a length of 10–30 μm from purified natural graphite by modified Hummers and Offeman's method [71]. The as-prepared GO nanosheets (0.5 mg/mL) were dispersed in water by bath ultrasonication, which resulted in no sedimentation for 5 weeks. After obtaining GO from Hummers' method, min et al. [72] prepared fluorinated GO (FGO) through 12 h hydrothermal treatment at 160 °C with the presence of 0.5 mL nitric acid and 9.5 mL hydrofluoric acid. The produced FGO was dispersed in water with 0.1–1% concentration by ultrasonication, and the as-prepared lubricants exhibited excellent stability for 12 days. Fan et al. [73] prepared 0.5 mg/mL FGO aqueous solution by dispersing 5 mg FGO in 10 mL distilled water under 30 min ultrasonication (300 W), which led to good stability for a week.

Other graphene derivatives, such as reduced GO (RGO) and pH-dependent GO, have been emerging to further enhance the dispersion stability of GO in water, according to the fact that the strong oxygen functionality and flatness together with possible defects of GO may prompt its agglomeration in water. Liu et al. [74] dispersed GO (100 mg) in DW (80 mL) under ultrasonication for 60 min, followed by addition of PEI (5 g) that preliminarily suffered a magnetic stirring with water (100 mL) for 30 min. The mixture was then stirred for 12 h at 80 °C until GO was transformed to RGO, during which the transformation was recognised by the colour change of mixture (from yellowish brown to black). The as-synthesised PEI-RGO was finally dispersed into DW to obtain water-based nanolubricants with concentrations of 0.03–0.1 wt.%, which showed no deposition for over 60 days. Hu et al. [75] proposed a facile process to modify RGO using β-Lactoglobulin (BLG). They dissolved GO (50 mg) with DW (50 mL) via mechanical stirring for 30 min to attain GO aqueous solution which was then mixed with BLG (12.5 mg) and hydrazine hydrate (2 mL). The as-synthesised BLG-RGO was diluted in water to obtain water-based nanolubricants with concentrations of 0.05–1.0 mg/mL, which showed outstanding stability without apparent precipitation for 8 months—the longest stability period as reported in GO-based aqueous lubricants.

In the case of pH-dependent GO as water-based nanoadditives, two studies were reported by He et al. [17] and Meng et al. [76], which proposed two different preparation processes. A mechanical de-agglomeration method was used by He et al. [17] to prepare GO suspension with 0.06 wt.% GO and 0.1 M NaOH in DW under high intensity ultrasonic agitation (400 W) for 10 min, resulting in good dispersion stability for a week. The pH value was adjusted by NaOH and varied from 3.1 to 9.7, which had insignificant influence on agglomeration of GO sheets. In contrast, Meng et al. [76] mixed triethanolamine (TEA) with 0.1 wt.% GO in DIW to adjust the pH value from acidic (pH 2.8) to alkaline (pH 9). They found that the lubricant with pH 9 was the most homogeneous with no precipitation for 50 days.

It has been documented that some researchers have investigated the carbon-based nanocomposites in the formulation of water-based lubricants, including GO/graphene [77], GO/nanodiamond [78,79], GO/carbon [80], and GO/g-C$_3$N$_4$ [18]. The detailed information regarding their preparation parameters as well as stability duration is listed in Table 1.

2.2.5. Composites

Nanocomposite which comprises two or more different nano-sized particles is becoming a significant part of nanotechnology and one of the fastest growing research areas in materials science and engineering [81]. The use of nanocomposite is to simultaneously combine physical and chemical properties of the constituent nanomaterials in an attempt to produce a homogeneous phase for better performance than single-component NPs [82]. The composite water-based nanolubricant, therefore, can be prepared by dispersing two or more nanoadditives in the base water.

Over the past few years, much research has been conducted on graphene-based composites including GO/SiO$_2$ [19,83], GO/TiO$_2$ [84], and GO/Al$_2$O$_3$ [21]. For example, Huang et al. [19] synthesised GO/SiO$_2$ water-based lubricants by mechanical stirring the aqueous suspension at 25 °C for 30 min, followed by ultrasonic processing (450 W) for 60 min (on/off interval of 5 s) in a chilled water bath. Similar preparation method was used to synthesis the water-based lubricants containing GO/Al$_2$O$_3$. Both results indicated that the formation of hybrid nanostructure enabled smaller particle size distribution in water, compared to that of constituent nanoadditives. Some other nanocomposites such as Cu/SiO$_2$, ZnO/Al$_2$O$_3$, Fe$_3$O$_4$/MoS$_2$, TiO$_2$/Ag, and Ag/C have also been used as nanoadditives in water. For example, Li et al. [85] used a two-step method to prepare 0.05–0.3 wt.% TiO$_2$/Ag in cooling water by 2 h magnetic stirring, followed by 12 h ultrasonication with 40 kHz frequency and 110 W power, which showed no sedimentation within a month. In contrast, Fe$_3$O$_4$/MoS$_2$ nanocomposite with 0.3–1.2 wt.% concentration was prepared by Zheng et al. [86] by dissolving 5.4 g FeCl$_3$·6H$_2$O, 1.27g FeCl$_2$, 0.48 g MoS$_2$ nanosheets in 50 mL DW, followed by drop-wise addition of 6 mL NH$_3$·H$_2$O in the solution under ultrasonic state. The precipitates were then centrifuged, washed, and dried for 24 h at 50 °C. The as-prepared nanocomposite Fe$_3$O$_4$/MoS$_2$ showed better dispersibility in water than only Fe$_3$O$_4$ NPs or MoS$_2$ nanosheets. Additional studies on the preparation of different nanocomposites are listed in Table 1.

2.2.6. Nitrides

Among all the nitrides, hexagonal boron nitride (h-BN), as a promising and an ideal alternative to other nanoadditives dispersed in water, has attracted extensive attention due to its exceptional performance, such as high-temperature stability, high thermal conductivity, high electrical resistivity, low coefficient of friction, strong inertness in a wide variety of chemical environments, and environmental friendliness [87]. Cho et al. [88] revealed superb stability of h-BN nanosheets in water without any precipitation within 1 month. The authors synthesised 30 nm thick and 300 nm wide h-BN nanosheets with 0.01, 0.05 and 1 wt.% concentration under 20 h bath sonication excluding any additional surfactants. Moreover, Bai et al. [89] recommended the use of thin hydroxylated BN nanosheets (HO-

BNNS) with 0.0125–0.20 wt.% concentration dispersed in water-glycol (55 wt.% DW and 45 wt.% glycol) under 30 min ultrasonication, which showed good dispersion for 5 days.

2.2.7. Carbides

Recently some research works have been conducted on carbides such as Nb_2C and Ti_3C_2. Cheng and Zhao [90] prepared Nb_2C nanofluid using three different degrees of oxidised Nb_2C nanosheets with the mass ratio of 0.25–1.0 mg/mL of pure water. The Nb_2C nanosheets were obtained by mixing 1 mg/mL accordion shaped Nb_2C powder into 100 mL aqueous solution, followed by 10 mL/50 wt.% TBAOH as intercalation agent, and 20 mg ascorbic acid as anti-oxidant under 12 h magnetic stirring in an ice bath. The pH level of the mixture was balanced by adding hydrochloric acid and 10 min centrifugation at 3000 rpm. The final Nb_2C solution was acquired by freeze-drying method and divided into three parts: Nb_2C of 20 nm (black coloured), after 6 h magnetic stirring at 60 °C water bath; moderately oxidised Nb_2C (MO-Nb_2C) of 12 nm (yellow-coloured), after 7 days magnetic stirring at room temperature; and completely oxidised Nb_2C (CO-Nb_2C) of 6 nm (transparent). The authors also mixed benzalkonium chloride (surfactant) to enhance stability and, among the three groups, MO-Nb_2C showed the best stability even after 1 month. Nguyen and Chung [91] prepared five solutions with 1 wt.%, 2 wt.%, 3 wt.%, 5 wt.%, and 7 wt.% Ti_3C_2 concentrations by adding 0.01, 0.02, 0.03, 0.05, and 0.07 g of Ti_3C_2 to 1 mL of DW respectively. Each solution was mixed under 1 h magnetic stirring at room temperature to ensure uniform dispersion of the NPs in water.

2.2.8. Others

The literature demonstrates that very little research has been conducted on some rare NPs, such as black phosphorus and hydroxides. Tang et al. [92] synthesised 3.9 nm black phosphorus quantum dots (BPQDs) by adding N-methyl-2-pyrrolidone (NMP) under ultrasonic treatment for 8 h. The as-prepared BPQDs with 0.1 wt.% concentration were dispersed into 2.0 wt.% Triethanolamine (TEA) aqueous solution by another 10 min ultrasonic treatment. The BPQDs dispersion showed good stability with no precipitates even after 2 weeks. Metal hydroxides have been demonstrated to perform well as lubricant additives, and layered double hydroxides (LDHs) stand out as particularly impressive examples of this. Wang et al. [93] presented the preparation of 19.42 nm wide and 8.59 nm thick oleylamine-modified Ni-Al LDH (NiAl-LDH/OAm) nanoplatelets as water-based lubricant additives with a concentration of 0.1–1.0 wt.%. The lubricants were synthesised by a microemulsion method under both stirring and ultrasonication in DW followed by 10 h drying at 80 °C. A transparent and stable solution was obtained without additional dispersion or surfactants. The authors also prepared aqueous polyalkylene glycol (PAG) solution by dispersing two different LDHs with 0.5 wt.% concentration under only ultrasonication, including ultrathin LDH nanosheets (ULDH-NS, 60 nm wide and 1 nm thick), and LDH NPs (19.73 nm wide and 8.68 nm thick) [94]. The ULDH-NS (0.5 wt.%) were uniformly dispersed in water, showing no agglomeration. While 0.5 wt.% LDH-NPs were dispersed in water, a transparent solution was obtained with no precipitates due to the addition of a surfactant 1-butanol and oleylamine under the reverse microemulsion reaction process.

In addition to this, some exceptional NPs such as chitosan [95] and stearic acid [96] were also explored recently. Li et al. [95] used 0.5 wt.% nanoscale liquid metal droplets wrapped by the chitosan (NLMWC) as water-based nanoadditives and compared with 0.5 wt.% gallium based liquid metal $Ga_{76}In_{24}$. The NLMWC was dispersed in water under 15 min ultrasonication in water bath, and the as-prepared aqueous solution remained stable for 1 month without any precipitation and agglomeration. In contrast, the precipitates were observed for $Ga_{76}In_{24}$ in water within 2 h.

The details of different types of nanoadditives used in water-based lubricants are listed in Table 1.

Table 1. List of different types of nanoadditives in water for preparation of various water-based nanolubricants.

Types	Nanoadditives	Size	Shape	Concentration	Stirring	Stability Duration	References
Metals	Cu	3 nm	Spherical	0.1–2.0 wt.%	Magnetic stirring for 10, 20 min	-	Zhao et al. [46]
		2 nm	-	0.5–5 wt.%		-	Zhang et al. [47]
	Ag	10–100 nm	-	1 g/L	-	5 to 19 days	Odzak et al. [97]
	Au	26.7 nm	-	0.018 vol%	Irradiation for 1–18 h	1 month	Kim et al. [98]
Metal and non-metal oxides	Al_2O_3	0.197 μm	Spherical	-	Ultrasonication for 10 min	Few minutes	Radice et al. [48]
		30, 150, and 500 nm	Spherical	0.2–8 wt.%	Ultrasonication for 10 min	3 days	He et al. [16]
	CeO_2	10–40 nm	-	0.05 wt.%	Ultrasonication for 2 min	4 days	Zhao et al. [53]
	CuO	60 nm wide and 230 nm long	Nanorod or spindle	0.1–2.0 wt.%	Magnetic stirring for 60 min	8 h	Zhao et al. [54]
	$\gamma\text{-}Fe_2O_3$	5 nm	-	0.1–1 wt.%	-	1 h	Pardue et al. [55]
	Fe_3O_4	-	Chain like	1 wt.%	Ultrasonication for 1 h	40 days	Lv et al. [99]
	MoO_3	80–100 nm	2D	0.1, 0.2, 0.3, 0.4, 0.5 wt.%	-	-	Sun et al. [56]
		100 nm	-	0.5 wt.%	-	1 h	Ding et al. [58]
	SiO_2	30 nm	Spherical	0.1–1 wt.%	Magnetic stirring for 10 min	-	Bao et al. [59,100]
		20 nm	Spherical	0.5 wt.%	-	30 days	Lv et al. [101]
		30 nm	-	0.1–2.0 wt.%	Stirring for 0.5 h	-	Gao et al. [102]
		20 nm	Spherical	0.1, 0.2, 0.4, 0.8, 1.6 wt.%	Mechanical stirring and ultrasonic	-	Gu et al. [103]
	TiO_2	0.2–0.4 μm	-	0.25, 0.5, 1.0 wt.$_{NaPa/TiO2}$ %	Stirring and ultrasonication	-	Ohenoja et al. [104]
		40 nm	Non-spherical	1.5 vol.%	Ultrasonication for 2 h	-	Najiha et al. [105]
		15 nm	Spherical	0.1, 0.5, 1.0, 1.5, 2.0 vol.%	Ultrasonic vibration	Several days	Kayhani et al. [106]

Table 1. *Cont.*

Types	Nanoadditives	Size	Shape	Concentration	Stirring	Stability Duration	References
		20 nm	Spherical	0.2–8.0 wt.%	Mechanical and ultrasonication for 10 min	7 days	Wu et al. [13–15,27,29]
		20 nm	Spherical	0.5–4.0 wt.%	Ultrasonic stirring	-	Huo et al. [26]
		300 nm	Spherical	2.0–4.0 wt.%	Mechanical stirring	48 h	Wu et al. [24]
	TiO$_2$	50 nm	Spherical	0, 0.25, 0.5, 0.75, 1.0 wt.%	Ultrasonication for 30 min	-	Kong et al. [49]
		~20 nm	-	0.03, 0.05, 0.07 wt.%	Magnetic stirring	-	Ukamanal et al. [107]
Metal and non-metal oxides		90 nm	-	0.5–1.5 wt.%	Magnetic stirring	7 days	Meng et al. [52]
		20 to 50 nm	-	0.1, 0.4, 0.7, 1.0, 2.0, 3.0, 4.0, 5.0 wt.%	-	-	Sun et al. [50]
		40 nm	-	0.1, 0.5, 1, 2, 4 wt.%	Ultrasonication	30 min	Wang et al. [108]
		20–25 nm	-	-	Magnetic heat for 1 h	7 days	Zhu et al. [51]
	SiO$_2$, TiO$_2$, ZnO	100 nm	Spherical	ASNPs 3 wt.%, AZNPs 1 wt.%, ATNPs 1 wt.%	Magnetic stirring for 6 h	-	Cui et al. [109]
	ZnO, CuO	ZnO (4.5 & 27 nm); CuO (7.5, 45 nm)	-	1 g/L	Ultrasonication for 30 min	19 days	Odzak et al. [97]
	WO$_3$	50 nm	Spherical	0–1 wt.%	Magnetic stirring for 2 h	5 days	Xiong et al. [57]
	Ag$_2$S	2–10 nm	-	-	Sonication for 10 min	1 h	Kuznetsova [61]
	CuS	4 nm	Uniform spherical	0.1 to 2.0 wt.%	Magnetic stirring for 10 min, 1 h	2 days	Zhao et al. [62]
Metal sulphides		-	-	0.1 g	-	10 days	Wu et al. [110]
		100–300 nm	layered	-	Stirring for 10, 20 min	-	Zhang et al. [111]
	MoS$_2$	height 3.5 nm	Chain like layered	0.05 and 0.1 wt.%	Ultrasonication for 1, 2, 3 h	10 days	Wang et al. [64]
		100 nm	-	0.3–0.5 wt.%	Magnetic stirring for 10 min	16 h	Meng et al. [63]

Table 1. Cont.

Types	Nanoadditives	Size	Shape	Concentration	Stirring	Stability Duration	References
	graphene-SiO$_2$	Graphene (5 nm thick, interlayer distance 0.34 nm); SiO$_2$ (30 nm)	SiO$_2$ spherical, graphene multi-layered sheet	Graphene:SiO$_2$ (0.4:0.1, 0.3:0.2, 0.2:0.3, and 0.1:0.4)	Stirring for 1hr, ultrasonic bathing for 2 h	-	Xie et al. [112]
	GO-SiO$_2$	GO (1–2 nm thick); SiO$_2$ (30–40 nm)	GO sheet wrinkled folded	0.03–0.5 wt.%	Magnetic stirring for 24 h	60 days	Guo et al. [83]
	GO-SiO$_2$	GO (4–6 nm); SiO$_2$ (25–30 nm)	GO lamellar wrinkled, SiO$_2$ spherical	0.04, 0.08, 0.12, 0.16 and 0.20 wt.%.	Mechanical stirring for 30 min, ultrasonication	-	Huang et al. [19]
	CNT-SiO$_2$	CNT (inner diameter 8 nm, outer diameter 15 nm); SiO$_2$ (30 nm)	SiO$_2$ spherical, CNT tubular	0.5 wt.%	Magnetic stirring for 1 h, ultrasonic bathing for 2 h	-	Xie et al. [113]
	GO-TiO$_2$	TiO$_2$ (25 nm)	TiO$_2$ spherical	0.5 wt.% (0.3 wt.% GO-0.2 wt.% TiO$_2$)	Stirring for 20 min, sonicating for 40 min	30 days	Du et al. [84]
Composites	GO-Al$_2$O$_3$	GO (4–6 nm thick, 10–50 μm lateral sizes), Al$_2$O$_3$ (15, 30 & 135 nm)	Layered	0.25, 0.5, 1.0 and 2.0 wt.%	Magnetic stirring for 30 min, ultrasonic probe for 30 min	1 h	Huang et al. [20]
	GO-Al$_2$O$_3$	GO (10–50 μm in diameter; 1–2 nm thick); Al$_2$O$_3$ (30 nm)	GO layered; Al$_2$O$_3$ near-spherical	0.04, 0.08, 0.12, 0.16, and 0.20 wt.%	Mechanical stirring for 10 min, ultrasonic agitation process	-	Huang et al. [21]
	GO-TiO$_2$/ZrO$_2$	TiO$_2$/ZrO$_2$ (25 nm); GO (3–5 nm thick, 1.5–5.5 μm lateral)	2D GO; zero dimension TiO$_2$/ZrO$_2$	0.5 wt.%	Magnetic stirring and ultrasonication for—30 min, 1 h	-	Huang et al. [12]
	GO-TiO$_2$-Ag	-	-	0.05 wt.%	Sonication for 4 h	-	Zayan et al. [114]
	PTEE-SiO$_2$	413.6 nm (SiO$_2$ layer 20–30 nm)	PTFE rod-like or spherical	0.2, 1 and 3 wt.% (PTFE:SiO$_2$_0.57:0.43)	Ultrasonication for 20 min	12 h	Wang et al. [115]
	Cu-SiO$_2$	20 nm average (Silica layer thick 2 nm)	network-like silica, Cu spherical	0, 0.5, 1.0, 1.5, 2.0 wt.%	Magnetic stirring	-	Zhang et al. [116]
		-	Sphere	0.4 wt.%	Magnetic stirring for 15 min	30 days	Liu et al. [117]

Table 1. Cont.

Types	Nanoadditives	Size	Shape	Concentration	Stirring	Stability Duration	References
	MoS$_2$-Al$_2$O$_3$	MoS$_2$-Al$_2$O$_3$ (144.8 nm), MoS$_2$ (178.6 nm), Al$_2$O$_3$ (35.4 nm)	Laminar	2.0 wt.%	Electro-magnetic stirring	168 h	He et al. [118]
	Al$_2$O$_3$, MoS$_2$, hBN, and WS$_2$	Al$_2$O$_3$ (<100 nm), hBN (70–80 nm), MoS$_2$ (80–100 nm), WS$_2$ (80–100 nm)	Al$_2$O$_3$ (spherical); hBN, MoS$_2$, and WS$_2$ (layered structure)	1% each	Ultrasonic bath for 1 h	24 h	Kumar et al. [119]
	Fe$_3$O$_4$-MoS$_2$	MoS$_2$ (100–400 nm), Fe$_3$O$_4$ (10 nm), FeO$_4$ (30–60 nm) on MoS$_2$	Laminated structure	0.3, 0.6. 0.9, 1.2 wt.%	Ultrasonication	-	Zheng et al. [86]
Composites	MWCNT-Fe$_2$O$_3$	Fe$_2$O$_3$ (20–30 nm); MWCNT (10–30 μm length, 10–20 nm outer diameter, 3–5 nm inner diameter)	Multi-walled carbon nanotube	0.1–1.5 vol.% (Fe$_2$O$_3$ 80%, MWCNT 20%)	Sonication for 120 min	1 month	Giwa et al. [120]
	Ag-C	350–400 nm (C shell 100–120 nm thick) Ag 130–180 nm	Core spherical, NPs elliptical (core like short rod)	Ag 28 wt.% in Ag-C	Magnetic stirring for 30 min, ultrasonication for 60 min	5 days	Song et al. [121]
	TiO$_2$-Ag	TiO$_2$ (40 nm)	Ellipsoidal	0.05, 0.1, 0.1, 0.25, 0.3 wt.%	Magnetic stirring for 2 h	1 month	Li et al. [85]
	ZnO-Al$_2$O$_3$	ZnO (70 nm), Al$_2$O$_3$ (45 nm)	ZnO elongated, Al$_2$O$_3$ spherical	0.1–23 wt.%	Ultrasonic bath for 30 min	-	Gara et al. [122]
	WO$_3$-Mn$_3$B$_7$O$_{13}$Cl	22.4 nm	Spherical	0.0, 0.1, 0.3, 0.5, 0.7 and 0.9 wt.%	Ultrasonic vibration for 1 h	48 h	Liang et al. [123]
Carbon-based materials	Carbon	outer diameter ~177 nm	Toroidal	2.0, 1.5, 1.2, 1.0, 0.5, and 0.1 wt.%	Magnetic stirring for 60 h	4 months	Peña-Parás et al. [65]
	Carbon	130, 170, 200 and 250 nm	Spherical	0.05, 0.1, 0.15, 0.2, 0.3 wt.%	Ultrasonication	5–10 h	Wang et al. [124]

Table 1. Cont.

Types	Nanoadditives	Size	Shape	Concentration	Stirring	Stability Duration	References
Carbon-based materials	Carbon nanotube	10–20 nm diameter; 1–2 μm axial dimension	Short and tube	0.1 wt.%	Sonication for 2 h	30 min	Peng et al. [125]
		90 nm diameter	Long rod like	0.1, 0.3, 0.5, 0.7, and 1.0 wt.%	Stirring for 0.5, 1, 3 h	-	Sun et al. [126]
		20–30 nm in outer diameter; 10–30 μm in length	Pentagonal and heptagonal	0.05, 0.10, 0.15, 0.20, and 0.25 wt.%	Proper stirring	12 days	min et al. [127]
		SWCNTs (2 nm diameter), MWCNTs (25 ± 10 nm diameter)	Sphere	50–100 μL	-	Few hours	Kristiansen et al. [128]
		8–50 nm in diameter, 0.5–30 μm in length	-	-	Magnetic stirring and ultrasonication for 2 h	168 h	Ye et al. [129]
	Carbon dots	CDs-IL 4.4 nm	Spherical	3, 12.2, 34.9, 19.4 wt.%	Magnetic stirring for 6 h	60 min	Tang et al. [130]
		Sulphur doped CQDs 4.8 nm	Spherical	0.25, 1.25, 2.5, 5, and 10 wt.%	Ultrasonication for 30 min	7 days	Xiao et al. [131]
		CDs-GO 3–4 nm	-	0.06, 0.08, 0.1, 0.2, 0.3 mg/mL	-	6 months	Hu et al. [66]
	Graphene	1 nm	-	23.8, 69.9, and 110 mg/mL	Magnetic stirring for 12 h	1 month	Liang et al. [67]
		100 nm	2D nanosheet	0.2 mg/mL	Stirring for 4 h	-	Fan et al. [132]
		Size several micrometres, interlayer spacing 0.63 nm	Crystal	0.5, 1.0, 1.5, 2.0, and 2.5 mg/mL	Ultrasonication for 30 min		Ma et al. [133]
		0.67–0.87 nm	Multiple layered	0, 0.5, 1, 2, and 4 mg/mL	Stirring for 4 h, ultrasonication 8 h	-	Ye et al. [134]

Table 1. Cont.

Types	Nanoadditives	Size	Shape	Concentration	Stirring	Stability Duration	References
	Graphene	2 nm	-	0.5, 1.5, 2.5, 4, 5, and 8 mg/mL	-	-	Qiang et al. [135]
		-	Flat flake	0.1, 1 wt.%	-	30 days	Piatkowska et al. [136]
	Diamond	3–10 nm	spherical	0.1, 0.5, 1, 2, 4, and 6 wt.%	Probe sonication, stirring	-	Mirzaamiri et al. [137]
		5–10 nm	-	0.01–0.07 wt.%	Simple stirring	-	Jiao et al. [68]
		1.20 & 1.45 nm	Sheet	0, 0.3, 0.5, and 1 mg/mL	Ultrasonication for 30 min	1 week	Fan et al. [73]
Carbon-based materials		10–50 µm thick, 0.335 nm high	Single monolayer	0.01 wt.%	Ultrasonication for 5 min	-	Kinoshita et al. [138]
		4 nm	-	0–2 wt.%	Ultrasonication	-	Elomaa et al. [139]
		200–1000 nm	Transparent nanosheet	0.1, 0.3, 0.5, 0.7, 1 wt.%	Ultrasonication	12 days	min et al. [72]
		0.5–5 µm diameter; 0.8–1.2 nm thick	-	0.01, 0.05, 0.1, and 0.5 wt.%	Sonication for 2 h	-	Singh et al. [140]
	Graphene oxide	500 nm–5 µm diameter; 0.8–1.2 nm thick	Ultra-thin	0.025, 0.05, 0.075, and 0.1 vol.%	Ultrasound, stirring	3 months	Bai et al. [141]
		20–30 nm outer diameters; 10–30 µm length	2D sheet	0.5 mg/mL	Stirring for 30 min	5 weeks	Song et al. [70]
		0.335 nm thick	Ultrathin and transparent	0.8, 1.2, and 1.6 mg/mL	Stirring for 24 h, ultrasonication	2 weeks	Gan et al. [142]
		10–50 µm lateral size; 1–2 nm thick	Spherical	0.06 wt.%, 0.5 wt.%	Stirring for 30 min, ultrasonic bath for 10 min	7 days	He et al. [17]
		2–5 nm thick 10–20 µm lateral size	-	0.1 wt.%	Stirring for 30 min ultrasonic bath for 20 min	50 days	Meng et al. [76]

14

Table 1. *Cont.*

Types	Nanoadditives	Size	Shape	Concentration	Stirring	Stability Duration	References
	Graphene oxide	1 nm thick	initially sheet shape, then parabolic shape	0.2 mg/mL	Sonication	-	Kim et al. [143]
		1.3 nm	Thin film	0.05 to 1.0 mg/mL	Mechanical stirring for 30 min	8 months	Hu et al. [75]
		0.8 μm lateral size 1.96 nm thick	Bathtub	0, 0.03, 0.05, 0.07, and 0.1 wt.%	Magnetic stirring for 12 h, ultrasonication for 1 h	90 days	Liu et al. [74]
		C (30–60 nm) and GO (30–60 nm)	C onion-like spherical; GO 2D nanosheet	C 0.06 wt.%; GO 0.02–0.06 wt.%	-	-	Su et al. [80]
	GO & carbon	oxidised wood-derived nano carbons 640–1300 nm and GO 50–200 nm	aggregated chain-like	0.001 and 1 wt.%	Ultrasonication for 30 min	1 month	Kinoshita et al. [144]
Carbon-based materials	GO & chitosan	GO 0.05–0.2 μm	GO optical 3D; copolymer brush-like	2 mg/mL	Stirring for 6, 12 h ultrasonication	30 days	Wei et al. [145]
	GO & 3-APS	3-APS (525.39 nm)	-	2 mg/mL	Stirring for a certain period	-	Li et al. [146]
	GO & graphene	GO 4.2 nm, graphene 5 nm	Multi-layered	0.2, 0.5, 0.7 and 1.0 wt.%	Stirring for 1 h, ultrasonic bath 2 h	-	Xie et al. [77]
		GO 2.5 nm and nanodiamond 2–10 nm	GO laminar	0, 0.2, 0.4, 0.6, 0.8, 1.0 wt.%	Magnetic stirring for 9 h	-	Wu et al. [78]
	GO & diamond	GO 30 nm, 2–3 nm thick; modified diamond 30 nm	GO lamellar and MD 3D structure	GO colloid (0.7 wt.%) and MD colloid (0.5 wt.%)	Ultrasonic ethanol bath for 5 min	2 months	Liu et al. [79]
	GO & graphitic CN	graphitic carbon nitride and GO 10–50 μm lateral size, 1–2 nm thick	unique one-layer	0.06 wt.% each	Stirring for 30 min, ultrasonic bath for 10 min	-	He et al. [18]
	PEGlated graphene	20 nm	laminar	0.005, 0.01, 0.03, 0.05, and 0.1 wt.%	Mechanical stirring for 3, 4 h	7 days	Hu et al. [147]

Table 1. *Cont.*

Types	Nanoadditives	Size	Shape	Concentration	Stirring	Stability Duration	References
Polymers	Cellulose	Length 200 ± 25 nm, Size 1–50 μm	Chain like, crystalline	1, 1.5, 2, 2.5, 3, and 4 wt.%	-	-	Shariatzadeh et al. [148,149]
	Fullerene–styrene and –acrylamide	3–40 nm	Ideal spherical	0.5 wt.%	-	-	Lei et al. [150]
		average 46 nm	Ideal spherical	0, 0.2, 0.4, 0.6 & 0.8 wt.%	-	-	Jiang et al. [151]
	Hydrogel	-	Fibrous-3D network	3, 4 & 5 wt.%	Stirring for 3–4 h, mechanical sheared	-	Wang et al. [152]
	Naphthalene	-	-	0.02, 0.04, 0.06, 0.08, 0.1, 0.15, 0.2 mol/L	Stirring for 24 h	-	Yang et al. [153]
Metal salts	LaF$_3$	LaDTP-10 (19.6 nm) and LaDTP-20 (8.5 nm)	LaDTP-10 polycrystalline; LaDTP-20 sphere	1 wt.%	Continuous magnetic stirring for 1 h	-	Zhang et al. [154]
		-	Chain like	0, 0.25, 0.5, 0.75 & 1 wt.%	Stirring for 2 h	-	Zheng et al. [155]
		-	Bilayered	1 wt.%	Stirring for 12 h	-	Dong et al. [156]
	Proton type-ionic liquids	-	brushy-like soft layer	0.1 & 1 wt.%	Magnetic stirring for 2 h, ultrasonication for 10 min	60 min	Khanmohammadi et al. [157]
		-	-	1 wt.%	Magnetic stirring for 10 min	-	Kreivaitis et al. [158]
Nitrides	Hydroxylated boron nitride (HO-BNNS)	0.6–0.8 nm	Thin flat	HO-BNNS/water-glycol (0.0125, 0.025, 0.05, 0.10, 0.20 wt.%)	Ultrasonic process for 30 min	5 days	Bai et al. [89]
	Hexagonal boron nitride	76.14 nm	-	0.2 to 1.0 wt.%	Ultrasonication	7 days	He et al. [159]
		-	-	0.1–5.0 vol.%	-	-	Abdollah [160]
		300 nm wide and 30 nm thick	-	1, 0.05 or 0.01 wt.%	Sonicator bath for 20 h	30 days	Cho et al. [88]
	Silicon nitride	Silica 20, 50, 100, 200 nm	-	-	-	-	Lin et al. [161]

Table 1. *Cont.*

Types	Nanoadditives	Size	Shape	Concentration	Stirring	Stability Duration	References
Carbides	Nb_2C	20 nm (Nb_2C), 12 nm ($MO-Nb_2C$), 6 nm ($CO-Nb_2C$)	Accordion like, Crystalline	1.0, 0.75, 0.5, and 0.25 mg/mL	Magnetic stirring for 6, 12 h, 7 days; ultrasonic stirring	$CO-Nb_2C$ 15 days; $MO-Nb_2C$ 30 days	Cheng et al. [90]
	Ti_3C_2	Lateral size 0.2–3 μm Layer thick 20 nm	Layered, Planar	1, 2, 3, 5 and 7 wt.%	Magnetic stirring for 1 h	-	Nguyen et al. [91]
	Black phosphorus	3.9 nm	Crystalline	0.001–0.02 wt.%	Ultrasonication for 8 h	2 weeks	Tang et al. [92]
		500 nm	Honey-comb	91.17% (wt.%)	Ultrasonication for 10 h	-	Wang et al. [162]
		100 nm wide; 7 nm thick	Multilayered	35, 70, and 200 mg/L	Stirring for 10 min, ultrasonication	-	Wang et al. [163]
Others	LDH	19.73 nm wide; 8.68 nm thick	-	0.5 wt.%	Ultrasonication	-	Wang et al. [94]
		19.42 nm wide; 8.59 nm thick	Layered	0.1–1.0 wt.%	Stirring for and ultrasonication	-	Wang et al. [93]
	Chitosan	70–145 nm	Crystalline	0–0.5 wt.%	Ultrasonication for 15 min	30 days	Li et al. [95]
	Stearic acid	-	2D layered	0.25, 0.5, 0.75 & 1.0 mg/mL	Ultrasonication	-	Ye et al. [96]

3. Dispersion Stability of Nanoadditives

3.1. Evaluation Methods

Research on dispersion stability of nanoadditives in water is of vital importance for the development of water-based nanolubricants, and it has become one of the key challenges in restricting their widespread practical application. Stable nanosuspension is usually considered as a prerequisite for the successful preparation of water-based nanolubricants. In general, there are some methods that can be used to effectively evaluate the dispersion stability, including microscopy, zeta potential, UV-vis spectral analysis, dynamic light scattering, and sedimentation.

3.1.1. Microscopy

Microscopic methods using optical microscope (OM), transmission electron microscope (TEM), and scanning electron microscope (SEM) are very useful to distinguish the size and shape of NPs, and the dispersion stability of NPs in water can be evaluated as per the distribution and agglomeration of NPs under microscopic observation. Among all the microscopic methods, the use of OM is the easiest and quickest technique to examine the agglomeration behaviour and trend of the nanolubricants even at micrometre scale. However, the limited resolution of OM is the main disadvantage of viewing the size and shape of NPs at nanometre scale.

Currently, electron microscopy is the most commonly used method to evaluate the NPs stability in a base lubricant due to its high resolution. TEM and SEM are two of the most popular methods for observing the morphology and size distribution of NPs. TEM samples are prepared by placing a nanolubricant drop on a carbon-coated copper grid until complete evaporation of the base lubricant. SEM samples are prepared by dropping a small amount of nanolubricant onto a tape that is attached to the top of a sample holder before heating and drying under vacuum [164,165]. During the sample preparation processes for both TEM and SEM, NPs may somewhat agglomerate, leading to inaccurate evaluation of NPs stability in base lubricant. In spite of this, it is still acceptable to evaluate the dispersion stability of NPs between different concentrations or formulations by comparing the NP size difference. Figure 4 shows TEM images of TiO_2 NPs with 2.0 wt.% and 4.0 wt.% concentrations in water. It can be found that the NPs were uniform and well dispersed in water, and no visible agglomeration was observed even at 4.0 wt.% concentration. With the increase of concentration, however, there is a trend of few NPs agglomeration.

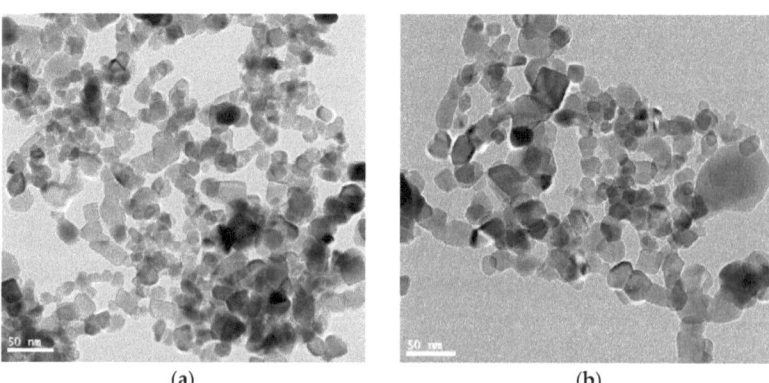

(a)   (b)

Figure 4. TEM image of TiO_2 NPs in water-based lubricants with concentrations (**a**) 2.0 wt.%; (**b**) 4.0 wt.% [15].

Given the shortcomings of using TEM and SEM, scientists recommend the use of freeze etching replication TEM (FERTEM) or cryogenic electron microscopy (cryo-TEM) [166–168] to observe the nanolubricants because each of them is appropriate to characterise wet samples, and the size and morphology of NPs can be kept the same as those in the original nanolubricants.

3.1.2. Zeta Potential Test

Zeta potential (ZP) test shows the potential difference between the static fluid layers adhered to the dispersed particles and the dispersion medium [166,169]. The nanofluid stability is evaluated by observing the fluid's electrophoretic behaviour, as a layer of charged particles is formed when the free charge in the base fluid is attracted by the surface opposite charges of dispersed particles [166]. ZP values range from a positive value at low pH to a negative value at high pH in any nanofluid. A nanofluid with relatively high ZP absolute value (>30 mV) is electrically stable due to a strong repulsive force between NPs. Instead, a nanofluid with relatively low ZP absolute value (<15 mV) tends to have NPs agglomerate because an attractive force dominates. In particular, a ZP absolute value above 60 mV indicates excellent dispersion stability of a nanofluid.

ZP test has been used in many studies to determine the size distribution and stability of water-based lubricants [63,90,93,95,111,170]. Figure 5 demonstrates a bar graph of the ZP (mV) vs. nanolubricant with 0.12 wt.% GO, 0.12 wt.% Al_2O_3, and 0.12 wt.% GO-Al_2O_3 (1:1) suspension [21]. It can be seen that the GO-Al_2O_3 lubricant exhibits comparatively higher ZP absolute value (40 mV) than Al_2O_3 (30 mV) and GO lubricants (35 mV), signifying a greater level of stability.

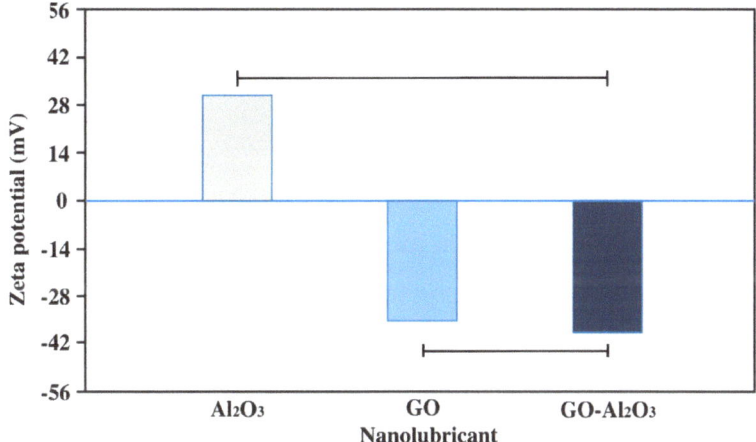

Figure 5. Zeta potential rate of GO, Al_2O_3, and GO-Al_2O_3 [21].

3.1.3. UV-Vis Spectral Analysis

One of the reliable methods to measure the dispersion stability and durability of nanofluids is spectral absorbency analysis against NPs concentration in nanofluids utilising an ultraviolet-visible (UV-vis) spectrophotometer which follows the Beer-Lambert law [171,172]. The stability of suspension is determined by calculating the volume of sediment relative to the time of sediment. The intensity of light different from the scattering and absorption of light passing through the fluid is used in the UV-vis spectrophotometer. It evaluates the absorbance of a fluid within a wavelength of 200–900 nm to analyse various dispersions in the fluid [173]. One of the unique aspects of this method is that it is capable of obtaining quantitative data of NP concentration in nanofluid and is applicable for all boundary fluids [174].

Research has been undertaken using UV-vis spectrometer to analyse the liquid absorbance, physical properties, and chemical state of NPs in water-based lubricants such as graphene [17,74,135], boron nitride [89], and copper [46,116]. For example, the absorbance of 0.05 wt.% hydroxylated boron nitride nanosheets (HO-BNNS) in water was measured within 5 days, as shown in Figure 6. A good dispersion of HO-BNNS in water can be observed from the slight drop of absorbance, probably because of the hydrogen bond existence between the hydroxyl groups in HO-BNNS and water-glycol. The NPs absorbency ratio is directly proportional to the nanofluid concentration. However, the use of UV-vis is limited by the dispersion of high concentration nanofluids, mainly for carbon nanotube CNT solutions [175] or when the colour of the base lubricant is too dark to distinguish the deposits of NPs.

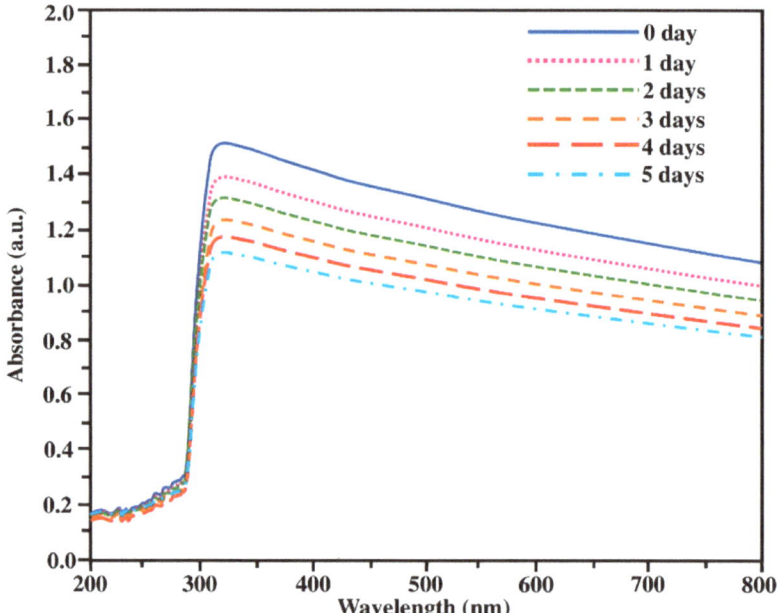

Figure 6. UV–vis absorption spectra of 0.05 wt.% HO-BNNS in water within five days [89].

Generally, a linear relationship exists between the NP concentration and the absorption intensity [166,176,177]. For instance, Liang et al. [123] assessed NP stability using an absorbance method for 48 h and noted that the dispersion stability was obtained for $Mn_3B_7O_{13}Cl$-WO_3:Eu^{3+} from a linear graph (absorbance vs. time), as shown in Figure 7a. The results also showed that WO_3: Eu^{3+} and $Mn_3B_7O_{13}Cl$ exhibited a decrease in relative absorbance over time, indicating that NPs settled more rapidly in acidity solution than that in alkalinity solution. Compared to $Mn_3B_7O_{13}Cl$, $Mn_3B_7O_{13}Cl$-WO_3:Eu^{3+} showed lower relative absorbance, thus presenting better stability. Similarly, Figure 7b revealed that the relative absorbance of the BN, BNNS, and HO-BNNS decreased with the increase in settling time [89]. The settling rate was different for these three additives. To be specific, the settling rate of BN was higher because of its larger mass. BNNS showed a better stability compared to BN due to fewer hydroxyl groups on its surface. While HO-BNNS exhibited the best dispersion stability due to the presence of more hydroxyl groups on its surface. It is worth noting that at higher nanofluid concentration this method might not work accurately because the absorption range is outside the system's highest limit [178].

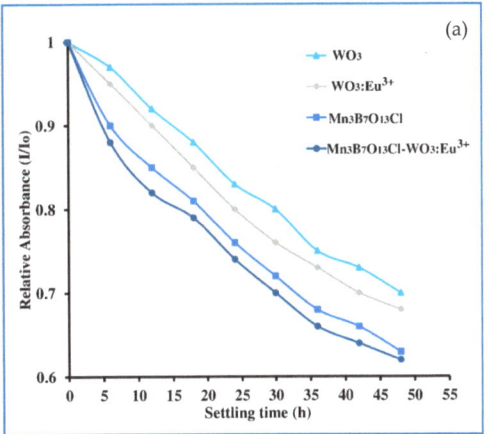
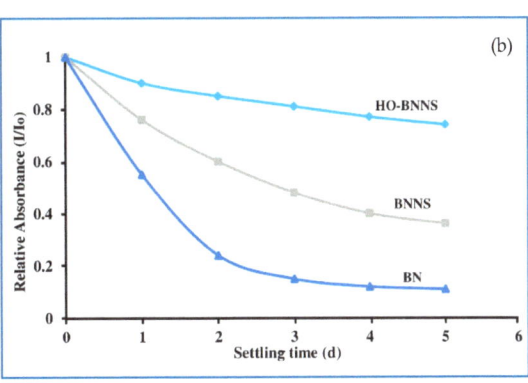

Figure 7. Relative absorbance under different time of (**a**) WO$_3$, WO$_3$:Eu^{3+}, Mn$_3$B$_7$O$_{13}$Cl, and Mn$_3$B$_7$O$_{13}$Cl-WO$_3$:Eu^{3+} NPs [123] and (**b**) BN, BNNS, and HO-BNNS [89].

3.1.4. Dynamic Light Scattering

The dispersion stability of lubricants over time can also be evaluated by measuring the NPs size distribution using the dynamic light scattering (DLS) method through analysing the autocorrelation index function [177,179,180]. This method evaluates the particle diffusion moving in Brownian motion and uses the Stokes-Einstein ratio to convert them into particle size ranging between 0.3 nm to 10 μm [181]. Numerous researchers use the DLS method in order to determine the diameter of NPs, although this method is also promising for estimating the size distribution of water-based nanoadditives such as graphene [132,141], ceria [53], and copper [65,117,118,182]. The more stable dispersion of NPs in base lubricant, the greater the impact is on the scattered light intensity, and vice versa. DLS was used to monitor colloidal stability of Cu@SiO$_2$ NPs in DW over time, as illustrated in Figure 8. After 30 days of storage at ambient conditions, very limited size variation was observed, and the hydrodynamic diameter (HD) stayed constant at 230 nm, demonstrating no agglomeration of Cu@SiO$_2$ NPs in DW.

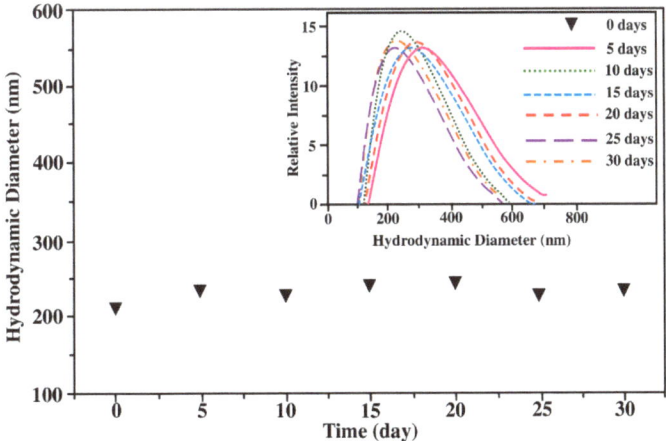

Figure 8. Dynamic light scattering (DLS) observation of Cu@SiO$_2$ NPs in water [117].

Basically, before measurement, thick nanolubricants must be diluted, which may change the microstructure nature of the NPs [168,171]. Therefore, a suitable method is needed to determine the particle size of nanolubricant without changing the particle's microstructure.

3.1.5. Sedimentation

The basic and low-cost method to analyse the dispersion stability of nanolubricants is the sedimentation method. It is a process in which NPs settle or precipitate as sludge at the lubricant base. Additionally, it is related to the balance of friction force and buoyancy [183]. In nanofluid, the NPs sediment volume or weight is an indicator of nanofluid stability under an external force field.

According to Stokes' law [171], the rate of NP sedimentation rate is calculated:

$$v_o = 54.5d^2 \frac{(\delta - \rho)}{\mu} \text{ (for spherical shape)} \tag{1}$$

where d is the diameter (cm) of NP. δ is the density (g/cm^3) of NP. ρ is the density (g/cm^3) of base lubricant and μ is the viscosity (Pa·s) of base lubricant.

Photography of nanofluid sedimentation using a camera over a time period is the simplest method to monitor the nanofluid stability and sedimentation coefficient [166,184]. Comparing the photos is widely used in nanofluid research, and many examples were mentioned for water-based nanoadditives such as TiO$_2$ [15,24,29,51]. The stability analysis using sedimentation is usually conducted immediately after the dispersion of NPs in the base lubricant, and the disturbance or movement should be avoided to ensure its reliability. Wu et al. [24] carried out the sedimentation experiment to compare the precipitation of TiO$_2$ NPs in different as-prepared water-based nanolubricants, as shown in Figure 9. They found that all the lubricants remained stable without apparent particle sedimentation at the bottom within 120 min as long as the dispersants such as sodium dodecyl benzene sulfonate and Snailcool were added into water. After standing the lubricants for 24 h, the TiO$_2$ NPs began to settle down slightly, and a shallow supernatant appeared.

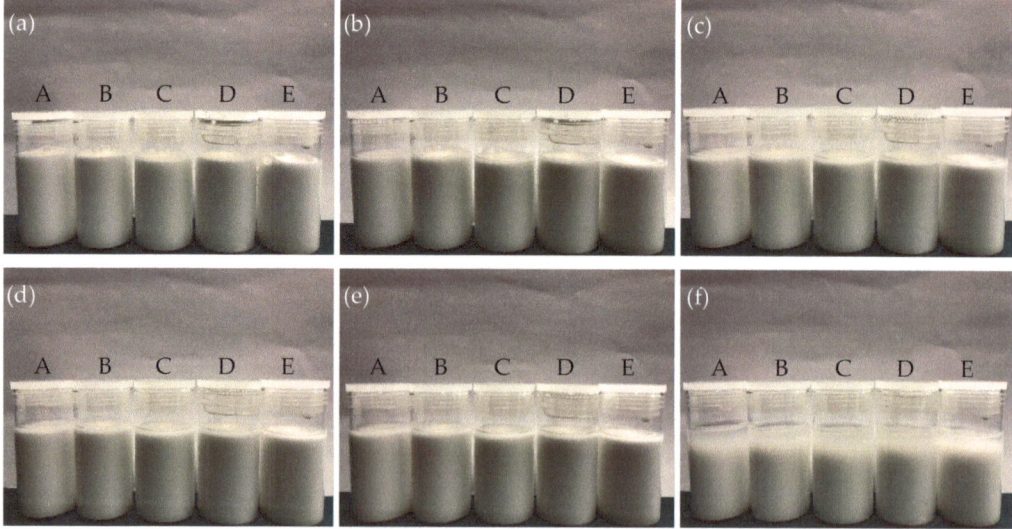

Figure 9. Sedimentation of TiO$_2$ NPs dispersed in different water-based nanolubricants at settling time of (**a**) 0; (**b**) 30 min; (**c**) 60 min; (**d**) 90 min; (**e**) 120 min and (**f**) 48 h [24].

Moreover, the sedimentation rate of NPs in a certain time is measured by the change in the mass/volume concentration in the base fluid using a highly sensitive analytical balance. According to the following formula, the NPs suspension fraction F at a specific time is calculated:

$$F = \frac{(W_o - W)}{W_o} \tag{2}$$

where W_o is the total mass of all NPs in the base fluid; W is the mass of the sediment NPs at a certain time [185].

The nanofluid is considered to be stable if the concentration of NPs in the base fluid stays constant over time, and vice versa [186]. This is a simple method that just requires an accurate and sensitive balance to quantify the volume of the deposited NPs. The only limitation of this method is that it cannot detect small NP aggregates that are not considered to settle at a reasonable amount [166,187].

3.1.6. Other Methods

There are some other methods applicable for evaluating dispersion stability of nanofluids, including three-omega, centrifugation, and density measurement, each of which has its own characteristics. For example, a densitometer can be used in the density measurement method to monitor NP concentration varying with time due to NPs aggregation and precipitation [188]. The three-omega method determines colloidal stability with large volume fractions of NPs through thermal conductivity [187,189]. In centrifugation method, analysis is carried out using the centrifugal force acting on the nanofluids at various speeds and time [164,177,180]. It is acknowledged that centrifugation is a faster way to determine nanofluids stability compared to sedimentation photography [164].

3.2. Factors Affecting Dispersion Stability
3.2.1. pH Control

Altering the pH of nanofluids affects the surface of NPs and enhances the stability of dispersed NPs [190]. Nanofluids are characterised by their electrokinetic properties, which influence their stability. Therefore, an increase or decrease in nanofluid pH is accompanied by an increase or decrease in zeta potential. As previously noted, NPs repel one another when their zeta potentials increase above +30 mV or decrease below −30 mV. With the objective to change the pH of a nanofluid, suitable acidic or alkaline solutions should be added [191].

In order to enhance the nanofluids' stability, it is necessary to modify its pH level. The pH of nanofluids must be equal to a specific critical value in order for them to have zero net charges. As the electrostatic force is diminished, the NPs begin to break apart. At this point, the system is at the point of zero charge (PZC) or an isoelectric point (IEP). Aggregation and sedimentation of NPs peaks at this stage [166]. At an IEP, both surface charge and zeta potential are zero. By enhancing repulsive forces among particles and zeta potential, a pH value far from the IEP demonstrates better durability. When a pH is far away from IEP and the zeta potential is high, it leads to a stable suspension with almost no particle coagulation [191]. The effect of pH on the stability of water-based lubricants has been investigated by many researchers using NPs such as graphene oxide (GO). For instance, Meng et al. [76] discovered that alkaline GO water-based lubricant (pH 9.0) presented the most effective lubricity, and an increase in lubricant pH value can modify the GO structure, improving the dispersibility of GO in water.

3.2.2. Ultrasonication

Generally, chemical or physical methods are used to generate NPs as dry powders that are then distributed into water for synthesising aqueous lubricants [192]. Every chemical treatment-based procedure changes the surface chemistry of the NPs distributed in boundary fluids, protecting them from agglomeration and sedimentation, and improving their long-term stability [166]. In contrast, physical methods such as magnetic stirring, shear homogenizer, and probe and bath ultrasonication are also widely used for producing stable nanofluids with a two-step method [188]. In comparison to bath ultrasonication, probe ultrasonication is expected to produce a nanofluid with better dispersion stability. Furthermore, it has been demonstrated that ultrasonication is more effective than mechanical stirring in reducing the agglomeration of NPs in water. In ultrasonication, an ultrasonic wave is spread through a periodic motion, leading to unstable cavitation bubbles bursting. Implosion occurs when unstable cavitation bubbles break up aggregates [193,194]. The produced ultrasonic waves can break up the attractive force between NPs to increase the stability of the nanofluid [195]. It should be noted that an excessive ultrasonication time may result in rapid sedimentation of NPs.

3.2.3. Surface Modification

In the surface modification method, the surface-modified NPs are straightforwardly added into water to attain stable water-based nanolubricants without the aid of surfactants. Surface modifiers are necessary to modify NPs' surface activity. Table 2 demonstrates a list of surface modifiers used for different types of NPs. For example, Tang and Cheng [196] modified ZnO NPs to enhance the stability in an aqueous solution using polymethacrylic acid (PMAA). Nano-ZnO has hydroxyl groups that interact with carboxyl groups of PMAA and develop zinc methacrylate complexes on nano-ZnO surfaces, which enables ZnO NPs to be stably dispersed in water. Zhao et al. [46] implemented an in situ modification of water soluble Cu NPs, which involves both the preparation and surface modification simultaneously. The two polar groups of Bis (2-hydroxyethyl) dithiocarbamic acid (HDA) that act as capping agents enhance the dispersibility of Cu NPs in water. Through in situ surface modification, they also synthesised water soluble CuO with different morphologies (nanorods, nanobelts, and spindle shapes) using polyethylene glycol (PEG, a guiding agent) and polyvinyl pyrrolidone (PVP, a stabilising agent), showing improved stability [54].

It is possible to obtain long-term nanofluid stability by functionalising the NP surface. Therefore, in order to prepare self-stabilised nanofluid, functionalised NPs are added to the base fluid. In this context, appropriate functional organic groups are selected that will most likely adhere to the NP surface, thereby assisting self-stabilisation. Functional groupings can be presented in two ways. The first technique requires bifunctional organic compounds to introduce all of the functional ligands in one step. The NP surface is attached with one functionality, and the other group is used to functionalise the NP. In the second method, bifunctional compounds are combined through reaction, in which a group acts as a bonding point, and can later be transformed to an ultimate functionality [197]. Kayhani et al. [106] functionalised spherical TiO_2 NPs (15 nm) by mixing with hexamethyldisilazane ($C_6H_{19}NSi_2$) at 2:1 mass fraction ratio to produce stable TiO_2/water nanofluid. The mixture was sonicated for 1 h at 30 °C to obtain soaked NPs which were dried and dispersed in distilled water for 3–5 h by ultrasonic vibration. No agglomeration was observed in the nanofluids for several days. The stable behaviour of TiO_2 NPs resulted from the hydrophilic ammonium group. In addition to this, Yang and Liu [198] succeeded in maintaining the stability of 10 wt.% SiO_2 water-based nanolubricant by functionalising with silanes of (3-glycidoxylpropyl) trimethoxysilane, and found no sedimentation existed for a year.

3.2.4. Surfactant Addition

In general, NPs will not agglomerate as long as the nanofluid surface tension is low. It is reported that surfactants can be used to reduce the surface tension of nanofluids [199]. The presence of surfactants decreases the interfacial tension between two liquids or between a liquid and a solid, supporting the spread of liquids [200]. In aqueous solutions, surfactants addition can significantly improve NP stability. It is not only necessary to investigate the influence of surfactants on base nanofluids, but also important to discover new surfactant candidates with the potential to enhance the stability while minimising the damage to native nanofluid properties [188].

Popular surfactants used in literature to maximise the dispersion stability have been listed in Table 2. It is essential to select the right surfactant, which is either cationic, anionic or non-ionic. For instance, the performance of sodium dodecyl sulfate (SDS) and polyvinylpyrrolidone (PVP) was compared by Xia and Jiang [201] to present the dispersion stability of Al_2O_3 NPs in DW. Based on this study, PVP demonstrated better dispersibility than SDS by improving stability at 0.5, 1.0, and 2.0 wt.% surfactant concentration. The nonionic PVP performed comparatively better because of its extended alkyl chain. Kakati et al. [202] prepared 0.1–0.8 wt.% Al_2O_3 and ZnO water-based nanolubricants with the aid of 0.03 wt.% SDS. The results showed that the nanofluid with SDS had 4–5 days stability while without SDS the NPs agglomerated within 1 h after preparation.

Table 2. A list of surface modifiers and surfactants used in various water-based nanolubricants.

NPs Type	Surface Modifier	Surfactant
Pure metals	Bis (2-hydroxyethyl) dithiocarbamic acid (HAD) [46], Methoxylpolyethyleneglycol xanthate potassium (MPEGOCS$_2$K) [47]	Polyvinylpyrrolidone (PVP) [46,97]
Metal and non-metal oxides	polyethylene glycol-200 [59], oleic acid (OA) [102], polyethyleneimine (PEI) [13,15,26–28], sodium hexametaphosphate (SHMP) [50], KH-570 [103]	(3-mercaptopropyl)trimethoxysilane (MPS) [85], sorbitan monostearate [53], polyvinylpyrrolidone (PVP) [54,97], polyethylene glycol (PEG) [54], cetrimonium bromide (CTAB), and sodium dodecylbenzene sulfonate (SDBS) [24,25,51,52,99], sodium silicate [51,52], snailcool [24], hexadecyl trimethyl ammonium bromide (CTAB) [101]
Metal sulphides	Bis (2-hydroxyethyl) dithiocarbamic acid (HDA) [62], sodium oleate soap, triethanolamine oleate, fatty alcohol polyethylene glycol ether (MOA), polyethylene glycol octyl phenyl ether (OP-4) [110],	(3-mercaptopropyl) trimethoxysilane (MPS) [61], cetrimonium bromide (CTAB), and sodium dodecylbenzene sulfonate (SDBS) [111], oleic acid, triethanolamine [111],
Carbon-based materials	Dopamine methacrylamide (DMA) 2-methacryloyloxyethyl phosphorylcholine (MPC) [68],	humic acid (HA) [128], sodium dodecyl sulfate (SDS) [119,120], Triton X-100 ($C_{34}H_{62}O_{11}$) [67]
Composites	hexadecyldithiophosphate (DDP) [117], 3-mercaptopropyl trimethoxysilane (MPTS) [116], polydopamine (PDA) [118],	sodium dodecyl sulfate (SDS) [124,125], polyvinylpyrrolidone (PVP) [119], Igepal CO-520 [117], cetrimonium bromide (CTAB), and sodium dodecylbenzene sulfonate (SDBS) [119,184]
Others	dialkyl polyoxyethylene glycol thiophosphate ester (DTP-10, DTP-20) [154], oleylamine [93]	benzalkonium chloride [90], sodium polyacrylate (PAAS) [159], SHMP (sodium hexametaphosphate), 1,4-butylene glycol [203], coconut diethanol amide (CDEA) [204]

3.3. Theories of Dispersion Stability

In addition to the above discussion, the dispersion stability of nanofluids can be explained by a number of theories including DLVO (Derjaguin and Landau 1941, Verwey and Overbeek 1948), depletion and steric stability theory. The DLVO theory describes the dispersion stability by electrostatic repulsive forces and van der Waals forces, which is only applicable for spherical NPs. When the van der Waals force is less than electrostatic repulsion, molecule agglomeration and collision are reduced significantly, resulting in more stable suspension [176,205]. The NPs tend to reassemble while the van der Waals force plays a main role. A DLVO potential is calculated as follows:

$$G_T(h) = G_e(h) + G_A(h) \qquad (3)$$

where $G_e(h)$ denotes the repulsive potential and $G_A(h)$ denotes the attractive potential. $G_T(h)$ is the total potential at varying values of h. Both are influenced by the distance between NPs [206], as shown in Figure 10.

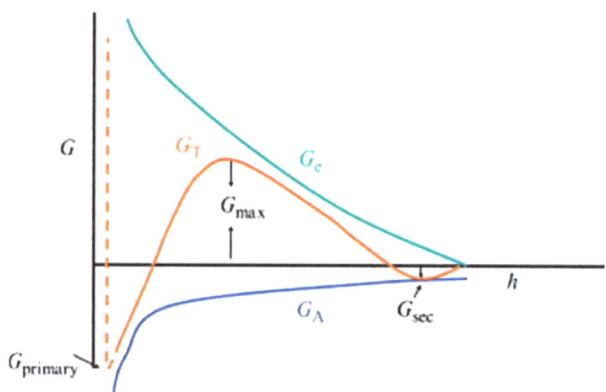

Figure 10. DLVO potential variation with a certain range of NPs distance [206].

Additionally, the electrostatic repulsive potential and van der Waals attraction potential can be determined using the following equations [207]:

$$\text{Repulsive potential } G_e(h) = 2\pi\varepsilon a\psi_o^2 \ln[1 + exp(-\kappa h)] \qquad (4)$$

$$\text{van der Waals attractive potential } G_A(h) = -\frac{A}{6}\left[\left(\frac{2a^2}{h^2 + 4ah}\right) + \left(\frac{2a^2}{h^2 + 4ah + 4a^2}\right) + \ln\left(\frac{h^2 + 4ah}{h^2 + 4ah + 4a^2}\right)\right] \qquad (5)$$

where a is particle radius; ψ_o is particle's surface potential; ε is medium's permittivity; κ is inverse Debye length that is influenced by the thickness of electrical double layer; h is the distance between NPs; and A is Hamaker constant. The electrostatic repulsion is influenced by three factors: size, distance, and surface potential of NPs, while the van der Waals force is influenced by the distance and radius of NPs. NPs aggregate when their distance is below a certain value due to the molecular attraction [207].

However, for NPs with platelet, rod, ellipse, or other shapes, both the depletion and steric stability theories have to be taken into account. The steric stability theory is explained by van der Waals force and elastic steric force in nanofluids with different NPs, ionic/non-ionic surfactants or absorbed polymers. The steric force is influenced by the chemical composition of the suspension and thickness, and the density of absorbed polymer layers [208–211]. The depletion theory contributes to nanofluids with free polymer additives, and depletion force is present between the NPs and non-absorbed polymers which results in depletion layer formation [212]. For depletion force, the concentration of free polymers is the influencer. When the polymer concentration is low, the NPs aggregate due to the attractive potential energy, whereas repulsive potential energy is strong with more free polymers developing more stable nanofluids.

Apart from the van der Waals, depletion, and steric stability forces, the Brownian force [213,214], buoyancy force [215], hydration force [216], and interphase resistance [217] may also contribute to the dispersion stability of nanofluids under specific conditions.

4. Tribo-Testing Methods

The American Society of Lubrication Engineers (ASLE) has compiled 234 apparatus for tribo-testing which are classified by their geometries. Tribometers that have been designed with advanced instrumentation are equipped with instruments that enable the measurement of coefficients of friction, friction forces, wear rates, noise, vibrations, and temperature of a system [218]. Several tribo-testing methods, including four-ball, pin-on-disk, ball-on-disk, ball-on-plate, ball-on-three-plates, and block-on-ring have been primarily used to evaluate the tribological performance of as-prepared water-based lubricants.

4.1. Four-Ball

A four-ball tribometer is one of the most common type of tribometers for analysing lubricant performance. Figure 11 shows a diagram of the four-ball tribometer. This structure consists of three balls fixed in a lubricant bath. The rotating ball is placed above the three fixed balls [218,219]. During the tribo-testing, a normal force is exerted pneumatically to the rotating ball at a constant speed, and the force is increased until the balls are welded together under the frictional heat. This tribometer can not only be used for measurement of COF and wear scar diameter (WSD), but also be applicable for assessment of load-carrying capacity and extreme pressure (EP) property. ASTM standard (D2783) is followed while conducting the tests. The EP characteristics can be evaluated by measurement of P_B (last non-seizure load, the last load at which the measured scar diameter is not more than 5% greater than the compensation value at that load) and P_D (weld point, the lowest applied load at which sliding surfaces seize and then weld). In principle, higher P_B and P_D values indicate better EP properties. It should be noted that the feasibility of evaluating EP properties is the unique feature of the four-ball test, which is independent of other types of tribo-testing configurations.

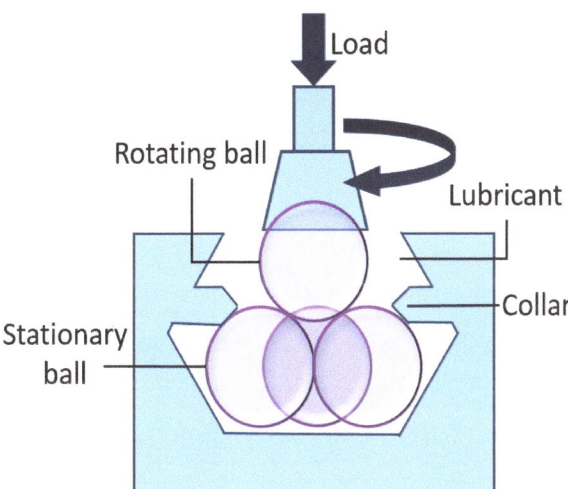

Figure 11. Four-ball tribometer.

The studies found in the literature using four-ball tribometer to evaluate lubricant properties are listed in Table 3. For instance, Zhang et al. used this tribometer to investigate the tribological properties of surface capped Cu [47] and Cu/SiO$_2$ [116] nanocomposite as additives in DW, and found that the water-based nanolubricants exhibited excellent load carrying capacity and reduced wear and friction. Furthermore, multiple studies have been conducted with TiO$_2$ NPs using four-ball tribometer to evaluate the tribological performance [49,52,102,103]. For example, Sun et al. [50] investigated the tribological behaviour of nano-TiO$_2$ water-based lubricant using the four-ball tribometer under 196 N load at a speed of 60 rpm for 30 min, which resulted in a significant reduction of wear by 30.6%, COF by 64%, with an optimal concentration of 0.7 wt.%. Additionally, with the increase of nano-TiO$_2$ concentration, the P_B increased. To be specific, for 0.1 wt.% nano-TiO$_2$ the P_B raised by 6.5%, ultimately reached 784 N at 5.0 wt.%. Meng et al. [52] also conducted tribo-testing with nano-TiO$_2$ using four-ball tribometer for 30 min at 1200–1760 r/min under 392 N. The test result indicated that adding nano-TiO_2 in lubricant assisted with the increase in P_B by 62.5%, along with the reduction in COF and wear by 33.8% and 47.4%, respectively. Many factors such as load, speed, optimal concentration, and temperature can be modified for testing tribological characteristics with various simulated circumstances.

4.2. Pin-on-Disk

The pin-on-disk tribometer shown in Figure 12a is used according to standard testing procedures. The arrangement consists of a fixed pin and a rotating disk. A circular sliding path is established on the rotating disk, while the load cells and sensors measure the frictional and tangential force generated by the fixed pin. The loading force is applied on the pin which is placed on the disk surface with a distance away from the disk centre. The disk is driven by a servo motor with a certain rotation speed (rpm). The applied pins are generally cylindrical in form, with flat, truncated, or spherical conical ends. The wear of pin can be measured by monitoring its dimensional changes, such as the length of pin, the WSD of the contact face, or by measuring weight loss. If the weight loss is too little to be weighed, it can be calculated or evaluated using wear area obtained from surface profile under a 3D microscope.

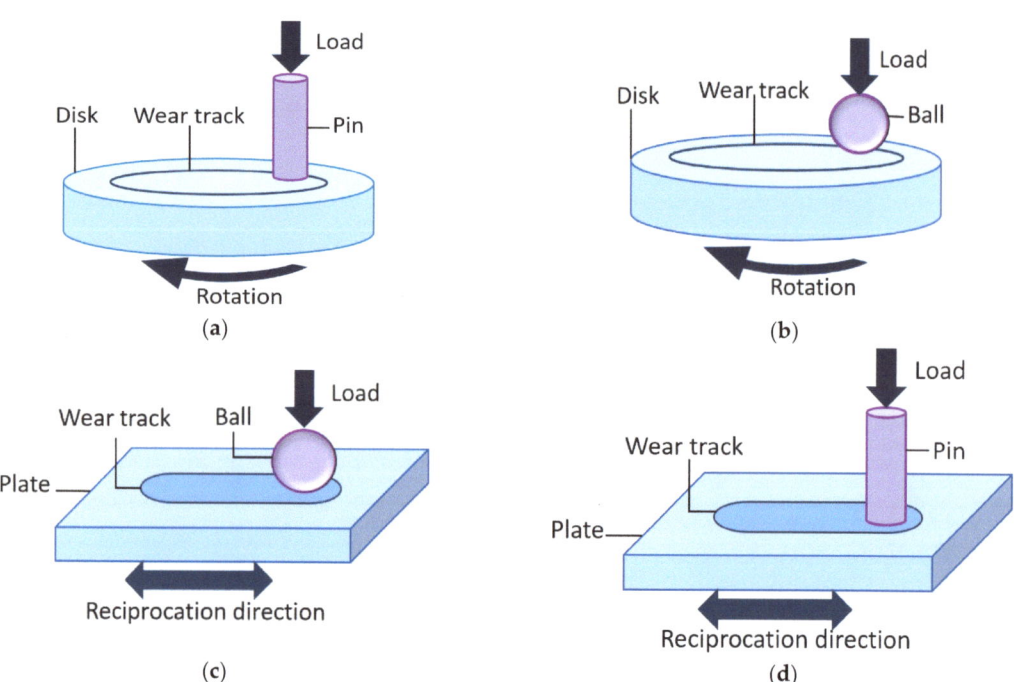

Figure 12. (**a**) Pin-on-disk, (**b**) ball-on-disk, (**c**) ball-on-plate, and (**d**) pin-on-plate.

In the field of tribology, this type of tribometers has been widely used in simulating practical working conditions. The pin-on-disk tribo-testing can be used for research on bearing systems, brake systems, train wheel systems, and manufacturing industries. This mechanism can be carried out under both the dry and lubricating conditions [218,219]. The pin-on-disk tribometer was used by He et al. [159] to evaluate the tribological properties of hexagonal boron nitride (h-BN) in base water at 300 rpm for 30 min under 400–600 N loads, which revealed an excellent reduction in WSD by 14.6% and COF by 29.1% with an optimal concentration of 0.7 wt.% h-BN. Zhao et al. [53] also used pin-on-disk tribometer to conduct 30 min tribological test at 50 mm/s using 0.05–0.2 wt.% cerium dioxide (CeO_2)

water-based nanolubricant. The results indicated 20% reduction in COF and 49% reduction in wear track depth. In a similar way, Elomaa et al. [139] conducted tribo-testing with graphene oxide (GO) water-based nanolubricant and found that the optimal parameters for significant friction and wear reduction can be achieved using 1 wt.% GO under 10 N.

4.3. Ball-on-Disk

The term pin-on-disk is substituted by ball-on-disk in the presence of the spherical cap end, which is also used extensively in tribo-testing configurations due to massive application of ball bearings in engineering. In this case, the parameters of the ball including diameter, Poisson's ratio, and elastic modulus are supposed to influence the Hertzian contact stress when contacting the disk. The initial experimental conditions, therefore, are quite different, which affects the final results of the tribological test. Normally, the working principles of the pin-on-disk and ball-on-disk arrangement are quite similar. The only difference is that the contact type is set by pin or ball. The ball-on-disk tribometer includes a ball and a circular flat disk, and a contact is established between these two surfaces of contact. The ball usually remains stationary when the flat disk rotates around its central axis, as shown in Figure 12b. A pre-set normal load is applied on the fixed ball, and a circular sliding path is generated on the disk after testing. Load cells and sensors can be used to measure the friction or tangential force generated between the fixed ball and the rotating disk. Usually, the tests are conducted in accordance with the standard test procedure (ASTM G99) [219].

There is a great demand for ball-on-disk tribometers to investigate the performance of water-based nanolubricants. For example, ball-on-disk tribometer was used by Radice and Mischler [48] to inspect the effect of Al_2O_3 NPs in aqueous solutions on the tribocorrosion behaviour of steel/alumina sliding surface under 4–10 N and 10–40 mm/s, which resulted in COF and wear rate reduction by 40–50%. Ball-on-disk tribometer has also drawn the attention of many researchers to analyse the tribological properties of nanoadditives including nanocomposite, graphene based, and ionic liquids. For example, tribological properties of carbon dots (CDs) [66] and poly ethylene glycol-graphene (PEG-G) [147] nanoadditives were evaluated under 10 N and a sliding velocity of 5 Hz. Superior friction-reduction and anti-wear properties were obtained with 0.05 wt.% PEG-G in water, indicating 39.04% wear rate reduction and 81.23% COF reduction. The use of CDs in water-based lubricants revealed outstanding tribological behaviour compared to pure water, with 39.66% COF reduction and 38% wear rate reduction. Table 3 contains a list of studies related to the use of ball-on-disk tribometers.

4.4. Ball-on-Plate

A ball-on-plate configuration is designed for sliding reciprocating motions (Figure 12c). Under certain conditions, the ball slides linearly along the plate. When compared to a ball-on-disk configuration, where the ball moves unidirectionally over a circular track, a ball-on-plate apparatus performs an alternating linear movement of the ball back and forth over the stationary plate at a constant speed [219]. The normal load is applied on the ball, and COF can be recorded against time during the sliding process. The wear track produced by reciprocating motion is much shorter than that produced by rotating motion for the same dimension of specimen under the same linear speed and testing duration. Therefore, ball-on-disk test is used more often than ball-on-plate test due to its higher efficiency for wear loss generation and more spots yielded for wear track observation.

There have been quite a few researchers who have used a ball-on-plate tribometer to characterise the tribological properties of graphene-based NPs. A reciprocating ball-on-plate tribometer was used by Xie et al. to evaluate the tribological properties of graphene/GO [77] and SiO_2/graphene [112] as nanoadditives in water-based lubricant for magnesium alloy sheets under 3 N normal load at a speed of 0.08 m/s for 30 min. Compared to pure water, 0.5 wt.% GO enabled 77.5% COF reduction and 90% wear rate reduction, while 0.5 wt.% graphene showed only 21.9% COF reduction and 13.5%

wear rate reduction. Furthermore, the best tribological properties were obtained using SiO_2/graphene (0.1:0.4) combination in water compared to only 0.5 wt.% graphene in water, presenting decreases in COF and wear volume by 48.5% and 79%, respectively. Moreover, black phosphorus quantum dots served as a high-efficient nanoadditive in water-based lubricant with an ultra-low concentration of 0.005 wt.%, not only exhibiting remarkable wear reduction by 56.4%, COF reduction by 32.3%, but also indicating an increment in load-carrying capacity from 120N to 300 N [92]. In recent years, ball-on-plate tribometer has been used consistently for various nanoadditivies including GO, nanodiamond, copper, ionic liquids, and naphthalene (listed in Table 3).

4.5. Ball-on-Three-Plates

The ball-on-three-plates tribometer is another device that researchers can use to examine the tribological properties of lubricants. Such a configuration is also called ball-on-pyramid. This device is composed of a spherical shaped ball and three plates that move independently in all directions. In order to efficiently distribute a normal load on the upper ball's three points of contact, the bottom plates need to be placed at 45° along the loaded axis. Results can be inaccurate when normal loads are unevenly distributed on three plates. It is possible to adapt the system to the desired material combinations since the balls and the plates are interchangeable [218,219]. Figure 13a shows a ball-on-three-plates tribometer that is commonly used for testing lubricants. He et al. [16] conducted studies to investigate the tribological behaviour of water-based Al_2O_3 nanosuspensions using a ball-on-three-plates tribometer under 10–40 N load at 20–100 mm/s sliding speed. Results indicated that, compared to water/glycerol solution, 1–2 wt.% Al_2O_3 (30 nm) nanolubricant presented the highest reductions in COF and wear mark by 27% and 22%, respectively. The authors also used this tribometer at a sliding speed of 50 mm/s under a normal load of 20 N to assess the tribological characteristics of GO sheets in water. By using 0.06 wt.% GO nanolubricant, vibration and noise in tribo-testing were minimised simultaneously, and the COF and WSD were reduced by 44.4% and 17.1%, respectively [17]. He et al. [18] also examined the tribological properties of pure GO and g-C_3N_4 nanosheets in water applying 10–35 N normal load with varying sliding speeds between 25–125 mm/s at 25 °C. With optimal concentration of 0.06 wt.% GO, g-C_3N_4 and g-C_3N_4/GO (1:1), reductions in COF by 37%, 26%, and 37% and WSD by 19.1%, 16.0%, and 19.6%, respectively, were observed. Thus, g-C_3N_4/GO presented better tribological performance than only GO and g-C_3N_4 in water under varying loads and speeds.

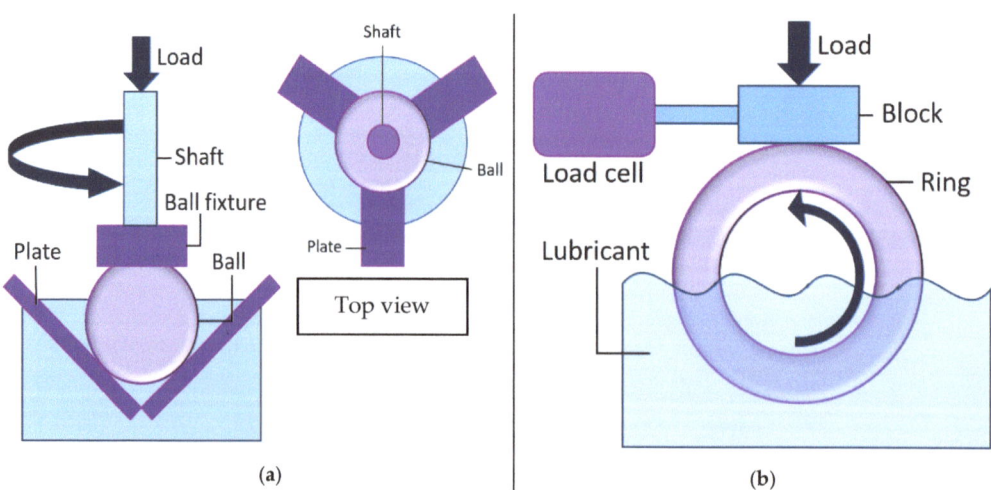

(a) (b)

Figure 13. (**a**) Ball-on-three-plates configuration and (**b**) block-on-ring configuration.

4.6. Block-on-Ring

In the block-on-ring setup, as shown in Figure 13b, the block is placed against the rotating ring under a predetermined load. The COF between the block and the ring is recorded when the ring rotates at a certain rate. Block-on-ring tribometers have been extensively used to study nanolubricants, coatings, and polymers. This type of tribometer configuration can evaluate the wear and COF between sliding surfaces in contact such as lubricating films, rings, and bearings.

Huang et al. [21] used a block-on-ring tribometer to study the role of GO, Al_2O_3, and GO-Al_2O_3 as nanoadditives in water-based lubricants under varying loads from 10 to 30 N and varying sliding speeds from 100 to 400 mm/s at 20–25 °C. Compared to individual 0.06 wt.% GO and 0.06 wt.% Al_2O_3 solutions, 0.12 wt.% GO-Al_2O_3 (1:1) lubricant showed significant decreases in COF by 47% and 64%, respectively, and surface roughness was improved by 60% and 63%, respectively. The authors conducted tribo-testing with 0.16 wt.% GO/SiO_2 water-based slurry under 20 N and 109 rpm, leading to a surface roughness (R_a) reduction by 35% and an increase in material removal rate (MRR) by 28% [19]. Huang et al. [12] conducted another study using a block-on-ring tribometer for testing water-based nanosuspension containing ZrO_2/TiO_2 NPs and GO with a normal load of 100 N and a sliding speed of 400 mm/s. The results demonstrated that the use of GO- ZrO_2/TiO_2 hybrid nanosuspension resulted in 65% surface roughness improvement and 25% reduction in COF. The studies using block-on-ring tribometers are summarised in Table 3.

4.7. Others

There are several tribometers with a wide range of testing capabilities used for different applications. For example, piston ring-on-cylinder line is used to analyse wear and friction phenomena in ring-piston pairs in combustion engines, compressors, and pumps, as shown in Figure 14a. Two-disk tribometers are used to measure relative displacement among two cylindrical surfaces to determine the wear on gears, rollers, bearings, and wheel systems [218,220], as shown in Figure 14b. Furthermore, a ring-on-ring tribometer is used to study the wear and friction of cylindrical tribological pairs, such as camshafts, clutches, and bearings, as shown in Figure 14c. Contact areas between rings typically vary based on their topology, which can be either tangential or concentric [218].

In recent months, a novel ultrahigh speed ball-on-disk tribometer with a sliding speed up to 50 m/s has been developed [221], which may be used to characterise the lubricants in a large speed range. This ultrahigh sliding speed can be achieved through a combined solution of on-line precision cutting and in situ dynamic.

In the tribological studies related to water-based lubricants, some researchers used other tribometers including ring-on-plate [222], ball-on-block, pin-on-cylinder [148,149], and 2-ball-plate tribometers [126] according to the testing parameters. In recent years, Ye et al. used ball-on-block tribometer to evaluate the tribological properties of multi-walled carbon nanotubes [129], stearic acid [96], and urea-modified fluorinated graphene [134] as water-based nanoadditives by analysing their anti-wear and friction-reduction properties.

Table 3 demonstrates a list of tribological studies summarised in the field of water-based nanolubrication using different tribometers.

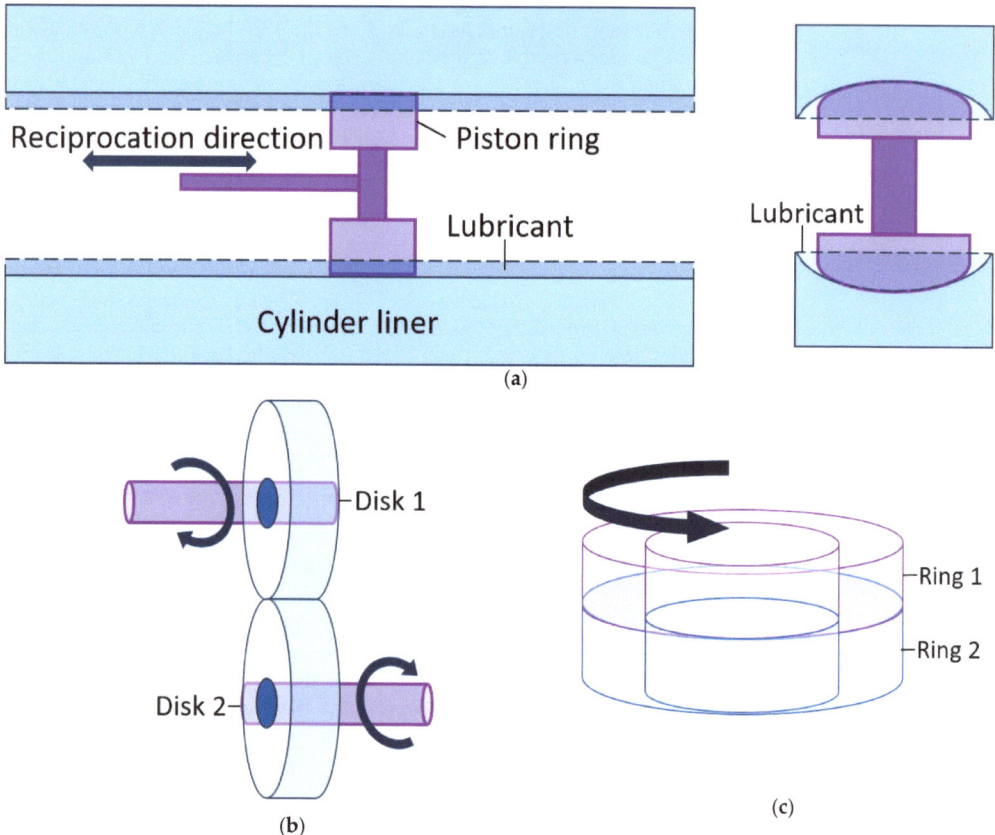

Figure 14. (**a**) piston ring-on-cylinder line, (**b**) two-disk, and (**c**) ring-on-ring configurations.

Table 3. A summary of different tribometers used for testing water-based nanolubricants.

Tribometer	Nanoparticle	Test Parameters			Wear Reduction	Testing Results		Reference
		Force	Speed	Temp. & Duration		Friction Reduction	Optimum Concentration	
	hBN	100 N	120–440 rpm	Room temp., 30 min	95.73%	60%	0.05 wt.%	Bai et al. [89]
	Capped Cu	-	1450 rpm	25 °C, 30 min	-	-	-	Zhang et al. [47]
	Cu-SiO$_2$	50 N	1450 rpm	25 °C, 30 min	37%	-	1 wt.%	Zhang et al. [116]
	hBN	392 N	1200 rpm	25 °C, 30 min	14.6%	29.1%	0.7 wt.%	He et al. [159]
	Novel C	0–7200 N	500 rpm	25 °C, 18 s	96%	76%	1.2–2.0 wt.%	Peña-Parás [65]
	MWCNT	-	1450 rpm	Room temp., 30 min	-	-	-	Peng et al. [125]
	CDs-IL	30–80 N	600 rpm	Room temp., 30 min	64%	57.5%	0.015 wt.%	Tang et al. [130]
	Fe$_3$O$_4$-MoS$_2$	294 N	0.479 m/s	30 min	29.7%	34.6%	-	Zheng et al. [86]
	LaF$_3$	100–900 N	1450 rpm	20 °C, 30 min	-	-	0.75–1 wt.%	Zhang et al. [154]
Four-balls	fullerene–styrene	200 N	1450 rpm	20 °C, 30 min	-	-	-	Lei et al. [150]
	fullerene–acrylamide	200 N	1450 rpm	20 °C, 30 min	-	-	-	Jiang et al. [151]
	GO-TiO$_2$	392 N	1200 rpm	20 °C, 30 min	-	-	0.5 wt.%	Du et al. [84]
	MoO$_3$	392 N	1200–1760 rpm	1800 s	-	-	0.4 wt.%	Sun et al. [56]
	MoS$_2$ and MoO$_3$	392 N	1200–1760 rpm	1800 s	-	-	0.3–0.5 wt.%	Meng et al. [63]
	Multilayer-MoS$_2$	588 N	1200 rpm	30 days	-	-	-	Zhang et al. [111]
	SiO$_2$	-	1760 rpm	Room temp., 10 s	-	-	0.3 wt.%	Bao et al. [59]
	Dual-Coated TiO$_2$	147 N	1440 rpm	-	34.8%	0.17%	1.6 wt.%	Gu et al. [103]
	OA–TiO$_2$	-	1450 rpm	25 °C, 30 min	-	-	0.5 wt.%	Gao et al. [102]
	Nano-TiO$_2$	196 N	60 rpm	30 min	30.6%	64.9%	0.7 wt.%	Sun et al. [50]
	Nano-TiO$_2$	200 N	1200–1450 rpm	Room temp., 30 min	-	-	0.5 wt.%	Kong et al. [49]

Table 3. *Cont.*

Tribometer	Nanoparticle	Test Parameters			Testing Results			Reference
		Force	Speed	Temp. & Duration	Wear Reduction	Friction Reduction	Optimum Concentration	
Four-balls	Nano-TiO$_2$	392 N	1200–1760 rpm	30 min	47.4%	33.8%	-	Meng et al. [52]
	Eu doped	392 N	-	60 min.	0.62–0.37 mm	0.083–0.065	0.5 wt.%	Liang et al. [123]
	Eu	-	-	45–55 °C, 2 h	0.62–0.35 mm	0.083–0.055	0.6 wt.%	Xiong et al. [57]
	hBN	400–600 N	300 rpm	25 °C, 30 min	14.6%	29.1%	0.7 wt.%	He et al. [159]
Pin-on-disk	Ceria	-	50 mm/s	Room temp., 30 min	49%	20%	0.05–0.2%	Zhao et al. [53]
	Cr$_2$O$_3$	20–150 N	50 mm/s	-	-	-	-	Cheng et al. [223]
	GO	10 N	0.02 m/s	21–23 °C, 30 min	-	57%	1 wt.%	Elomaa et al. [139]
	Two phase fluids	20 N	100 rpm	22 °C	-	~0.05	-	Pawlak et al. [224]
	Ag-C	1–9 N	100–500 rpm	Room temp., 30 min	40.4%	80.6%	1.0 wt.%	Song et al. [121]
	Polyalkylene Glycol	3 N	24 mm/s	Room temp.	-	Around 20%	0.5 wt.%	Wang et al. [94]
	Al$_2$O$_3$ (also disk on ball)	4–10 N	10–40 mm/s	-	40–50%	40–50%	-	Radice and Mischler [48]
	C dots	10 N	-	Room temp., 1 h	38%	39.66%	-	Hu et al. [66]
Ball-on-disk	CQD	2 N	150 cycles/min	Room temp., 12 min	-	30%	-	HuaPing et al. [131]
	Urea modified C	3–7 N	200–400 rpm	30 min	96.70%	80.86%	0.15 wt.%	min et al. [127]
	Hexagonal BN	5.64 N	10.2 mm/s	Room temp, 30 days	-	-	-	Cho et al. [88]
	DDP-Cu	1–4 N	-	25 °C, 30 min	60.5%	45.5%	0.2–0.4 wt.%	Liu et al. [117]
	Diamond	-	80 mm/s	30 °C	88%	70%	2 wt.%	Mirzaamiri [137]
	γ-Fe$_2$O$_3$	4 N	0.20 m/s	Room temp.	-	-	0.6 wt.%	Pardue et al. [55]
	GO/Chitosan	100 N	-	-	47%	84%	-	Wei et al. [145]

Table 3. *Cont.*

Tribometer	Nanoparticle	Test Parameters				Testing Results		Reference
		Force	Speed	Temp. & Duration	Wear Reduction	Friction Reduction	Optimum Concentration	
	Graphene quantum dots	100 N	-	Room temp., 60 min	58.5%	42.5%	-	Qiang et al. [135]
	GO-MoS$_2$	0.5–3 N	60 rpm	25 °C	-	50%	-	Liu et al. [225]
	FGO	5 N	300 r/min	30 min	88.1%	41.4%	0.7 wt.%	min et al. [72]
	GO	5–20 N	0.005–0.1 m/s	Room temp	68%	78.5%	0.1 wt.%	Singh et al. [140]
	GO-OLC	2–10 N	200 rpm	Room temp	-	-	0.06 wt.%	Su et al. [80]
	Nanofilm GO	2 N	12 mm/s	25 °C, 60 min	79.7%	43.6%	-	Li et al. [146]
	SiO$_2$-GO	10 N	-	25–35 °C	78.3%	-	0.05 wt.%	Guo et al. [83]
	PEGlated graphene	10 N	-	30 min	81.23%	39.04%	0.05 wt.%	Hu et al. [147]
	Hydroxide	2N	0.024 m/s	25 °C, 45 min	43.2%	83.1%	0.5 wt.%	Wang et al. [93]
	Al$_2$O$_3$-WS$_2$-MoS$_2$	10 N	320 rpm	Amb. Temp.	23.4%	53.89%	-	Kumar et al. [119]
Ball-on-disk	Black phosphorus	10–70 N	-	-	97.1%	25%	-	Wang et al. [162]
	BP	8–15 N	150 r/min	30 min	61.1%	32.4%	-	Wang et al. [163]
	Si$_3$N$_4$	15, 30, 60 N	0.25 m/s, 0.5 m/s	27 °C, 3600 s	-	-	-	Lin et al. [161]
	Ti$_3$C$_2$	3–10 N	120 rpm, 0.126 m/s	24–26 °C, 1 h	48%	20%	5 wt.%	Nguyen and Chung [91]
	TiO$_2$	5 N	50 mm/s	25 °C, 30 min	-	16.3%	0.4–8.0 wt.%	Wu et al. [14]
	TiO$_2$	20–80 N	50 mm/s	10 min	70.5%	84.3%	4 wt.%	Wu et al. [25]
	NaCl saline	10–100	50 mm/s	1 h	-	-	3.5 wt.%	Wu et al. [226]
	ZnO and Al$_2$O$_3$	10 N	100 mm/s	-	-	56.9%	-	Gara and Zou [122]
	Ceramics	30 N	0.5 m/s	Room temp., 3600 s	54.0%	78.8%	-	Cui et al. [109]
	Chitosan	5–30 N	12–36 mm/s	25 °C, 30 min	69%	40%	0.3 wt.%	Li et al. [95]

Table 3. *Cont.*

Tribometer	Nanoparticle	Test Parameters				Testing Results			Reference
		Force	Speed	Temp. & Duration	Wear Reduction	Friction Reduction	Optimum Concentration		
Ball-on-disk	Individual additives	3 N	20 mm/s	Room temp. 1 h	-	12%, 30%	0.05%, 0.1%		Tomala et al. [203]
	Hard C microsphere	100–300 mN	10 mm/s	30 min	-	-	0.1 wt.%		Wang et al. [124]
	Cu	1–4 N	0.02 m/s	22 °C, 30 min	85–99.9%	80.6%	0.6 wt.%		Zhao et al. [46]
	CuO	-	20 mm/s	22 °C, 30 min	72.6–89.1%	43.2–52.2%	0.8 wt.%		Zhao et al. [54]
	Nano diamond	1 N	360 rpm	25 °C, 30 min	-	40%	-		Jiao et al. [68]
	Graphene and GO	1–8 N	0.08 m/s	30 min	13.5%	21.9%	0.5 wt.%		Xie et al. [77]
	Fluorinated GO	20 N	4 mm/s	Room temp., 2000 s	47%				Fan et al. [73]
	MGO	5–25 N	-	Room temp., 3000 s	74%				Gan et al. [142]
Ball-on-plate	GO-ND	0–1 N	0.4 mm/s	25 °C, 1800 s	-	-	0.1 wt.% GO, 0.5 wt.% ND		Wu et al. [78]
	GO-MD	-	-	250 s	-	0.6–0.01	0.7 wt.% GO, 0.5 wt.% MD		Liu et al. [79]
	Graphene water-based	10 N	0.01 m/s	-	-	-	0.1 wt.% graphene flakes. 1% wt.% graphite		Piątkowska et al. [136]
	Graphene-SiO$_2$	3 N	0.08 m/s	Room temp., 30 min	79%	48.5%	0.5 wt.%		Xie et al. [112]
	Monolayer GO	1.88 N	0.5 mm/s	-	Marginal after 60,000 cycles	~0.05 after 60,000 cycles	0.01 wt.%		Kinoshita et al. [138]
	Oxide graphene	10 N	120 rpm	10 min	-	-	<0.1 wt.%		Song and Li [70]
	Metal doped CDs	40–500 N	-	20–120 min	Up to 43.1%	Up to 73.5%	1.0 wt.%		Tang et al. [227]
	Reduced GO	50–200 N	4mm/s	-	70 μm after 100,000 cycles	Around 0.1 after 100,000 cycles	0.01 wt.%		Kim and Kim [143]

Table 3. *Cont.*

Tribometer	Nanoparticle	Test Parameters			Testing Results			Reference
		Force	Speed	Temp. & Duration	Wear Reduction	Friction Reduction	Optimum Concentration	
	PEI-RGO	-	9000 r/min	-	45%	54.6%	0.05 wt.%	Liu et al. [74]
	Protic ionic (PILs)	2-4 N	-	30 °C, 30 min	85%	80%	1 wt.%	Kreivaitis et al. [158]
Ball-on-plate	MoS$_2$	20 N	-	25 °C	-	-	0.1 wt.%	Wang et al. [64]
	Naphthalene	100 N	1475 rpm	30 min	-	-	-	Yang et al. [153]
	BPQDs	40–300 N	10 mm/s	30 °C, 20–120 min	56.4%	32.3%	0.005 wt.%	Tang et al. [92]
	CNT/SiO$_2$	5 N	120 rpm	10 min	66.4%	66.4%	0.5 wt.%	Xie et al. [113]
	rGO	20 mN	4 mm/s	-	-	12 times	5 µL/min	Kim et al. [228]
Ball-on-three-plates	Alumina	10–40 N	20 to 100 mm/s	10 min	22%	27%	2 wt.%	He et al. [16]
	g-C$_3$N$_4$/GO	10–35 N	25 to 125 mm/s	25 °C	19.6%	37%	0.06 wt.%	He et al. [18]
	pH-GO	20 N	50 mm/s	-	17.1%	44.4%	0.06 wt.%	He et al. [17]
	MR fluid	0.5 N	1.18 m/s	2–10 min	-	-	1 vol%	Rosa et al. [229]
	Novel C	245 N	300 rpm	1200 s	96%	76%	2 wt.%	Peña-Parás [65]
Block-on-ring	GO-Al$_2$O$_3$	10 to 30 N	100 to 400 mm/s	20–25 °C, last 7 min	-	47–64%	0.06 wt.%	Huang et al. [21]
	ZrO$_2$/TiO$_2$	100 N	400 mm/s	-	65%	25%	-	Huang et al. [12]
	GO-SiO$_2$	20 N	109 rpm	Ambient Temp.	-	-	0.16 wt.%	Huang et al. [19]
Ring-on-plate	Alkyl glucopyranosides (AGPs)	50 N	0.1 m/s	Room temp, 1 h	-	>95%	-	Chen et al. [222]
Ball-on-block	MWCNT	50 N	-	30 min	66%	-	-	Ye et al. [129]
	urea-modified FG		-	-	-	-	-	Ye et al. [134]
	Stearic acid	-	-	30–500 °C	57–90%	68–83%	-	Ye et al. [96]
Piston ring-on-cylinder	Cellulose	50 N	130–300 rpm	Room temp.	>50%	~75%	2 wt.%	Shariatzadeh and Grecov [148,149]
2 ball-plate	Px-CNTs	5 N	120 rpm	10 min	-	66.4%	0.5 wt.%	Sun et al. [126]

5. Lubrication Mechanism

Due to small sizes and high surface areas, NPs have unique properties that are different from bulk materials. Hence, research groups have focused on the addition of NPs to lubricant dispersions for enhancing the thermophysical and tribological properties [230,231]. Scientists have been committed to understanding the roles of nanoadditives in base lubricants and the lubrication mechanism of synthesised nanolubricants in terms of friction-reduction and anti-wear properties. Numerous mechanisms, including rolling effect, protection film, mending effect, polishing effect, and synergistic effect [232], have been proposed to explain the enhancement of the lubricity.

5.1. Rolling/Ball Bearing Effect

According to the theory of rolling/ball-bearing effect, nanolubricants have outstanding lubricity due to two factors. First, NPs are spherical and they can act as ball bearings under friction. Second, the NPs flatten and create a sliding system between the two friction surfaces, eventually reducing friction and wear [233]. Figure 15a illustrates the rolling/ball bearing mechanism. NPs with a sphere, quasi-spherical, or hemispherical shape play this role. Generally, spherical NPs reduce the COF by converting sliding friction to rolling friction as a result of their morphological properties [233–235]. Some scholars have revealed that these NPs themselves might roll or embed into surfaces [236–238]. Moreover, many scholars have mentioned the ball bearing effect in lubrication mechanism analysis of the water-based nanoadditives including metal and non-metal oxide [54,59,100], composite [86,119], and many more [68,78,92,124,137,159]. In the past few years, it has been noted that due to their spherical shape, metal and non-metal oxides such as SiO_2 [59,100] and TiO_2 [15,25,29] generate a ball bearing effect between sliding surfaces, converting the friction mechanism from sliding to rolling and thus causing friction reduction. NPs will maintain their shape and stiffness under mild conditions [239–241].

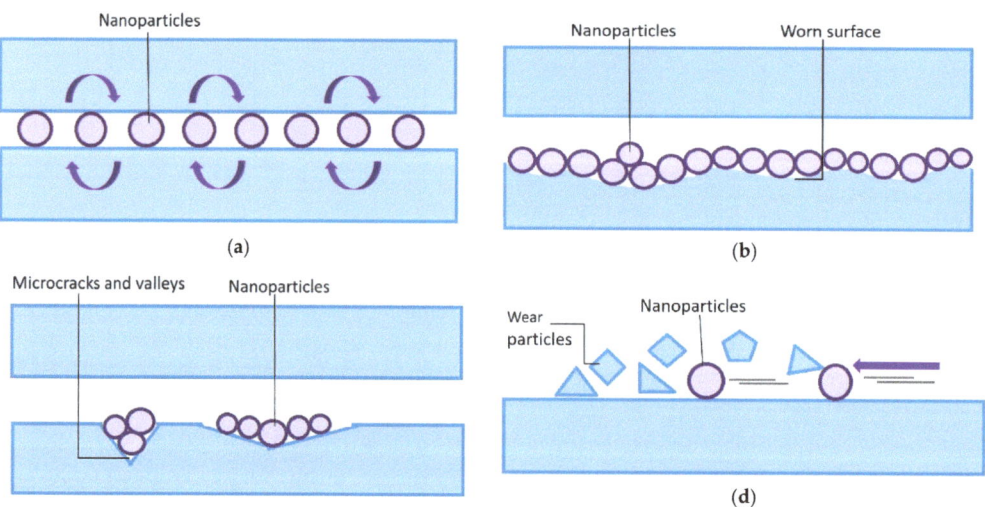

Figure 15. (a) Rolling/ball bearing mechanism; (b) protective film/tribo-film formation; (c) mending effect; and (d) polishing/smoothing effect.

On the other hand, stable spherical NPs can enhance bearing capacity and performance under maximum pressure [242,243]. The ball bearing effect of spherical NPs is influenced by the film thickness, and NPs with a diameter close to the film thickness will retain their shape. A transfer film is generated when the film's thickness is less than the NPs' diameter [59]. During the hot rolling test, Wu et al. [25,29] found that the rolling effect of spherical shaped nano-TiO_2 assisted in reducing wear and friction to a large extent, thereby decreasing the roll roughness and rolling force.

5.2. Protective Film/Tribo-Film

The idea behind this method is the formation of a thin protective film consisting of NPs on the area of frictional contact [236]. The protective film refers to coating the friction surfaces by separating them from direct contact, and it is derived from three aspects. First, the NPs are able to be deposited or adsorbed on the rubbing surfaces to form a boundary film. Second, the NPs are melted as a protective film under friction heat to cover the friction surfaces. Third, the formation of the protective film is produced by tribo-sintering of the NPs, and the compacted and smooth tribo-film is conducive to the decreases in friction and wear. Figure 15b shows the schematics of protective film/tribo-film formation.

It is believed by scientists that the nanofluids have better lubricity than the base lubricant because the NPs can form a protective film over friction surfaces and thereby prevent the friction surfaces from sliding [244–246]. Multiple studies have mentioned the significance of protective film formation using NPs such as graphene based [73,75,77,134,145,147], ionic liquids [155–157], and phosphorus [162]. Some other NPs such as metal oxides [27–29,56–58], diamond [137], and composites [19,113,115,123] are also able to form a protective film on friction surfaces. Based on previous studies, it is concluded that the protective film not only reduces the friction and wear between friction pairs but also helps in isolating air, thus leading to reduced oxide scale thickness during hot steel rolling [27–29]. It is also found that the enhanced wettability of nanofluids caused by the addition of NPs promotes the formation of protective film.

5.3. Mending Effect

Mending effect is very important in moving mechanical parts that are repeatedly loaded and might fail from the crack, as there is the possibility that cracks can spread outside the contact area to other parts of the metal body. Using this technology, the metal surface can become hardened by sintering and filling NPs into macroscopic cracks [236], as shown in Figure 15c. A mending effect has been observed in numerous studies [16,21,27,66,92,117,121]. Many results reveal that NPs act with a mending effect by filling in the surface defects, thereby improving the surface quality [247–249]. Bao et al. [59,100] used SiO_2 NPs to improve the tribological performance of water-based lubricants through the mending effect, which contributed to the decreases in scratches and pits on friction surface with enhancement of anti-wear and friction-reduction properties. To compensate for weight loss, it is particularly important to deposit NPs on the interaction surface [11,59,171,234,249].

Scientists proposed the roles of NPs to self-heal or repair surfaces by depositing them in grooves or scars, which can prevent further wear [250]. Tribo-sintering may occur in the mending process when the frictional heat generated is strong enough to melt the scattered NPs, thus causing permanent deposition of the NPs on the worn surface [219,233]. It is reported that soft metals with a face-centred cubic structure are generally capable of self-repairing [233]. However, the mending effect does not simply include the deposition of the NPs on the friction surface. The melting point of NPs falls significantly as their size decreases. These NPs are thus easy to melt at a high friction surface temperature, thus forming a uniform filler and tightly bonding to the friction surface [233].

5.4. Polishing/Smoothing Effect

The polishing effect can be observed prior to and after using lubricants that contain nanoadditives. The surface roughness is decreased because of NPs deposition on the surface profile [251]. In the polishing/smoothing effect, NPs are used to minimise the roughness of the lubricated surface by abrasion treatment [252], as shown in Figure 15d. NPs deposited on the hollow contour provides smoothness to the metal surface. A number of studies demonstrated that the polishing effect can be contributed by some types of NPs such as phosphorus [92], metal oxide [27–29,100], and nitride [159]. Under high pressure and speed, NPs flatten the surface peaks to make a smooth tool surface, which in turn greatly improves the surface quality of workpiece.

Nanofluid is an ideal polishing substance that mechanically polishes the friction surface. The tribo-pair contact area increases when harder nanoparticles are used to polish a smoother surface, which reduces friction and meanwhile increases the load-carrying capacity of the nanolubricant. However, the mechanical properties of NPs are poor on a very rough surface. Due to their small size, NPs can only restore rough surfaces at an atomic scale. In other words, a mechanical polishing effect is more evident when the surface of tribo-pair is relatively smooth. In addition to the polishing effect, NPs can be deposited easily in microcracks and valleys as per the mending effect, which also produces smooth surfaces [233].

5.5. Synergistic Effect

Lubricants and modified friction surfaces are often combined with nanoadditives to generate a synergistic effect. Most of the time, NPs perform their lubrication effects through not just one mechanism, but rather a combination of various mechanisms. Furthermore, lubrication conditions can change when various mechanisms are modified. Research on lubrication mechanism analysis has always been focused on the roles of NPs in base lubricants. It is also important to consider the interaction among the NPs, the dispersant, and the base lubricant [233].

Synergistic lubrication effect may improve the tribological performance of nanolubricants by integrating two or more mechanisms, and this effect is usually contributed by composite nanoadditives [21,78,84,115,118,121,225]. A synergistic effect is also reported for some other nanoadditives such as metal carbide [90], nitrides [159,160], and oxides [73,96,108,130,146,147,154], and these nanoadditives themselves play multiple roles in combining two or more lubrication effects together for enhanced lubricity.

5.6. Exfoliation

For the NPs with layered structure, the lubrication mechanisms are different from those of spherical NPs. Tevet et al. [253] pointed out three main lubrication mechanisms in terms of fullerene-like (IF) NPs which have layered hollow polyhedral structure. First, the IF NPs may act as a spacer between the friction pairs, which supplies low frictional force during sliding process, as shown in Figure 16a. Second, the IF NPs can also behave as ball bearings that roll on the friction surface, as shown in Figure 16b. Third, the IF NPs can be exfoliated into nanosheets under certain shear force. The evolved nanosheets are deposited on the asperities of friction surface and thus supply an easy shearing for the subsequent friction process, as shown in Figure 16c. The "exfoliation" is typical lubrication mechanism for layer-structure NPs dispersed in base lubricant.

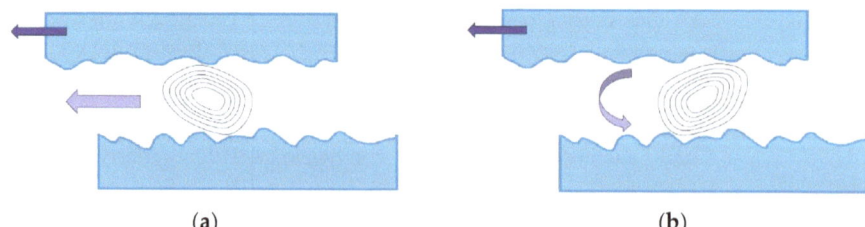

(a)   **(b)**

Figure 16. *Cont.*

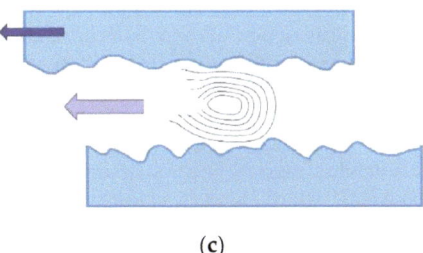

(c)

Figure 16. Schematic of lubrication mechanism for the lubricants containing nanoadditives with layered structure: (**a**) sliding; (**b**) rolling; and (**c**) exfoliation [253].

IF NPs are inorganic and layered compounds such as fullerenes and carbon nanotubes, with spherical or tubular structures. These concentric layered materials are often solid, though hollow sometimes, and are arranged in an onion shape [254]. This onion morphology is spherical outside and lamellar inside. It is more likely that the onion may have spherical morphology if it is stable, otherwise it may exfoliate and turn into a sheet [255]. One of the benefits of onion structure is the sphere-like shape and the lack of dangling bonds [256]. Su et al. [80] evaluated the lubrication performance of onion-like carbon (OLC) NPs and graphene oxide (GO) sheets, and reported that due to its onion-like structure, OLC may reduce friction and wear in water, producing rolling motion and tribo-film formation during sliding. Due to the protective film formation between adjacent wear surfaces, the 2D structure of GO provided better sliding and shear between adjacent wear surfaces, exhibiting excellent lubricity. Another layer-lattice-structured NPs, hexagonal boron nitride, has also exhibited superb friction-reduction properties, probably through a similar mechanism [246,257].

5.7. Hydration Lubrication

In water-based lubricants, regardless of whether NPs are added or not, water is absolutely the largest component. The water molecule, due to its positive (H atoms) and negative charges (O atom), interacts strongly with charged ions or zwitterions to form stable hydration layers (thin water films) in aqueous media [258]. The hydration layers form hydrated charges between sliding surfaces, which can sustain large pressures due to the reluctance of the hydration water to be squeezed out and meanwhile generate a fluid response to shear [259]. The hydration lubrication mechanism enables a significant friction reduction between surfaces which expose or slide across such hydrated layers under low shear stresses. This striking lubrication mechanism is expected to provide new insight into the boundary lubrication processes in water-based lubricant.

Although these mechanisms are well established, it has always been a matter of debate among researchers. In fact, it has been difficult for any single theory to fully explain the lubrication mechanism in water-based nanolubricants. In this case, further research is needed to establish a more accurate theory of nanolubrication.

6. Application of Water-Based Nanolubricants in Metal Rolling

6.1. Physicochemical Properties of Applied Lubricants

Traditional water-based lubricants are often restricted in practical metal rolling due to their low viscosity, poor wettability on roll surface, and harsh corrosion of base water, despite their eco-friendliness, low-cost, noninflammability, and outstanding cooling performance. The physicochemical properties of applied lubricants including viscosity, wettability, and corrosivity are crucial to the lubrication performance.

Viscosity is a measurement of the tendency of liquid to resist flow, and it is defined as the ratio of the shear stress to the shear rate [82]. The liquid is known to be Newtonian or non-Newtonian when the viscosity is constant or varies, respectively, at different values of shear rate [260]. The viscosity of lubricant is dependent on many factors, and it influences lubrication performance in most cases. It has been reported that the viscosity of base lubricant increases with the addition of NPs [25,261,262], and it also increases with the increase of their mass fraction [173]. In addition, the viscosity decreases with the increase of temperature [263], and it is a function of pressure throughout the inlet zone in metal rolling [264]. Most importantly, a higher viscosity leads to a lower COF due to a transition of lubrication regime from boundary lubrication into mixed or hydrodynamic lubrication [241], which indicates an increase in the thickness of lubricating film as per the well-known Stribeck curve [4]. This increased film thickness restrains the work rolls and workpiece from direct contact, leading to decreased friction in the contact zone.

Wettability is one of the most important lubricant characteristics, which reveals how well a lubricant can wet a solid surface [265]. Wetting of surfaces in contact is of vital importance to decrease friction and wear, which relates directly to the decreases of rolling force and roll wear in metal rolling [29,266]. This result is ascribed to the phenomena that excellent wettability facilitates the formation of a lubricating film [112], and also helps retain the effective lubricants on work roll surfaces [28]. In general, wettability can be characterised by the measurement of contact angle on a solid surface, and a smaller contact angle indicates a better wettability [267]. It has been reported that the wettability of base lubricant can be significantly improved by adding NPs, and the addition of dispersant or surfactant also enables improved wettability due to their functional groups [25].

Corrosivity of water-based lubricants often plays a negative role in aggravating the surface quality of rolled products, especially for ferrous materials. Water itself and pH values of the lubricants both determine the corrosion effect of as-prepared water-based lubricants on metal surfaces. It has been proved that the corrosion resistance of water can be enhanced by adding water-soluble additives. For example, nanoadditives such as fluorinated graphene [72], surface-modified CuS NPs [62], and GO-TiO$_2$ [84] are able to retard the corrosion of steel by forming a protective film on the steel surface. Corrosion inhibitors such as Triethanolamine (TEA) [92] and sodium dodecyl benzene sulfonate (SDBS) [63], have also been used to minimise the corrosivity of water-based lubricants. On the other hand, water-based lubricant with a pH value of 7 and above leads to insignificant corrosion on the rolled surface [76]. Nevertheless, the optimal pH values in the lubricants should be determined with the consideration of their effects on the dispersion stability of the lubricants. In some cases, the corrosion resistance of water-based lubricants may be negligible, especially in hot rolling of steels and cold rolling of non-ferrous metals such as aluminium and magnesium. However, special attention should be given to the corrosion effect of water-based lubricants on cold rolling of ferrous metals.

6.2. Hot Rolling of Steels

Hot rolling of steels is applied to obtain not only the required dimensions and mechanical properties, but also satisfying surface finish [268]. Friction and wear are unavoidably generated between the work rolls and the workpiece, which results in increased consumption of energy and damage of work rolls [269,270]. Water-based nanolubricants are emerging to substitute the traditional oil-containing lubricants to resolve these issues with environmental concern. The use of water-based nanolubricants in hot steel rolling also exhibits a great potential in reducing the thickness of oxide scale, improving the surface quality of rolled products, and refining the grains in rolled steel. The water-based nanolubricants can also act as better coolants of work rolls than conventional cooling water due to their enhanced thermal conductivity [271], which further prolongs the roll service life by replacing the cooling water using a lubricant supply system. In view of these, energy consumption of the rolling mill can be lowered, and rolling high-strength steels with heavy reduction can thus be readily achieved within the limits of mill load. Meanwhile, the roll changing frequency would be reduced, and the mill configuration would be simplified, thus leading to increased productivity and decreased operation cost. The overall quality and yield together with the properties of rolled products are therefore greatly enhanced.

As the hot rolling process involves harsh working conditions such as high temperature and pressure, special requirements are placed on the physicochemical properties of the nanoadditives and other chemical additives. The applications of different types of water-based nanolubricants in hot rolling are comprehensively reviewed in the subsequent sub-sections in terms of their effects on rolling force, surface morphology, oxidation behaviour of steel, and microstructure obtained after rolling.

6.2.1. Rolling Force

In general, rolling force data during hot steel rolling are measured through two individual transducers placed at both the drive and the operation sides over the bearing blocks of the top work roll. The presence of lubricants between the work rolls and the workpiece enables a decrease in friction and hence the rolling force is lowered by up to 25% [272]. The value of rolling force is one of the key indicators that evaluates the energy expenditure during hot steel rolling.

TiO_2 water-based nanolubricants used in hot steel rolling were first reported in [273]. The rolling force obtained using the lubricant with 2% anatase TiO_2 decreased in each pass, and the total decrease reached up to 20%, in comparison to that obtained under dry condition. However, there was a lack of experimental evidence to underpin the analysis of lubrication performance, and the role of TiO_2 NPs was not well understood. The effects of TiO_2 concentration and rolling parameters on the rolling force were not discussed either. In light of this, single-pass hot steel rolling tests were conducted to investigate the influences of TiO_2 concentration and rolling temperature on the rolling force [29]. It was found that the use of 4% TiO_2 lubricant led to the lowest rolling force at rolling temperatures of 850 and 950 °C, while the rolling force obtained under 1050 °C did not vary significantly under all the lubrication conditions (see Figure 17a). The mechanisms of the decrease in rolling force were dominated by rolling and mending effects together with the formation of protective film, and they were demonstrated through cross-sectional SEM-EDS analysis. The hot rolled steel samples observed under SEM are, in fact, inevitably involved in grinding and polishing processes, which may affect the distribution of TiO_2 NPs and therefore the understanding of corresponding lubrication mechanisms. To overcome this drawback, a focused ion beam (FIB) foil was cut from the surface of rolled steel, and then observed under TEM to identify the NP distribution through EDS mapping [28]. The synergistic effect of lubricating film, rolling, polishing, and mending was thus confirmed to contribute to the predominant lubrication mechanisms. The effect of work roll roughness on the rolling force was also examined, which has been neglected by the majority of researchers in the field of hot steel rolling [25]. This revealed a significant research finding that the continuous use of TiO_2 water-based nanolubricants was inclined to enable a successive decrease in rolling force up to 8.3% due to the polishing on work roll surface (see Figure 17b). Some other researchers applied composite nanomaterials such as MoS_2-Al_2O_3 as the nanoadditives in water, and the average rolling force within five rolling passes was reduced by 26.9% compared with the base fluid [118]. Another notable finding was that the effect of MoS_2-Al_2O_3 lubricant on the decrease in rolling force was superior to that of the lubricant with individual nanoadditive, owing to the synergistic effect of the composite nanoadditives.

Although significant decrease in rolling force has been achieved in [25,28,29,118,273], the nanoadditives used in these lubricants all had particle sizes less than 100 nm, which brought forth extremely high material cost especially when large quantities of lubricants were applied, let alone the combination of two or more nanoadditives. Beyond this, the nanolubricants have always been prepared using complex chemical agents, followed by subsequent processes including ultrasonic treatment, which led to extra production cost. All these disadvantages have greatly restricted the application and popularisation of water-based nanolubricants in industrial-scale hot steel rolling. Accordingly, relatively coarse TiO_2 NPs (~300 nm in diameter) were adopted to replace the expensive nanosized particles in water-based lubrication formula [24]. Novel dispersant (SDBS with a linear structure) and extreme pressure agent (Snailcool) were added to compensate the degradation of lubrication performance caused by coarsened TiO$_2$ NPs. These TiO$_2$ water-based nanolubricants were prepared using mechanical agitation without applying ultrasonic treatment. Nevertheless, the rolling force could be decreased up to 8.1% in a single-pass hot rolling at 850 °C when using 4% TiO$_2$ lubricant.

An overview of the typical representatives of the use of water-based nanolubricants in decreasing rolling force is presented in Table 4. The total decrease in rolling force is expected to be more significant upon the use of lubricants in multi-pass hot steel rolling.

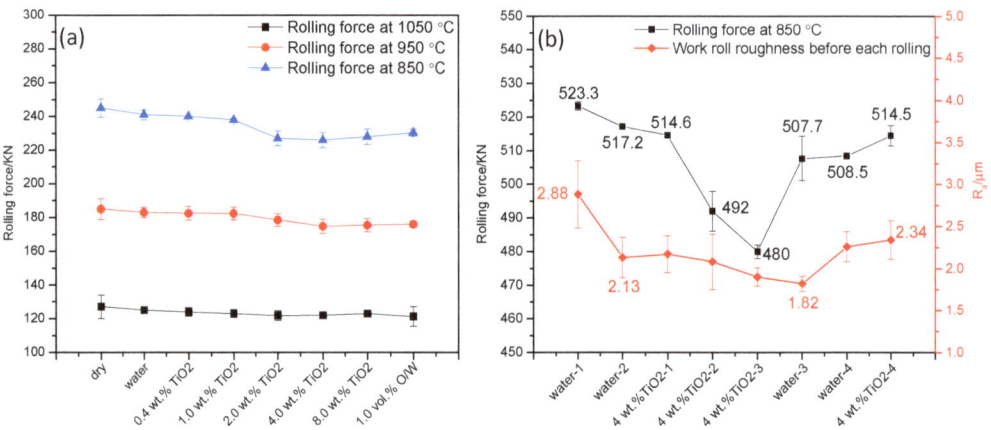

Figure 17. (**a**) The rolling force obtained in each single pass under different lubrication conditions at 1050, 950, and 850 °C [29]; (**b**) the rolling force varying with the work roll roughness at 850 °C under water and 4 wt.% TiO$_2$ lubrication [25]. (Rolling reduction of 30% in each single pass; rolling speed of 0.35 m/s).

Table 4. The decrease in rolling force using different types of water-based nanolubricants under certain rolling parameters.

Workpiece & Dimensions	Lubricant	Benchmark	Rolling Temp.	Rolling Reduction	Rolling Speed	Decrease in %	Ref.
Mild steel with 30 mm in thickness	2% anatase TiO$_2$, SHMP	Dry condition & Water	~950–750 °C	~81.8% in five passes	-	Up to 20% in the final pass	[273]
Mild steel 300 × 50 × 8 mm^3	0.4–8% TiO$_2$, 0.004–0.08% PEI, 10% glycerol	Dry condition & Water	~1050–850 °C	~30% in one pass	0.35 m/s	Up to 8%	[29]
Mild steel 300 × 100 × 8 mm^3	1–8% TiO$_2$, 0.01–0.08% PEI, 10% glycerol	Dry condition & Water	~850 °C	~30% in one pass	0.35 m/s	Up to 6.8%	[28]
Mild steel 300 × 91 × 8.5 mm^3	2% & 4% TiO$_2$, 0.2% & 0.4% SDBS, 10% glycerol	Water	~850 °C	~30% in one pass	0.35 m/s	Up to 8.3%	[25]
Mild steel 100 × 70 × 30 mm^3	MoS$_2$-Al$_2$O$_3$, glycerol, TEOA, SDBS, and SHMP	Base fluid	~1000–800 °C	~86.7% in five passes	1 m/s	Up to 26.9%	[118]
Mild steel 300 × 100 × 12 mm^3	2% & 4% TiO$_2$, 0.1% & 0.2% SDBS, 10% glycerol, 1% Snailcool	Water	~850 °C	~27% in one pass	0.35 m/s	Up to 8.1%	[24]

6.2.2. Surface Morphology of Rolled Steel

Surface morphology of hot rolled steel strips, i.e., the surface topography of oxide scale, can be characterised in terms of surface roughness measurement and microscopic observation, which is directly related to the assessment of surface quality. In the case of pickle-free as-hot-rolled steel strip, in particular, the surface roughness plays an important role in the downstream processing such as sheet metal forming, coating, and stamping [274,275]. It has been of great interest in recent years to use TiO_2 water-based lubricants to improve the surface morphology of as-hot-rolled steel strips [24,25,28,29,52]. Among these studies, a typical example is the use of 4% TiO_2 lubricant dispersed with PEI and glycerol, which yielded the smoothest strip surface after rolling according to the 3D surface morphologies (see Figure 18) [28]. The surface roughness of the rolled strip under dry condition can thus be improved by up to 19.5% (see Figure 19). Some other researchers used SiO_2 NPs (<0.5%) as nanoadditives in the base lubricant to improve the surface morphology of hot rolled strips [100]. The main mechanisms of the decrease in surface roughness were derived from mending and polishing effects of SiO_2 NPs. These SiO_2 NPs not only filled the surface defects such as pores and cracks, but also removed the peaks protruded from the surface [29]. The other mechanism that contributed to the decrease in surface roughness was the formation of tribofilm, and hence the direct contact between the work roll and the workpiece was relieved [276]. When the concentration of nanoadditive exceeded the optimal one, the agglomeration of NPs might aggravate the friction and wear in the contact zone, thereby increasing the surface roughness. On the contrary, a nanoadditive concentration lower than the optimal one resulted in insufficient lubrication, and therefore an insignificant effect on the decrease in surface roughness.

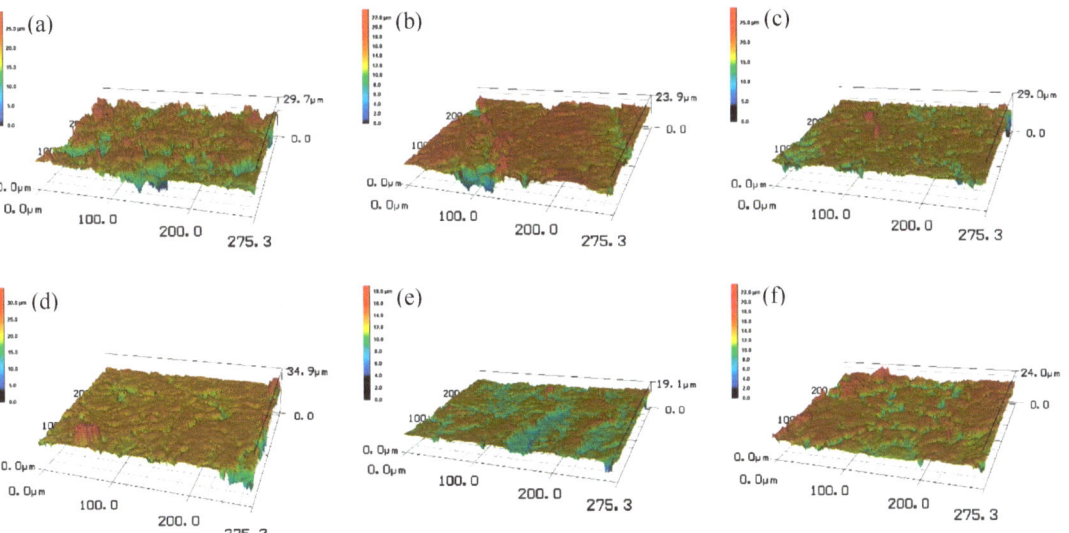

Figure 18. 3D profiles of surface morphologies of rolled steels at 850 °C under different lubrication conditions of (**a**) dry, (**b**) water, (**c**) 1.0 wt.% TiO_2, (**d**) 2.0 wt.% TiO_2, (**e**) 4.0 wt.% TiO_2, and (**f**) 8.0 wt.% TiO_2 [28]. (Rolling reduction of 30% in each single pass; rolling speed of 0.35 m/s).

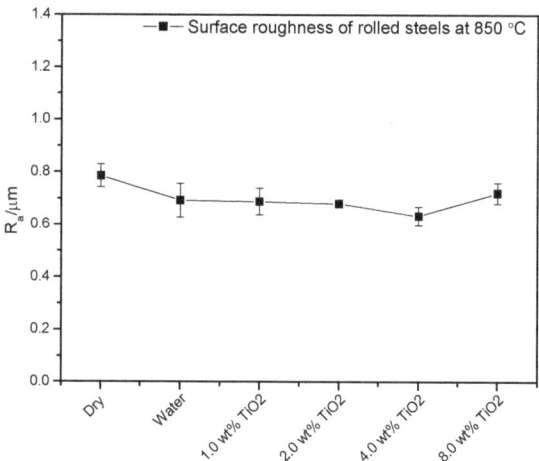

Figure 19. Surface roughness of rolled steels at 850 °C under different lubrication conditions as shown in Figure 18 [28]. (Rolling reduction of 30% in each single pass; rolling speed of 0.35 m/s).

6.2.3. Oxidation Behaviour of Steel

As hot steel rolling is normally conducted at the temperatures ranging from 800 to 1050 °C, significant oxidation occurs instantaneously in air [268]. The oxide scale formed on a hot rolled steel strip plays a prominent role in the subsequent industrial processes, including pickling, cold rolling, heat treatment, and electrolytic tinning [277]. In most cases, as-hot-rolled steel strip needs to suffer pickling, during which the oxide scale is removed by the use of acid treatment to obtain a high-quality surface for upcoming cold rolling. The oxides descaled as such have a dramatic impact on the consumption of acid and the yield of finished products. Decreasing oxide scale thickness is thus a highly desirable target in practical hot rolling production line. In some other cases, it is required to produce pickle-free as-hot-rolled steel strip that has 'tight oxide scale' formed on the strip surface prior to downstream forming [278]. Besides the roughness of the strip surface, the constitution of oxide phases in the scale also has a significant effect on the tribological feature during metal working [279]. Taking the oxide scale formed in mild steel as an example, hematite increases friction and wear as abrasive behaviour due to its high hardness, while magnetite and wustite are more ductile and hence resistant to wear [280]. As a whole, thin and tight oxide scale with considerable amounts of magnetite and wustite is always preferred after hot steel rolling.

Several studies were devoted to decreasing the oxide scale thickness during hot steel rolling by the use of various nanoadditives in water, including Eu-doped $CaWO_4$ NPs [276], SiO_2 NPs [100], MoS_2-Al_2O_3 nanocomposite [118], and TiO_2 NPs [24,28,29,52]. Some possible mechanisms of the decrease in oxide scale thickness have been proposed. First, the increase in strip deformation leads to the decrease in oxide scale thickness [275]. Second, the NPs fill in the voids of oxide scales and then prevent oxygen from penetrating into steel matrix for further oxidation [52]. Third, the NPs deposit on the strip surface to form a protective film that isolates oxygen and thus reduces the diffusion of O^{2-} into oxide layers [118]. These so-called 'mending effect' and 'protective film' can be characterised using SEM or TEM, in which the distribution of NPs can be clearly identified in the oxide layers [28,29]. The effects of using water-based nanolubricants on steel oxidation in these studies, however, were only confined to the entire thickness of oxide scale. It is also very important to systematically examine the formation and evolution mechanisms of different oxide phases during hot steel rolling. Given this point, Wu et al. [27] detailed the effect of water-based nanolubrication on the oxidation behaviour of steel through quantitative analyses of the oxide phases by the use of Raman microscope and SEM along with image processing software Image J. The schematic illustration of oxide scale formed on steel surface under no deformation (sampled from the tapered edge of steel workpiece), dry or water condition, and TiO_2 nanolubrication is presented in Figure 20. For non-deformed steels (see Figure 20a), a typical three-layered oxide scale was formed with a dominant inner layer of wustite, intermediate layer of magnetite, and top layer of hematite, as well as the magnetite seam at the scale/substrate interface and proeutectoid magnetite precipitated inside the wustite layer. For the steel under dry or water condition (see Figure 20b), considerable amounts of cracks and pores were generated, thereby causing fast conversion of magnetite to hematite near the scale surface and wustite to magnetite close to the scale/substrate interface. When TiO_2 water-based nanolubricants were used (see Figure 20c), TiO_2 NPs not only reduced the friction and the rolling force due to rolling effect, but also enabled the formation of protective film which was a barrier to inhibit oxygen diffusion. These effects decreased the extent of oxide deformation and therefore the porosity and cracks in oxide scale. As a result, the channels for oxygen penetration were reduced, which thus slowed down the conversions of wustite to magnetite and magnetite to hematite.

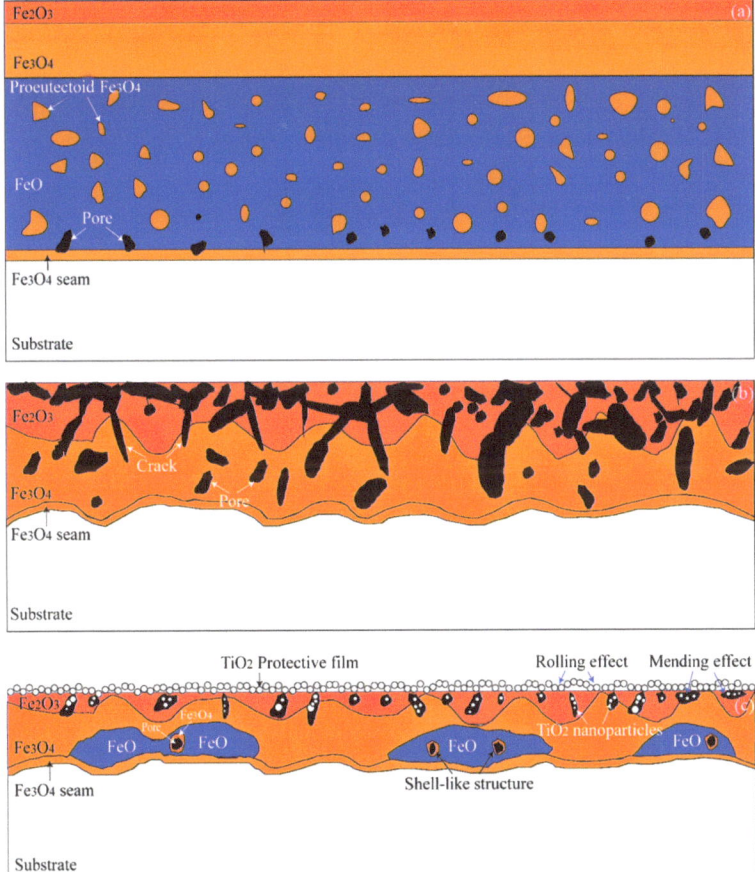

Figure 20. Schematic illustration of oxide scale formed on steel surface under (**a**) no deformation, (**b**) dry or water condition, and (**c**) TiO_2-containing nanolubrication condition during hot steel rolling [27].

For the steel experiencing hot rolling, in general, the conversions between different oxide phases follow the reactions below [27]:

$$6FeO + O_2 = 2Fe_3O_4 \tag{6}$$

$$4Fe_3O_4 + O_2 = 6Fe_2O_3 \tag{7}$$

$$4FeO = Fe_3O_4 + \alpha\text{-Fe} \tag{8}$$

Equations (6) and (7) exist at a temperature above 570 °C, while Equation (8) exists during cooling process at temperatures below 570 °C. The formation of proeutectoid magnetite and magnetite seam is related to the eutectoid reaction in Equation (8) [281,282].

6.2.4. Microstructure of Rolled Steel

It has been widely accepted that the steel's microstructure largely determines its mechanical properties which can be greatly enhanced by grain refinement and phase transformation [283]. For a steel with specified chemical compositions, controls of deformation during rolling and cooling rate after rolling are the dominating strategies to attain desired microstructure and targeted mechanical properties [284]. During hot steel rolling, heavy reduction and accelerated cooling are two of the most effective ways to refine the grains in steels [285]. In recent years, some researchers have been committed to applying water-based nanolubricants during hot steel rolling, instead of conventional ways, to achieve grain refinement. Nanolubricants containing 0.1–0.5 wt.% SiO_2, for instance, were used to refine the grain size in microstructure during hot rolling of ASTM 1045 steel within 5 passes from 1000 to 750 °C [100]. Nano-TiO_2 lubricants were also used to refine the grain size of rolled strips by around 50% after 5-pass hot rolling [52]. In particular, Wu et al. [28] conducted single-pass hot steel rolling at 850 °C using TiO_2 water-based nanolubricant, and found that the grain size of ferrite was refined up to 50.5% through statistical analysis of size distribution in the surface microstructure of rolled steels (see Figure 21). The possible factors that contribute to the decrease in grain size during hot steel rolling include the cooling rate of nanolubricant and the actual rolling deformation. On one hand, adding NPs into water increases the lubricant's thermal conductivity, and therefore the enhancement in cooling rate [185,261,286,287]. At an instant contact of work rolls with the workpiece, the lubricant in between acts as a coolant that offers a higher cooling rate than that of non-lubrication conditions. In this case, the increase in cooling rate leads to the increase in the number of nucleation sites for ferrite grains, and meanwhile a higher cooling rate enables the ferrite grains to pass through the high temperature region faster, thus shortening the time for ferrite growth [288]. Therefore, both the increased nucleation of ferrite and the retarded grain growth prompt the grain refinement due to enhanced cooling rate. On the other hand, the decrease in rolling force enabled by the use of nanolubricants gives rise to the decrease in exit thickness due to reduced spring-back value according to the Gaugemeter equation [289], where the exit thickness is the sum of the height of roll gap and the spring-back value. Given the increase in rolling deformation, the lattice distortion energy increases, and the dislocation density also increases, which respectively provides the driving force and nucleation sites for grain nucleation and therefore the grain refinement [290]. In some other cases, however, the grain size of surface microstructure would not change significantly after hot rolling with water-based nanolubricants [118,276], which might be ascribed to the slight differences in the deformation rate and the cooling rate of the lubricants being used.

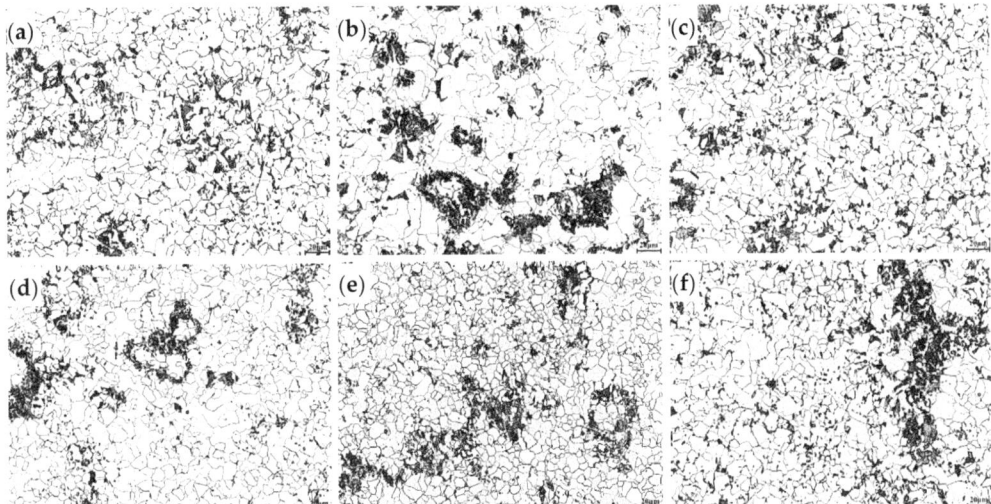

Figure 21. Surface microstructures of rolled steels at 850 °C under different lubrication conditions of (**a**) dry, (**b**) water, (**c**) 1.0 wt.% TiO_2, (**d**) 2.0 wt.% TiO_2, (**e**) 4.0 wt.% TiO_2, and (**f**) 8.0 wt.% TiO_2 [28]. (Rolling reduction of 30% in each single pass; rolling speed of 0.35 m/s).

6.3. Cold Rolling of Steels

In most cases, hot rolled steel strips experience subsequent cold rolling which is conducted at ambient temperature to obtain higher surface quality, special surface textures, or dimensional accuracy [291]. Due to the plastic deformation of workpiece in cold rolling, similar to hot rolling, friction exists between the work roll and the workpiece, and therefore the wear of work rolls is generated. To relieve the friction and wear occurred in the contact zone, rolling lubricants are also extensively applied in cold steel rolling. The requirements for cold rolling lubricants are a bit different from those for hot rolling ones in terms of rolling temperature and corrosive property of the lubricants. Nevertheless, the lubrication mechanisms, in general, are at least similar for both rolling processes.

It has been reported that the use of 0.7 wt.% nano-TiO_2 water-based rolling lubricant would largely decrease the surface scratches and adhesion defects on rolled steel surface [50]. However, the corresponding lubrication mechanisms were not well understood. Furthermore, 2D materials such as MoS_2, MoO_3, and graphene have also presented great potential in cold steel rolling due to their weak interlayer interaction and easy sliding nature between neighbouring atomic layers [292]. For example, water-based MoS_2 nanolubricant was applied in cold rolling of a low-carbon steel, and the results revealed that both the rolling force and the minimum rolling thickness obtained under base lubricant were decreased even using the recycled MoS_2 nanolubricant [63]. As nano-MoS_2 used as a lubricant additive would be readily oxidised and then converted to nano-MoO_3 under cold rolling conditions, they also applied MoO_3 water-based lubricant in a four-high cold rolling mill [56]. The rolling force and power were found to be significantly reduced in each rolling pass, and the minimum rolling thickness was apparently thinned, owing to a lubricating film composed of MoO_3, MoO_2, and $Fe(OH)_3$ formed on strip surface. Graphene, as another promising nanoadditive in water, has also been extensively studied over the past five years [67,132]. As a matter of fact, graphene oxide (GO) instead of graphene has been more commonly used in water due to the presence of hydrophilic groups such as hydroxyl, carboxyl, and epoxy which relieve the agglomeration caused by π-π stacking interaction [142,145]. However, the pristine GO suspension without pH modification is acidic, which is detrimental to the strip quality [17]. In light of this, Meng et al. [76] used triethanolamine (TEA) to prepare alkaline GO aqueous lubricants that were feasibly applied in cold steel rolling. The results demonstrated that the use of alkaline lubricant with pH 9 resulted in the best lubrication performance by distinctly reducing the rolling force in each pass, and the minimum rolling thickness after 7-pass rolling was 21.36% lower than that without using lubricant. The lubrication mechanisms were proposed as the combination of mending effect and formation of physically adsorbed layers.

Given the excellent tribological characteristics of 0D and 2D materials as individual additives in water, GO-based composites are attracting considerable research interest in the area of water-based nanolubrication, which exhibits improved tribological properties by combining two or more components through a synergistic effect [12,19–21]. In light of this, Du et al. [84] conducted cold steel rolling by means of 0.5 wt.% GO-TiO$_2$ nanolubricants in comparison with 0.5 wt.% GO and 0.5 wt.% TiO$_2$ nanolubricants. The results indicated that the hybrid nanolubricant exhibited better tribological performance than the lubricant with individual nanoadditive. Beyond that, the strip rolled with GO-TiO$_2$ nanolubricant had the minimum rolling thickness and surface roughness, and meanwhile the rolling force was the lowest among all the lubricants being used. The exceptional lubricating properties of GO-TiO$_2$ nanocomposite were caused by the excellent dispersion stability and the formation of absorption films, carbonaceous protective films and transfer films.

6.4. Cold Rolling of Non-Ferrous Metals

Magnesium (Mg) alloys, owing to their exceptional physicochemical properties such as low density, high strength/weight ratio, good damping performance, superb biocompatibility, excellent machinability, and castability, have been widely used in many engineering applications including transportation, electronics, aerospace, biomedical, and energy sectors [293]. During the forming of Mg alloys, generating severe friction and wear between the tool and the workpiece is inevitable, which may result in inhomogeneous deformations, shortened tool life, and poor surface quality of workpiece [294]. To overcome these disadvantages, the use of suitable lubricants is essential in the forming process. Although some eco-friendly lubricant additives, for instance, N-containing compounds [295], borates [296], and ionic liquids [297] have been developed for forming Mg alloys, they still raise some issues such as poor extreme pressure property, complex synthesis process, and high material cost. To date, there have been few choices of lubricants for the forming process of Mg alloys, while water-based nanolubricants are becoming increasingly popular due to their minimum impact on the environment, low cost, desirable recyclability, and exceptional lubrication performance. For example, graphene and graphene oxide (GO) have been adopted as water-based lubricant additives for steel-Mg alloy contacts through ball-on-plate tribo-testing, and 0.5 wt.% GO lubricant has been proposed for the forming of Mg alloys due to its prominent tribological performance in the reduction of COF and wear rate of Mg alloy plate [77]. It has been always anticipated that hybrid nanolubricants provide possibilities to further improve the tribological performance of the lubricant with individual nanoadditive through a synergistic effect of two or more components. Based on this, SiO$_2$/graphene composite was dispersed in water to evaluate their tribological performance for Mg alloy rolling [112]. It was found that the combinations of 0.1 wt.% nano-SiO$_2$ and 0.4 wt.% graphene displayed the best friction-reduction and anti-wear properties among all the SiO$_2$/graphene mixing ratios. During the cold rolling of Mg alloy under a single-pass reduction of 10%, the rolling force and rolled surface roughness obtained with dry rolling were decreased by up to 12% and 42.4%, respectively, when using the nanolubricant with the optimal SiO$_2$/graphene combinations. It has also been documented that SiO$_2$/nanotube hybrid water-based nanolubricant with a mass ratio of 0.3:0.2 can be used for Mg alloy forming according to its excellent tribological properties and load-carrying capacity as well as formation of stable lubricating film on the surface of Mg alloy [113]. These research findings have provided new insights into the potential of applying carbon-based aqueous nanolubricants in Mg alloy forming.

7. Conclusions and Outlook

In this paper, we provide a comprehensive overview of recent advances in water-based nanolubricants with particular emphasis on the preparation of different types of nanolubricants, the methods to evaluate their dispersion stability and tribological properties, the discussion on lubrication mechanisms, and their application in metal rolling. Among all the candidate nanoadditives used in water, carbon-based nanomaterials have attracted the most attention due to their exceptional physicochemical properties and unique chemical structures. Nanocomposites with carbon-based nanomaterials as constituent materials have shown better performance and greater potential than individual component in water-based nanolubricants. Although some water-based nanolubricants have presented remarkable performance in metal rolling, great efforts are still needed to well understand the behaviours of the nanoadditives and reduce their material and preparation costs for accelerating their applications on an industrial scale. Some research directions that deserve further exploration are provided as follows.

- Ensuring long-term dispersion stability of nanoadditives in water is still a big challenge. The interaction among different lubricant components needs to be investigated for the perfection of the theories of dispersion stability.
- For the application in metal rolling, the formulation of water-based nanolubricants needs to be optimised to further enhance their physicochemical properties in terms of dispersion stability, wettability, and extreme pressure property. Special attention should be given to the strategies for reducing material and preparation costs of the applied nanolubricants.
- The application of water-based nanolubricants in hot steel rolling has exhibited positive effects on the decreases in rolling force, rolled surface roughness, and oxide scale thickness, and also enabled refined grains in microstructure. However, the lubrication effects on controls of profile, flatness, and texture have been rarely involved. More studies are also needed to examine the grain refinement mechanism and attain maximally refined grains, which is a promising and economical technique to significantly promote the overall properties of hot rolled steels.
- For the case of application of cold steel rolling, it is of vital importance to have more focus on the study of the corrosive property of applied water-based nanolubricants. In addition to the lubrication effects on rolling force and surface quality, extra attention should be paid to those on rolling texture and shape control.
- Although certain water-based nanolubrication mechanisms in rolling of steels have been proposed through analysis of post-rolling specimen by means of electron microscopy, in situ observation of NPs and demonstration of their motion behaviour have not been specifically conducted. To have a systematic and comprehensive understanding of the lubrication mechanisms, varying rolling parameters such as rolling temperature, rolling reduction, and speed should be employed, and corresponding multi-scale numerical simulation can be carried out.
- As pointed out earlier, work roll service life can be prolonged using water-based nanolubricants, which largely reduces the roll changing frequency and thus enhances the productivity of rolling mill. However, no research has been conducted to quantitatively evaluate the wear of work rolls under water-based nanolubrication conditions.
- The use of green lubricant is becoming mainstream in sustainable manufacturing. It is of vital importance to develop a cost-effective recycling technology for waste water-based nanolubricants.

Author Contributions: Conceptualization, Z.J. and H.W.; writing, A.M. and H.W.; proofreading, Z.J., H.W. and A.M.; supervision, Z.J. and H.W. All authors have read and agreed to the published version of the manuscript.

Funding: This research was funded by the Australian Research Council (ARC, Grant Nos. DP190100738 and DP190100408).

Acknowledgments: The authors acknowledge the financial support from the Australian Research Council (ARC, Grant Nos. DP190100738 and DP190100408).

Conflicts of Interest: The authors declare no conflict of interest.

References

1. Bhushan, B.; Israelachvili, J.N.; Landman, U. Nanotribology-Friction, Wear and Lubrication at the Atomic-Scale. *Nature* **1995**, *374*, 607–616. [CrossRef]
2. Fu, Y.; Batchelor, A.W.; Loh, N.K.; Tan, K.W. Effect of lubrication by mineral and synthetic oils on the sliding wear of plasma nitrided AISI 410 stainless steel. *Wear* **1998**, *219*, 169–176. [CrossRef]
3. Haus, F.; German, J.; Junter, G.-A. Primary biodegradability of mineral base oils in relation to their chemical and physical characteristics. *Chemosphere* **2001**, *45*, 983–990. [CrossRef]
4. Sotres, J.; Arnebrant, T. Experimental Investigations of biological lubrication at the nanoscale: The cases of synovial joints and the oral cavity. *Lubricants* **2013**, *1*, 102–131. [CrossRef]
5. Dubey, S.; Sharma, G.; Shishodia, K.; Sekhon, G. Study on the performance of oil-in-water emulsions during cold rolling of steel strip. *Tribol. Trans.* **2005**, *48*, 499–504. [CrossRef]
6. Hu, X.; Wang, Y.; Jing, H. Application of oil-in-water emulsion in hot rolling process of brass sheet. *Ind. Lubr. Tribol.* **2010**, *62*, 224–231. [CrossRef]
7. Xia, W.; Zhao, J.; Wu, H.; Jiao, S.; Zhao, X.; Zhang, X.; Xu, J.; Jiang, Z. Analysis of oil-in-water based nanolubricants with varying mass fractions of oil and TiO$_2$ nanoparticles. *Wear* **2018**, *396–397*, 162–171. [CrossRef]
8. Xia, W.Z.; Zhao, J.W.; Wu, H.; Jiao, S.H.; Jiang, Z.Y. Effects of oil-in-water based nanolubricant containing TiO2 nanoparticles on the tribological behaviour of oxidised high-speed steel. *Tribol. Int.* **2017**, *110*, 77–85. [CrossRef]
9. Xia, W.Z.; Zhao, J.W.; Wu, H.; Zhao, X.M.; Zhang, X.M.; Xu, J.Z.; Jiao, S.H.; Wang, X.G.; Zhou, C.L.; Jiang, Z.Y. Effects of oil-in-water based nanolubricant containing TiO2 nanoparticles in hot rolling of 304 stainless steel. *J. Mater. Process. Technol.* **2018**, *262*, 149–156. [CrossRef]
10. Chen, Y.; Renner, P.; Liang, H. Dispersion of Nanoparticles in Lubricating Oil: A Critical Review. *Lubricants* **2019**, *7*, 7. [CrossRef]
11. Gulzar, M.; Masjuki, H.H.; Kalam, M.A.; Varman, M.; Zulkifli, N.W.M.; Mufti, R.A.; Zahid, R. Tribological performance of nanoparticles as lubricating oil additives. *J. Nanopart. Res.* **2016**, *18*, 223. [CrossRef]
12. Huang, S.Q.; Lin, W.K.; Li, X.L.; Fan, Z.Q.; Wu, H.; Jiang, Z.Y.; Huang, H. Roughness-dependent tribological characteristics of water-based GO suspensions with ZrO$_2$ and TiO$_2$ nanoparticles as additives. *Tribol. Int.* **2021**, *161*, 107073. [CrossRef]
13. Wu, H.; Jia, F.H.; Zhao, J.W.; Huang, S.Q.; Wang, L.Z.; Jiao, S.H.; Huang, H.; Jiang, Z.Y. Effect of water-based nanolubricant containing nano-TiO$_2$ on friction and wear behaviour of chrome steel at ambient and elevated temperatures. *Wear* **2019**, *426*, 792–804. [CrossRef]
14. Wu, H.; Zhao, J.W.; Cheng, X.W.; Xia, W.Z.; He, A.S.; Yun, J.H.; Huang, S.Q.; Wang, L.Z.; Huang, H.; Jiao, S.H.; et al. Friction and wear characteristics of TiO$_2$ nano-additive water-based lubricant on ferritic stainless steel. *Tribol. Int.* **2018**, *117*, 24–38. [CrossRef]
15. Wu, H.; Zhao, J.W.; Xia, W.Z.; Cheng, X.W.; He, A.S.; Yun, J.H.; Wang, L.Z.; Huang, H.; Jiao, S.H.; Huang, L.; et al. A study of the tribological behaviour of TiO$_2$ nano-additive water-based lubricants. *Tribol. Int.* **2017**, *109*, 398–408. [CrossRef]
16. He, A.S.; Huang, S.Q.; Yun, J.H.; Wu, H.; Jiang, Z.Y.; Stokes, J.; Jiao, S.H.; Wang, L.Z.; Huang, H. Tribological Performance and Lubrication Mechanism of Alumina Nanoparticle Water-Based Suspensions in Ball-on-Three-Plate Testing. *Tribol. Lett.* **2017**, *65*, 40. [CrossRef]
17. He, A.S.; Huang, S.Q.; Yun, J.H.; Jiang, Z.Y.; Stokes, J.; Jiao, S.H.; Wang, L.Z.; Huang, H. The pH-dependent structural and tribological behaviour of aqueous graphene oxide suspensions. *Tribol. Int.* **2017**, *116*, 460–469. [CrossRef]
18. He, A.S.; Huang, S.Q.; Yun, J.H.; Jiang, Z.Y.; Stokes, J.R.; Jiao, S.H.; Wang, L.Z.; Huang, H. Tribological Characteristics of Aqueous Graphene Oxide, Graphitic Carbon Nitride, and Their Mixed Suspensions. *Tribol. Lett.* **2018**, *66*, 42. [CrossRef]
19. Huang, S.Q.; Li, X.; Yu, B.; Jiang, Z.; Huang, H. Machining characteristics and mechanism of GO/SiO$_2$ nanoslurries in fixed abrasive lapping. *J. Mater. Process. Technol.* **2020**, *277*, 116444. [CrossRef]
20. Huang, S.Q.; Li, X.L.; Zhao, Y.T.; Sun, Q.; Huang, H. A novel lapping process for single-crystal sapphire using hybrid nanoparticle suspensions. *Int. J. Mech. Sci.* **2021**, *191*, 106099. [CrossRef]
21. Huang, S.Q.; He, A.S.; Yun, J.H.; Xu, X.F.; Jiang, Z.Y.; Jiao, S.H.; Huang, H. Synergistic tribological performance of a water based lubricant using graphene oxide and alumina hybrid nanoparticles as additives. *Tribol. Int.* **2019**, *135*, 170–180. [CrossRef]
22. Lin, W.; Kampf, N.; Klein, J. Designer Nanoparticles as Robust Superlubrication Vectors. *ACS Nano* **2020**, *14*, 7008–7017. [CrossRef]
23. Lin, W.; Kampf, N.; Goldberg, R.; Driver, M.J.; Klein, J. Poly-phosphocholinated Liposomes Form Stable Superlubrication Vectors. *Langmuir* **2019**, *35*, 6048–6054. [CrossRef] [PubMed]
24. Wu, H.; Kamali, H.; Huo, M.; Lin, F.; Huang, S.; Huang, H.; Jiao, S.; Xing, Z.; Jiang, Z. Eco-Friendly Water-Based Nanolubricants for Industrial-Scale Hot Steel Rolling. *Lubricants* **2020**, *8*, 96. [CrossRef]
25. Wu, H.; Jia, F.; Li, Z.; Lin, F.; Huo, M.; Huang, S.; Sayyar, S.; Jiao, S.; Huang, H.; Jiang, Z. Novel water-based nanolubricant with superior tribological performance in hot steel rolling. *Int. J. Extrem. Manuf.* **2020**, *2*, 025002. [CrossRef]
26. Huo, M.; Wu, H.; Xie, H.; Zhao, J.; Su, G.; Jia, F.; Li, Z.; Lin, F.; Li, S.; Zhang, H.; et al. Understanding the role of water-based nanolubricants in micro flexible rolling of aluminium. *Tribol. Int.* **2020**, *151*, 106378. [CrossRef]

27. Wu, H.; Jiang, C.Y.; Zhang, J.Q.; Huang, S.Q.; Wang, L.Z.; Jiao, S.H.; Huang, H.; Jiang, Z.Y. Oxidation Behaviour of Steel During hot Rolling by Using TiO$_2$-Containing Water-Based Nanolubricant. *Oxid. Met.* **2019**, *92*, 315–335. [CrossRef]
28. Wu, H.; Zhao, J.W.; Luo, L.; Huang, S.Q.; Wang, L.Z.; Zhang, S.Q.; Jiao, S.H.; Huang, H.; Jiang, Z.Y. Performance Evaluation and Lubrication Mechanism of Water-Based Nanolubricants Containing Nano-TiO$_2$ in Hot Steel Rolling. *Lubricants* **2018**, *6*, 57. [CrossRef]
29. Wu, H.; Zhao, J.W.; Xia, W.Z.; Cheng, X.W.; He, A.S.; Yun, J.H.; Wang, L.Z.; Huang, H.; Jiao, S.H.; Huang, L.; et al. Analysis of TiO$_2$ nano-additive water-based lubricants in hot rolling of microalloyed steel. *J. Manuf. Process.* **2017**, *27*, 26–36. [CrossRef]
30. Yu, X.; Zhou, J.; Jiang, Z. Developments and Possibilities for Nanoparticles in Water-Based Lubrication During Metal Processing. *Rev. Nanosci. Nanotechnol.* **2016**, *5*, 136–163. [CrossRef]
31. Rahman, M.H.; Warneke, H.; Webbert, H.; Rodriguez, J.; Austin, E.; Tokunaga, K.; Rajak, D.K.; Menezes, P.L. Water-Based Lubricants: Development, Properties, and Performances. *Lubricants* **2021**, *9*, 73. [CrossRef]
32. Canter, N. Special Report: Trends in extreme pressure additives. *Tribol. Lubr. Technol.* **2007**, *63*, 10.
33. Choi, S.U.; Eastman, J.A. Enhanced Heat Transfer Using Nanofluids. U.S. Patent 6221275, 1 January 2001.
34. Akoh, H.; Tsukasaki, Y.; Yatsuya, S.; Tasaki, A. Magnetic properties of ferromagnetic ultrafine particles prepared by vacuum evaporation on running oil substrate. *J. Cryst. Growth* **1978**, *45*, 495–500. [CrossRef]
35. Wang, H.; Xu, J.-Z.; Zhu, J.-J.; Chen, H.-Y. Preparation of CuO nanoparticles by microwave irradiation. *J. Cryst. Growth* **2002**, *244*, 88–94. [CrossRef]
36. Sandhya, S.U.; Nityananda, S.A. A Facile One Step Solution Route to Synthesize Cuprous Oxide Nanofluid. *Nanomater. Nanotechnol.* **2013**, *3*, 5. [CrossRef]
37. Han, S.Y.; Shin, S.Y.; Lee, H.-J.; Lee, B.-J.; Lee, S.; Kim, N.J.; Kwak, J.-H. Effects of annealing temperature on microstructure and tensile properties in ferritic lightweight steels. *Metall. Mater. Trans. A Phys. Metall. Mater. Sci.* **2012**, *43*, 843–853. [CrossRef]
38. Sun, Y.; Mayers, B.; Herricks, T.; Xia, Y. Polyol Synthesis of Uniform Silver Nanowires: A Plausible Growth Mechanism and the Supporting Evidence. *Nano Lett.* **2003**, *3*, 955–960. [CrossRef]
39. Eastman, J.A.; Choi, S.U.S.; Li, S.; Yu, W.; Thompson, L.J. Anomalously increased effective thermal conductivities of ethylene glycol-based nanofluids containing copper nanoparticles. *Appl. Phys. Lett.* **2001**, *78*, 718–720. [CrossRef]
40. Hsin, Y.L.; Hwang, K.C.; Chen, F.-R.; Kai, J.-J. Production and in-situ Metal Filling of Carbon Nanotubes in Water. *Adv. Mater.* **2001**, *13*, 830–833. [CrossRef]
41. Lo, C.-H.; Tsung, T.-T.; Chen, L.-C. Shape-controlled synthesis of Cu-based nanofluid using submerged arc nanoparticle synthesis system (SANSS). *J. Cryst. Growth* **2005**, *277*, 636–642. [CrossRef]
42. Angayarkanni, S.A.; Philip, J. Review on thermal properties of nanofluids: Recent developments. *Adv. Colloid. Interface Sci.* **2015**, *225*, 146–176. [CrossRef] [PubMed]
43. Haddad, Z.; Abid, C.; Oztop, H.F.; Mataoui, A. A review on how the researchers prepare their nanofluids. *Int. J. Sci.* **2014**, *76*, 168–189. [CrossRef]
44. Shahnazar, S.; Bagheri, S.; Abd Hamid, S.B. Enhancing lubricant properties by nanoparticle additives. *Int. J. hydrogen Energy* **2016**, *41*, 3153–3170. [CrossRef]
45. Liu, G.; Li, X.; Qin, B.; Xing, D.; Guo, Y.; Fan, R. Investigation of the mending effect and mechanism of copper nano-particles on a tribologically stressed surface. *Tribol. Lett.* **2004**, *17*, 961–966. [CrossRef]
46. Zhao, J.; Yang, G.; Zhang, C.; Zhang, Y.; Zhang, S.; Zhang, P. Synthesis of water-soluble Cu nanoparticles and evaluation of their tribological properties and thermal conductivity as a water-based additive. *Friction* **2019**, *7*, 246–259. [CrossRef]
47. Zhang, C.; Zhang, S.; Song, S.; Yang, G.; Yu, L.; Wu, Z.; Li, X.; Zhang, P. Preparation and Tribological Properties of Surface-Capped Copper Nanoparticle as a Water-Based Lubricant Additive. *Tribol. Lett.* **2014**, *54*, 25–33. [CrossRef]
48. Radice, S.; Mischler, S. Effect of electrochemical and mechanical parameters on the lubrication behaviour of Al$_2$O$_3$ nanoparticles in aqueous suspensions. *Wear* **2006**, *261*, 1032–1041. [CrossRef]
49. Kong, L.; Sun, J.; Bao, Y.; Meng, Y. Effect of TiO$_2$ nanoparticles on wettability and tribological performance of aqueous suspension. *Wear* **2017**, *376–377*, 786–791. [CrossRef]
50. Sun, J.; Li, Y.; Xu, P.; Zhu, Z. Study on the lubricating performance of nano-TiO$_2$ in water-based cold rolling fluid. *Mater. Sci. Forum* **2015**, *3*, 3988–3992.
51. Zhu, Z.; Sun, J.; Wei, H.; Niu, T.; Zhu, Z. Research on Lubrication Behaviors of Nano-TiO$_2$ in Water-Based Hot Rolling Liquid. *Adv. Mater. Res.* **2013**, *643*, 139–143. [CrossRef]
52. Meng, Y.; Sun, J.; Wu, P.; Dong, C.; Yan, X. The Role of Nano-TiO$_2$ Lubricating Fluid on the Hot Rolled Surface and Metallographic Structure of SS41 Steel. *Nanomaterials* **2018**, *8*, 111. [CrossRef]
53. Zhao, C.; Chen, Y.K.; Ren, G. A Study of Tribological Properties of Water-Based Ceria Nanofluids. *Tribol. Trans.* **2013**, *56*, 275–283. [CrossRef]
54. Zhao, J.; Yang, G.; Zhang, Y.; Zhang, S.; Zhang, C.; Gao, C.; Zhang, P. Controllable synthesis of different morphologies of CuO nanostructures for tribological evaluation as water-based lubricant additives. *Friction* **2020**, *9*, 963–977. [CrossRef]
55. Pardue, T.; Acharya, B.; Curtis, C.; Krim, J. A Tribological Study of γ-Fe$_2$O$_3$ Nanoparticles in Aqueous Suspension. *Tribol. Lett.* **2018**, *66*, 130. [CrossRef]
56. Sun, J.; Meng, Y.; Zhang, B. Tribological Behaviors and Lubrication Mechanism of Water-based MoO$_3$ Nanofluid during Cold Rolling Process. *J. Manuf. Process.* **2021**, *61*, 518–526. [CrossRef]

57. Xiong, S.; Liang, D.; Kong, F. Effect of pH on the Tribological Behavior of Eu-Doped WO$_3$ Nanoparticle in Water-Based Fluid. *Tribol. Lett.* **2020**, *68*, 126. [CrossRef]
58. Ding, M.; Lin, B.; Sui, T.; Wang, A.; Yan, S.; Yang, Q. The excellent anti-wear and friction reduction properties of silica nanoparticles as ceramic water lubrication additives. *Ceram. Int.* **2018**, *44*, 14901–14906. [CrossRef]
59. Bao, Y.; Sun, J.; Kong, L. Tribological properties and lubricating mechanism of SiO$_2$ nanoparticles in water-based fluid. In *IOP Conference Series: Materials Science and Engineering*; IOP Publishing: Bristol, UK, 2017; Volume 182, p. 012025. [CrossRef]
60. Kogovšek, J.; Remškar, M.; Mrzel, A.; Kalin, M. Influence of surface roughness and running-in on the lubrication of steel surfaces with oil containing MoS2 nanotubes in all lubrication regimes. *Tribol. Int.* **2013**, *61*, 40–47. [CrossRef]
61. Kuznetsova, Y.; Rempel, S.V.; Popov, I.D.; Gerasimov, E.; Rempel, A. Stabilization of Ag$_2$S nanoparticles in aqueous solution by MPS. *Colloids Surf. Phys. Eng. Asp.* **2017**, *520*, 369–377. [CrossRef]
62. Zhao, J.; Yang, G.; Zhang, Y.; Zhang, S.; Zhang, P. A Simple Preparation of HDA-CuS Nanoparticles and Their Tribological Properties as a Water-Based Lubrication Additive. *Tribol. Lett.* **2019**, *67*, 88. [CrossRef]
63. Yanan, M.; Jianlin, S.; Jiaqi, H.; Xudong, Y.; Yu, P. Recycling prospect and sustainable lubrication mechanism of water-based MoS$_2$ nano-lubricant for steel cold rolling process. *J. Clean. Prod.* **2020**, *277*, 123991. [CrossRef]
64. Wang, Y.; Du, Y.; Deng, J.; Wang, Z. Friction reduction of water based lubricant with highly dispersed functional MoS$_2$ nanosheets. *Colloids Surf. A Phys. Eng Asp.* **2019**, *562*, 321–328. [CrossRef]
65. Peña-Parás, L.; Maldonado-Cortés, D.; Kharissova, O.V.; Saldívar, K.I.; Contreras, L.; Arquieta, P.; Castaños, B. Novel carbon nanotori additives for lubricants with superior anti-wear and extreme pressure properties. *Tribol. Int.* **2019**, *131*, 488–495. [CrossRef]
66. Hu, Y.; Wang, Y.; Wang, C.; Ye, Y.; Zhao, H.; Li, J.; Lu, X.; Mao, C.; Chen, S.; Mao, J.; et al. One-pot pyrolysis preparation of carbon dots as eco-friendly nanoadditives of water-based lubricants. *Carbon* **2019**, *152*, 511–520. [CrossRef]
67. Liang, S.; Shen, Z.; Yi, M.; Liu, L.; Zhang, X.; Ma, S. In-situ exfoliated graphene for high-performance water-based lubricants. *Carbon* **2016**, *96*, 1181–1190. [CrossRef]
68. Jiao, Y.; Liu, S.; Sun, Y.; Yue, W.; Zhang, H. Bioinspired Surface Functionalization of Nanodiamonds for Enhanced Lubrication. *Langmuir* **2018**, *34*, 12436–12444. [CrossRef]
69. Si, Y.; Samulski, E.T. Synthesis of Water Soluble Graphene. *Nano Lett.* **2008**, *8*, 1679–1682. [CrossRef] [PubMed]
70. Song, H.-J.; Li, N. Frictional behavior of oxide graphene nanosheets as water-base lubricant additive. *Appl. Phys. A* **2011**, *105*, 827–832. [CrossRef]
71. Hummers, W.S.; Offeman, R.E. Preparation of Graphitic Oxide. *J. Am. Chem. Soc.* **1958**, *80*, 1339. [CrossRef]
72. Min, C.; He, Z.; Song, H.; Liang, H.; Liu, D.; Dong, C.; Jia, W. Fluorinated graphene oxide nanosheet: A highly efficient water-based lubricated additive. *Tribol. Int.* **2019**, *140*, 105867. [CrossRef]
73. Fan, K.; Liu, J.; Wang, X.; Liu, Y.; Lai, W.; Gao, S.; Qin, J.; Liu, X. Towards enhanced tribological performance as water-based lubricant additive: Selective fluorination of graphene oxide at mild temperature. *J. Colloid Interface Sci.* **2018**, *531*, 138–147. [CrossRef] [PubMed]
74. Liu, C.; Guo, Y.; Wang, D. PEI-RGO nanosheets as a nanoadditive for enhancing the tribological properties of water-based lubricants. *Tribol. Int.* **2019**, *140*, 105851. [CrossRef]
75. Hu, Y.; Wang, Y.; Zeng, Z.; Zhao, H.; Li, J.; Ge, X.; Wang, L.; Xue, Q.; Mao, C.; Chen, S. BLG-RGO: A novel nanoadditive for water-based lubricant. *Tribol. Int.* **2019**, *135*, 277–286. [CrossRef]
76. Meng, W.; Sun, J.; Wang, C.; Wu, P. pH-dependent lubrication mechanism of graphene oxide aqueous lubricants on the strip surface during cold rolling. *Surf. Interface Anal.* **2021**, *53*, 406–417. [CrossRef]
77. Xie, H.; Jiang, B.; Dai, J.; Peng, C.; Li, C.; Li, Q.; Pan, F. Tribological Behaviors of Graphene and Graphene Oxide as Water-Based Lubricant Additives for Magnesium Alloy/Steel Contacts. *Materials* **2018**, *11*, 206. [CrossRef] [PubMed]
78. Wu, P.; Chen, X.; Zhang, C.; Luo, J. Synergistic tribological behaviors of graphene oxide and nanodiamond as lubricating additives in water. *Tribol. Int.* **2019**, *132*, 177–184. [CrossRef]
79. Liu, Y.; Wang, X.; Pan, G.; Luo, J. A comparative study between graphene oxide and diamond nanoparticles as water-based lubricating additives. *Sci. China Technol. Sci.* **2013**, *56*, 152–157. [CrossRef]
80. Su, F.; Chen, G.; Huang, P. Lubricating performances of graphene oxide and onion-like carbon as water-based lubricant additives for smooth and sand-blasted steel discs. *Friction* **2020**, *8*, 47–57. [CrossRef]
81. Oh, S.-T.; Lee, J.-S.; Sekino, T.; Niihara, K. Fabrication of Cu dispersed Al$_2$O$_3$ nanocomposites using Al$_2$O$_3$/CuO and Al$_2$O$_3$/Cu-nitrate mixtures. *Scr. Mater.* **2001**, *44*, 2117–2120. [CrossRef]
82. Devendiran, D.K.; Amirtham, V.A. A review on preparation, characterization, properties and applications of nanofluids. *Renew. Sustain. Energy Rev.* **2016**, *60*, 21–40. [CrossRef]
83. Guo, P.; Chen, L.; Wang, J.; Geng, Z.; Lu, Z.; Zhang, G. Enhanced Tribological Performance of Aminated Nano-Silica Modified Graphene Oxide as Water-Based Lubricant Additive. *ACS Appl. Nano Mater.* **2018**, *1*, 6444–6453. [CrossRef]
84. Du, S.; Sun, J.; Wu, P. Preparation, characterization and lubrication performances of graphene oxide-TiO$_2$ nanofluid in rolling strips. *Carbon* **2018**, *140*, 338–351. [CrossRef]
85. Li, W.; Zou, C.; Li, X. Thermo-physical properties of cooling water-based nanofluids containing TiO$_2$ nanoparticles modified by Ag elementary substance for crystallizer cooling system. *Powder Technol.* **2018**, *329*, 434–444. [CrossRef]

86. Zheng, X.; Xu, Y.; Geng, J.; Peng, Y.; Olson, D.; Hu, X. Tribological behavior of Fe_3O_4/MoS_2 nanocomposites additives in aqueous and oil phase media. *Tribol. Int.* **2016**, *102*, 79–87. [CrossRef]
87. Lelonis, D.A.; Tereshko, J.W.; Andersen, C.M. Boron Nitride Powder—A High-Performance Alternative for Solid Lubrication. *GE Adv. Ceram.* **2003**, *4*, 81506.
88. Cho, D.-H.; Kim, J.-S.; Kwon, S.-H.; Lee, C.; Lee, Y.-Z. Evaluation of hexagonal boron nitride nano-sheets as a lubricant additive in water. *Wear* **2013**, *302*, 981–986. [CrossRef]
89. Bai, Y.; Wang, L.; Ge, C.; Liu, R.; Guan, H.; Zhang, X. Atomically Thin Hydroxylation Boron Nitride Nanosheets for Excellent Water-Based Lubricant Additives. *J. Am. Ceram. Soc.* **2020**, *103*, 6951–6960. [CrossRef]
90. Cheng, H.; Zhao, W. Regulating the Nb_2C nanosheets with different degrees of oxidation in water lubricated sliding toward an excellent tribological performance. *Friction* **2021**, 1–13. [CrossRef]
91. Nguyen, H.; Chung, K. Assessment of Tribological Properties of Ti_3C_2 as a Water-Based Lubricant Additive. *Materials* **2020**, *13*, 5545. [CrossRef]
92. Tang, W.; Jiang, Z.; Wang, B.; Li, Y. Black phosphorus quantum dots: A new-type of water-based high-efficiency lubricant additive. *Friction* **2021**, *9*, 1528–1542. [CrossRef]
93. Wang, H.; Liu, Y.; Chen, Z.; Wu, B.; Xu, S.; Luo, J. Layered Double Hydroxide Nanoplatelets with Excellent Tribological Properties under High Contact Pressure as Water-Based Lubricant Additives. *Sci. Rep.* **2016**, *6*, 22748. [CrossRef]
94. Wang, H.; Liu, Y.; Liu, W.; Liu, Y.; Wang, K.; Li, J.; Ma, T.; Eryilmaz, O.L.; Shi, Y.; Erdemir, A.; et al. Superlubricity of Polyalkylene Glycol Aqueous Solutions Enabled by Ultrathin Layered Double Hydroxide Nanosheets. *ACS Appl. Mater. Interfaces* **2019**, *11*, 20249–20256. [CrossRef] [PubMed]
95. Li, X.; Wang, Z.; Dong, G. Preparation of nanoscale liquid metal droplet wrapped with chitosan and its tribological properties as water-based lubricant additive. *Tribol. Int.* **2020**, *148*, 106349. [CrossRef]
96. Ye, X.; Wang, J.; Fan, M. Evaluating tribological properties of the stearic acid-based organic nanomaterials as additives for aqueous lubricants. *Tribol. Int.* **2019**, *140*, 105848. [CrossRef]
97. Odzak, N.; Kistler, D.; Behra, R.; Sigg, L. Dissolution of metal and metal oxide nanoparticles in aqueous media. *Environ. Pollut.* **2014**, *191*, 132–138. [CrossRef] [PubMed]
98. Kim, H.J.; Bang, I.C.; Onoe, J. Characteristic stability of bare Au-water nanofluids fabricated by pulsed laser ablation in liquids. *Opt. Lasers Eng.* **2009**, *47*, 532–538. [CrossRef]
99. Lv, T.; Xu, X.; Yu, A.; Niu, C.; Hu, X. Ambient air quantity and cutting performances of water-based Fe_3O_4 nanofluid in magnetic minimum quantity lubrication. *Int. J. Adv. Manuf. Technol.* **2021**, *115*, 1711–1722. [CrossRef]
100. Bao, Y.; Sun, J.; Kong, L. Effects of nano-SiO_2 as water-based lubricant additive on surface qualities of strips after hot rolling. *Tribol. Int.* **2017**, *114*, 257–263. [CrossRef]
101. Lv, T.; Xu, X.; Yu, A.; Hu, X. Oil mist concentration and machining characteristics of SiO_2 water-based nano-lubricants in electrostatic minimum quantity lubrication-EMQL milling. *J. Mater. Process. Technol.* **2021**, *290*, 116964. [CrossRef]
102. Gao, Y.; Chen, G.; Oli, Y.; Zhang, Z.; Xue, Q. Study on tribological properties of oleic acid-modified TiO_2 nanoparticle in water. *Wear* **2002**, *252*, 454–458. [CrossRef]
103. Gu, Y.; Zhao, X.; Liu, Y.; Lv, Y. Preparation and Tribological Properties of Dual-Coated TiO_2 Nanoparticles as Water-Based Lubricant Additives. *J. Nanomater.* **2014**, *2014*, 785680. [CrossRef]
104. Ohenoja, K.; Saari, J.; Illikainen, M.; Niinimäki, J. Effect of molecular weight of sodium polyacrylates on the particle size distribution and stability of a TiO_2 suspension in aqueous stirred media milling. *Powder Technol.* **2014**, *262*, 188–193. [CrossRef]
105. Najiha, M.S.; Rahman, M.M. Experimental investigation of flank wear in end milling of aluminum alloy with water-based TiO_2 nanofluid lubricant in minimum quantity lubrication technique. *Int. J. Adv. Manuf. Technol.* **2016**, *86*, 2527–2537. [CrossRef]
106. Kayhani, M.H.; Soltanzadeh, H.; Heyhat, M.M.; Nazari, M.; Kowsary, F. Experimental study of convective heat transfer and pressure drop of TiO_2/water nanofluid. *Int. Commun. Heat Mass Transf.* **2012**, *39*, 456–462. [CrossRef]
107. Ukamanal, M.; Chandra Mishra, P.; Kumar Sahoo, A. Temperature distribution during AISI 316 steel turning under TiO_2-water based nanofluid spray environments. *Mater. Today Proc.* **2018**, *5*, 20741–20749. [CrossRef]
108. Wang, L.; Tieu, A.K.; Zhu, H.; Deng, G.; Cui, S.; Zhu, Q. A study of water-based lubricant with a mixture of polyphosphate and nano-TiO2 as additives for hot rolling process. *Wear* **2021**, *477*, 203895. [CrossRef]
109. Cui, Y.; Ding, M.; Sui, T.; Zheng, W.; Qiao, G.; Yan, S.; Liu, X. Role of nanoparticle materials as water-based lubricant additives for ceramics. *Tribol. Int.* **2020**, *142*, 105978. [CrossRef]
110. Wu, C.; Hou, S.X.; Zhang, H.Q.; Jia, X.M. Study and Evaluation on Dispersion of Molybdenum Disulfide in Aqueous Solution. *Adv. Mater. Res.* **2013**, *750–752*, 2175–2178. [CrossRef]
111. Zhang, B.; Sun, J. Tribological performances of multilayer-MoS_2 nanoparticles in water-based lubricating fluid. In *IOP Conference Series: Materials Science and Engineering*; IOP Publishing: Bristol, UK, 2017; Volume 182, p. 012023. [CrossRef]
112. Xie, H.; Dang, S.; Jiang, B.; Xiang, L.; Zhou, S.; Sheng, H.; Yang, T.; Pan, F. Tribological performances of SiO_2/graphene combinations as water-based lubricant additives for magnesium alloy rolling. *Appl. Surf. Sci.* **2019**, *475*, 847–856. [CrossRef]
113. Xie, H.; Wei, Y.; Jiang, B.; Tang, C.; Nie, C. Tribological properties of carbon nanotube/SiO2 combinations as water-based lubricant additives for magnesium alloy. *J. Mater. Res. Technol.* **2021**, *12*, 138–149. [CrossRef]
114. Zayan, M.; Rasheed, A.K.; John, A.; Khalid, M.; Ismail, A. Experimental Investigation on Rheological Properties of Water Based Novel Ternary Hybrid Nanofluids. *Nanoscience* **2021**. [CrossRef]

115. Wang, N.; Wang, H.; Ren, J.; Gao, G.; Chen, S.; Zhao, G.; Yang, Y.; Wang, J. Novel additive of PTFE@SiO₂ core-shell nanoparticles with superior water lubricating properties. *Mater. Des.* **2020**, *195*, 109069. [CrossRef]
116. Zhang, C.; Zhang, S.; Yu, L.; Zhang, Z.; Wu, Z.; Zhang, P. Preparation and tribological properties of water-soluble copper/silica nanocomposite as a water-based lubricant additive. *Appl. Surf. Sci.* **2012**, *259*, 824–830. [CrossRef]
117. Liu, T.; Zhou, C.; Gao, C.; Zhang, Y.; Yang, G.; Zhang, P.; Zhang, S. Preparation of Cu@SiO₂ composite nanoparticle and its tribological properties as water-based lubricant additive. *Lubr. Sci.* **2020**, *32*, 69–79. [CrossRef]
118. He, J.; Sun, J.; Meng, Y.; Pei, Y. Superior lubrication performance of MoS₂-Al₂O₃ composite nanofluid in strips hot rolling. *J. Manuf. Process.* **2020**, *57*, 312–323. [CrossRef]
119. Kumar, A.S.; Deb, S.; Paul, S. Tribological characteristics and micromilling performance of nanoparticle enhanced water based cutting fluids in minimum quantity lubrication. *J. Manuf. Process.* **2020**, *56*, 766–776. [CrossRef]
120. Giwa, S.O.; Sharifpur, M.; Ahmadi, M.H.; Sohel Murshed, S.M.; Meyer, J.P. Experimental Investigation on Stability, Viscosity, and Electrical Conductivity of Water-Based Hybrid Nanofluid of MWCNT-Fe(2)O(3). *Nanomaterials* **2021**, *11*, 136. [CrossRef] [PubMed]
121. Song, H.; Huang, J.; Jia, X.; Sheng, W. Facile synthesis of core–shell Ag@C nanospheres with improved tribological properties for water-based additives. *New J. Chem.* **2018**, *42*, 8773–8782. [CrossRef]
122. Gara, L.; Zou, Q. Friction and Wear Characteristics of Water-Based ZnO and Al₂O₃ Nanofluids. *Tribol. Trans.* **2012**, *55*, 345–350. [CrossRef]
123. Liang, D.; Ling, X.; Xiong, S. Preparation, characterisation and lubrication performances of Eu doped WO₃ nanoparticle reinforce Mn₃B₇O₁₃ Cl as water-based lubricant additive for laminated Cu-Fe composite sheet during hot rolling. *Lubr. Sci.* **2021**, *33*, 142–152. [CrossRef]
124. Wang, Y.; Cui, L.; Cheng, G.; Yuan, N.; Ding, J.; Pesika, N.S. Water-Based Lubrication of Hard Carbon Microspheres as Lubricating Additives. *Tribol. Lett.* **2018**, *66*, 148. [CrossRef]
125. Peng, Y.; Hu, Y.; Wang, H. Tribological behaviors of surfactant-functionalized carbon nanotubes as lubricant additive in water. *Tribol. Lett.* **2007**, *25*, 247–253. [CrossRef]
126. Sun, X.; Han, B.; Kang, H.; Fan, Z.; Liu, Y.; Umar, A.; Guo, Z. Frictional Reduction with Partially Exfoliated Multi-Walled Carbon Nanotubes as Water-Based Lubricant Additives. *J. Nanosci. Nanotechnol.* **2018**, *18*, 3427–3432. [CrossRef] [PubMed]
127. Min, C.; He, Z.; Liu, D.; Zhang, K.; Dong, C. Urea Modified Fluorinated Carbon Nanotubes: Unique Self-Dispersed Characteristic in Water and High Tribological Performance as Water-Based Lubricant Additives. *New J. Chem.* **2019**, *43*, 14684–14693. [CrossRef]
128. Kristiansen, K.; Zeng, H.; Wang, P.; Israelachvili, J. Microtribology of Aqueous Carbon Nanotube Dispersions. *Adv. Funct. Mater.* **2011**, *21*, 4555–4564. [CrossRef]
129. Ye, X.; E, S.; Fan, M. The influences of functionalized carbon nanotubes as lubricating additives: Length and diameter. *Diam. Relat. Mater.* **2019**, *100*, 107548. [CrossRef]
130. Tang, W.; Wang, B.; Li, J.; Li, Y.; Zhang, Y.; Quan, H.; Huang, Z. Facile pyrolysis synthesis of ionic liquid capped carbon dots and subsequent application as the water-based lubricant additives. *J. Mater. Sci.* **2019**, *54*, 1171–1183. [CrossRef]
131. Xiao, H.; Liu, S.; Xu, Q.; Zhang, H. Carbon quantum dots: An innovative additive for water lubrication. *Sci. China Technol. Sci.* **2019**, *62*, 587–596. [CrossRef]
132. Fan, K.; Liu, X.; Liu, Y.; Li, Y.; Chen, Y.; Meng, Y.; Liu, X.; Feng, W.; Luo, L. Covalent functionalization of fluorinated graphene through activation of dormant radicals for water-based lubricants. *Carbon* **2020**, *167*, 826–834. [CrossRef]
133. Ma, L.; Li, Z.; Jia, W.; Hou, K.; Wang, J.; Yang, S. Microwave-assisted synthesis of hydroxyl modified fluorinated graphene with high fluorine content and its high load-bearing capacity as water lubricant additive for ceramic/steel contact. *Colloids Surf. A* **2021**, *610*, 125931. [CrossRef]
134. Ye, X.; Ma, L.; Yang, Z.; Wang, J.; Wang, H.; Yang, S. Covalent Functionalization of Fluorinated Graphene and Subsequent Application as Water-based Lubricant Additive. *ACS Appl. Mater. Interfaces* **2016**, *8*, 7483–7488. [CrossRef] [PubMed]
135. Qiang, R.; Hu, L.; Hou, K.; Wang, J.; Yang, S. Water-Soluble Graphene Quantum Dots as High-Performance Water-Based Lubricant Additive for Steel/Steel Contact. *Tribol. Lett.* **2019**, *67*, 64. [CrossRef]
136. Piatkowska, A.; Romaniec, M.; Grzybek, D.; Mozdzonek, M.; Rojek, A.; Diduszko, R. A study on antiwear properties of graphene water-based lubricant and its contact with metallic materials. *Tribologia* **2018**, *281*, 71–81. [CrossRef]
137. Mirzaamiri, R.; Akbarzadeh, S.; Ziaei-Rad, S.; Shin, D.-G.; Kim, D.-E. Molecular dynamics simulation and experimental investigation of tribological behavior of nanodiamonds in aqueous suspensions. *Tribol. Int.* **2021**, *156*, 106838. [CrossRef]
138. Kinoshita, H.; Nishina, Y.; Alias, A.A.; Fujii, M. Tribological properties of monolayer graphene oxide sheets as water-based lubricant additives. *Carbon* **2014**, *66*, 720–723. [CrossRef]
139. Elomaa, O.; Singh, V.K.; Iyer, A.; Hakala, T.J.; Koskinen, J. Graphene oxide in water lubrication on diamond-like carbon vs. stainless steel high-load contacts. *Diam. Relat. Mater.* **2015**, *52*, 43–48. [CrossRef]
140. Singh, S.; Chen, X.; Zhang, C.; Tyagi, R.; Luo, J. Investigation on the lubrication potential of graphene oxide aqueous dispersion for self-mated stainless steel tribo-pair. *Vacuum* **2019**, *166*, 307–315. [CrossRef]
141. Bai, M.; Liu, J.; He, J.; Li, W.; Wei, J.; Chen, L.; Miao, J.; Li, C. Heat transfer and mechanical friction reduction properties of graphene oxide nanofluids. *Diam. Relat. Mater.* **2020**, *108*, 107982. [CrossRef]
142. Gan, C.; Liang, T.; Li, X.; Li, W.; Li, H.; Fan, X.; Zhu, M. Ultra-dispersive monolayer graphene oxide as water-based lubricant additive: Preparation, characterization and lubricating mechanisms. *Tribol. Int.* **2021**, *155*, 106768. [CrossRef]

143. Kim, H.-J.; Kim, D.-E. Water Lubrication of Stainless Steel using Reduced Graphene Oxide Coating. *Sci. Rep.* **2015**, *5*, 17034. [CrossRef]
144. Kinoshita, H.; Suzuki, K.; Suzuki, T.; Nishina, Y. Tribological properties of oxidized wood-derived nanocarbons with same surface chemical composition as graphene oxide for additives in water-based lubricants. *Diam. Relat. Mater.* **2018**, *90*, 101–108. [CrossRef]
145. Wei, Q.; Fu, T.; Yue, Q.; Liu, H.; Ma, S.; Cai, M.; Zhou, F. Graphene oxide/brush-like polysaccharide copolymer nanohybrids as eco-friendly additives for water-based lubrication. *Tribol. Int.* **2021**, *157*, 106895. [CrossRef]
146. Li, X.; Lu, H.; Guo, J.; Tong, Z.; Dong, G. Synergistic water lubrication effect of self-assembled nanofilm and graphene oxide additive. *Appl. Surf. Sci.* **2018**, *455*, 1070–1077. [CrossRef]
147. Hu, Y.; Wang, Y.; Zeng, Z.; Zhao, H.; Ge, X.; Wang, K.; Wang, L.; Xue, Q. PEGlated graphene as nanoadditive for enhancing the tribological properties of water-based lubricants. *Carbon* **2018**, *137*, 41–48. [CrossRef]
148. Shariatzadeh, M.; Grecov, D. Cellulose Nanocrystals Suspensions as Water-Based Lubricants for Slurry Pump Gland Seals. *Int. J. Aerosp. Mech. Eng.* **2018**, *12*, 603–607. [CrossRef]
149. Shariatzadeh, M.; Grecov, D. Aqueous suspensions of cellulose nanocrystals as water-based lubricants. *Cellulose* **2019**, *26*, 4665–4677. [CrossRef]
150. Lei, H.; Guan, W.; Luo, J. Tribological behavior of fullerene–styrene sulfonic acid copolymer as water-based lubricant additive. *Wear* **2002**, *252*, 345–350. [CrossRef]
151. Jiang, G.; Guan, W.; Zheng, Q. A study on fullerene–acrylamide copolymer nanoball—A new type of water-based lubrication additive. *Wear* **2005**, *258*, 1625–1629. [CrossRef]
152. Wang, Y.; Yang, W.; Yu, Q.; Zhang, J.; Ma, Z.; Zhang, M.; Zhang, L.; Bai, Y.; Cai, M.; Zhou, F.; et al. Significantly Reducing Friction and Wear of Water-Based Fluids with Shear Thinning Bicomponent Supramolecular Hydrogels. *Adv. Mater. Interfaces* **2020**, *7*, 2001084. [CrossRef]
153. Yang, D.; Du, X.; Li, W.; Han, Y.; Ma, L.; Fan, M.; Zhou, F.; Liu, W. Facile Preparation and Tribological Properties of Water-Based Naphthalene Dicarboxylate Ionic Liquid Lubricating Additives. *Tribol. Lett.* **2020**, *68*, 84. [CrossRef]
154. Zhang, J.; Zhang, Y.; Zhang, S.; Yu, L.; Zhang, P.; Zhang, Z. Preparation of Water-Soluble Lanthanum Fluoride Nanoparticles and Evaluation of their Tribological Properties. *Tribol. Lett.* **2013**, *52*, 305–314. [CrossRef]
155. Zheng, D.; Wang, X.; Liu, Z.; Ju, C.; Xu, Z.; Xu, J.; Yang, C. Synergy between two protic ionic liquids for improving the antiwear property of glycerol aqueous solution. *Tribol. Int.* **2020**, *141*, 105731. [CrossRef]
156. Dong, R.; Yu, Q.; Bai, Y.; Wu, Y.; Ma, Z.; Zhang, J.; Zhang, C.; Yu, B.; Zhou, F.; Liu, W.; et al. Towards superior lubricity and anticorrosion performances of proton-type ionic liquids additives for water-based lubricating fluids. *Chem. Eng. J.* **2020**, *383*, 123201. [CrossRef]
157. Khanmohammadi, H.; Wijanarko, W.; Espallargas, N. Ionic Liquids as Additives in Water-Based Lubricants: From Surface Adsorption to Tribofilm Formation. *Tribol. Lett.* **2020**, *68*, 130. [CrossRef]
158. Kreivaitis, R.; Gumbytė, M.; Kupčinskas, A.; Kazancev, K.; Ta, T.N.; Horng, J.H. Investigation of tribological properties of two protic ionic liquids as additives in water for steel–steel and alumina–steel contacts. *Wear* **2020**, *456–457*, 203390. [CrossRef]
159. He, J.; Sun, J.; Meng, Y.; Yan, X. Preliminary investigations on the tribological performance of hexagonal boron nitride nanofluids as lubricant for steel/steel friction pairs. *Surf. Topogr. Metrol. Prop.* **2019**, *7*, 015022. [CrossRef]
160. Abdollah, M.F.B.; Amiruddin, H.; Azmi, M.A.; Tahir, N.A.M. Lubrication mechanisms of hexagonal boron nitride nano-additives water-based lubricant for steel–steel contact. *J. Eng. Tribol.* **2020**, *235*, 1038–1046. [CrossRef]
161. Lin, B.; Ding, M.; Sui, T.; Cui, Y.; Yan, S.; Liu, X. Excellent Water Lubrication Additives for Silicon Nitride to Achieve Superlubricity under Extreme Conditions. *Langmuir* **2019**, *35*, 14861–14869. [CrossRef] [PubMed]
162. Wang, W.; Xie, G.; Luo, J. Black phosphorus as a new lubricant. *Friction* **2018**, *6*, 116–142. [CrossRef]
163. Wang, Q.; Hou, T.; Wang, W.; Zhang, G.; Gao, Y.; Wang, K. Tribological behavior of black phosphorus nanosheets as water-based lubrication additives. *Friction* **2021**, 1–14. [CrossRef]
164. Ali, N.; Amaral Teixeira, J.; Addali, A. A Review on Nanofluids: Fabrication, Stability, and Thermophysical Properties. *J. Nanomater.* **2018**, *2018*, 6978130. [CrossRef]
165. Dey, D.; Kumar, P.; Samantaray, S. A review of nanofluid preparation, stability, and thermo-physical properties. *Heat Transf. Asian Res.* **2017**, *46*, 1413–1442. [CrossRef]
166. Ghadimi, A.; Saidur, R.; Metselaar, H.S.C. A review of nanofluid stability properties and characterization in stationary conditions. *Int. J. Heat Mass Transf.* **2011**, *54*, 4051–4068. [CrossRef]
167. Wu, D.; Zhu, H.; Wang, L.; Liu, L. Critical Issues in Nanofluids Preparation, Characterization and Thermal Conductivity. *Curr. Nanosci.* **2009**, *5*, 103–112. [CrossRef]
168. Zhu, H.; Li, C.; Wu, D.; Zhang, C.; Yin, Y. Preparation, characterization, viscosity and thermal conductivity of $CaCO_3$ aqueous nanofluids. *Sci. China Technol. Sci.* **2010**, *53*, 360–368. [CrossRef]
169. Urmi, W.T.; Rahman, M.M.; Kadirgama, K.; Ramasamy, D.; Maleque, M.A. An overview on synthesis, stability, opportunities and challenges of nanofluids. *Mater. Today Proc.* **2021**, *41*, 30–37. [CrossRef]
170. Fang, Y.; Ma, L.; Luo, J. Modelling for water-based liquid lubrication with ultra-low friction coefficient in rough surface point contact. *Tribol. Int.* **2020**, *141*, 105901. [CrossRef]
171. Azman, N.F.; Samion, S. Dispersion Stability and Lubrication Mechanism of Nanolubricants: A Review. *Int. J. Precis. Eng. Manuf.-Green Technol.* **2019**, *6*, 393–414. [CrossRef]

172. Cacua, K.; Murshed, S.M.S.; Pabón, E.; Buitrago, R. Dispersion and thermal conductivity of TiO_2/water nanofluid: Effects of ultrasonication, agitation and temperature. *J. Therm. Anal. Calorim.* **2019**, *140*, 109–114. [CrossRef]
173. Lee, K.; Hwang, Y.; Cheong, S.; Kwon, L.; Kim, S.; Lee, J. Performance evaluation of nano-lubricants of fullerene nanoparticles in refrigeration mineral oil. *Curr. Appl. Phys.* **2009**, *9*, e128–e131. [CrossRef]
174. Ghadimi, A.; Metselaar, H.; LotfizadehDehkordi, B. Nanofluid stability optimization based on UV-Vis spectrophotometer measurement. *J. Eng. Sci. Technol.* **2015**, *10*, 32–40.
175. Jiang, L.; Gao, L.; Sun, J. Production of aqueous colloidal dispersions of carbon nanotubes. *J. Colloid Interface Sci.* **2003**, *260*, 89–94. [CrossRef]
176. Yu, W.; Xie, H. A Review on Nanofluids: Preparation, Stability Mechanisms, and Applications. *J. Nanomater.* **2012**, *2012*, 128. [CrossRef]
177. Sajid, M.U.; Ali, H.M. Thermal conductivity of hybrid nanofluids: A critical review. *Int. J. Heat Mass Transf.* **2018**, *126*, 211–234. [CrossRef]
178. Alawi, O.A.; Mallah, A.R.; Kazi, S.N.; Sidik, N.A.C.; Najafi, G. Thermophysical properties and stability of carbon nanostructures and metallic oxides nanofluids. *J. Therm. Anal. Calorim.* **2019**, *135*, 1545–1562. [CrossRef]
179. Kumar, M.S.; Vasu, V.; Gopal, A.V. Thermal conductivity and rheological studies for Cu–Zn hybrid nanofluids with various basefluids. *J. Taiwan Inst. Chem. Eng.* **2016**, *66*, 321–327. [CrossRef]
180. Yu, F.; Chen, Y.; Liang, X.; Xu, J.; Lee, C.; Liang, Q.; Tao, P.; Deng, T. Dispersion stability of thermal nanofluids. *Prog. Nat. Sci. Mater. Int.* **2017**, *27*, 531–542. [CrossRef]
181. Cremaschi, L.; Bigi, A.; Wong, T.; Deokar, P. Thermodynamic properties of Al_2O_3 nanolubricants: Part 1—Effects on the two-phase pressure drop. *Sci. Technol. Built Environ.* **2015**, *21*, 607–620. [CrossRef]
182. Huminic, G.; Huminic, A.; Fleacă, C.; Dumitrache, F.; Morjan, I. Experimental study on viscosity of water based Fe–Si hybrid nanofluids. *J. Mol. Liq.* **2021**, *321*, 114938. [CrossRef]
183. Haghighi, E.; Nikkam, N.; Saleemi, M.; Behi, R.; Mirmohammadi, S.A.; Poth, H.; Khodabandeh, R.; Toprak, M.; Muhammed, M.; Palm, B. Shelf stability of nanofluids and its effect on thermal conductivity and viscosity Shelf stability of nanofluids and its effect on thermal conductivity and viscosity. *Meas. Sci. Technol.* **2013**, *24*, 105301–105311. [CrossRef]
184. Baghbanzadeh, M.; Rashidi, A.; Soleimanisalim, A.H.; Rashtchian, D. Investigating the rheological properties of nanofluids of water/hybrid nanostructure of spherical silica/MWCNT. *Thermochim. Acta* **2014**, *578*, 53–58. [CrossRef]
185. Zhu, H.; Zhang, C.; Tang, Y.; Wang, J.; Ren, B.; Yin, Y. Preparation and thermal conductivity of suspensions of graphite nanoparticles. *Carbon* **2007**, *45*, 226–228. [CrossRef]
186. Li, Y.; Zhou, J.; Tung, S.; Schneider, E.; Xi, S. A review on development of nanofluid preparation and characterization. *Powder Technol.* **2009**, *196*, 89–101. [CrossRef]
187. Mukherjee, S.; Paria, S. Preparation and Stability of Nanofluids-A Review. *IOSR J. Mech. Civ. Eng.* **2013**, *9*, 63–69. [CrossRef]
188. Qamar, A.; Anwar, Z.; Ali, H.; Shaukat, R.; Imran, S.; Arshad, A.; Ali, H.M.; Korakianitis, T. Preparation and dispersion stability of aqueous metal oxide nanofluids for potential heat transfer applications: A review of experimental studies. *J. Therm. Anal. Calorim.* **2020**, 1–24. [CrossRef]
189. Oh, D.-W.; Jain, A.; Eaton, J.K.; Goodson, K.E.; Lee, J.S. Thermal conductivity measurement and sedimentation detection of aluminum oxide nanofluids by using the 3ω method. *Int. J. Heat Fluid Flow* **2008**, *29*, 1456–1461. [CrossRef]
190. Peng, X.F.; Yu, X.; Xia, L.F.; Zhong, X. Influence factors on suspension stability of nanofluids. *J. Zhejiang Univ.* **2007**, *41*, 577–580.
191. Azizian, R.; Doroodchi, E.; Moghtaderi, B. Influence of Controlled Aggregation on Thermal Conductivity of Nanofluids. *J. Heat Transf.* **2016**, *138*, 021301. [CrossRef]
192. Jama, M.; Singh, T.; Mahmoud, S.; Koç, M.; Samara, A.; Isaifan, R.; Atieh, M. Critical Review on Nanofluids: Preparation, Characterization, and Applications. *J. Nanomater.* **2016**, *2016*, 6717624. [CrossRef]
193. Kumar Dubey, M.; Bijwe, J.; Ramakumar, S.S.V. PTFE based nano-lubricants. *Wear* **2013**, *306*, 80–88. [CrossRef]
194. Lee, G.-J.; Rhee, C.K. Enhanced thermal conductivity of nanofluids containing graphene nanoplatelets prepared by ultrasound irradiation. *J. Mater. Sci.* **2014**, *49*, 1506–1511. [CrossRef]
195. Mahbubul, I.M.; Elcioglu, E.B.; Saidur, R.; Amalina, M.A. Optimization of ultrasonication period for better dispersion and stability of TiO2–water nanofluid. *Ultrason. Sonochem.* **2017**, *37*, 360–367. [CrossRef]
196. Tang, E.; Cheng, G.; Ma, X.; Pang, X.; Zhao, Q. Surface modification of zinc oxide nanoparticle by PMAA and its dispersion in aqueous system. *Appl. Surf. Sci.* **2006**, *252*, 5227–5232. [CrossRef]
197. Neouze, M.-A.; Schubert, U. Surface Modification and Functionalization of Metal and Metal Oxide Nanoparticles by Organic Ligands. *Mon. Chem.-Chem. Mon.* **2008**, *139*, 183–195. [CrossRef]
198. Yang, X.-F.; Liu, Z.-H. Pool boiling heat transfer of functionalized nanofluid under sub-atmospheric pressures. *Int. J. Therm. Sci.* **2011**, *50*, 2402–2412. [CrossRef]
199. Sezer, N.; Atieh, M.A.; Koç, M. A comprehensive review on synthesis, stability, thermophysical properties, and characterization of nanofluids. *Powder Technol.* **2019**, *344*, 404–431. [CrossRef]
200. Narendar, G.; Gupta, A.V.S.S.K.S.; Krishnaiah, A.; Satyanarayana, M.G.V. Experimental investigation on the preparation and applications of Nano fluids. *Mater. Today Proc.* **2017**, *4*, 3926–3931. [CrossRef]
201. Xia, G.; Jiang, H.; Liu, R.; Zhai, Y. Effects of surfactant on the stability and thermal conductivity of Al_2O_3/de-ionized water nanofluids. *Int. J. Therm. Sci.* **2014**, *84*, 118–124. [CrossRef]

202. Kakati, H.; Mandal, A.; Laik, S. Promoting effect of Al₂O₃/ZnO-based nanofluids stabilized by SDS surfactant on CH₄+C₂H₆+C₃H₈ hydrate formation. *J. Ind. Eng. Chem.* **2016**, *35*, 357–368. [CrossRef]

203. Tomala, A.; Karpinska, A.; Werner, W.S.M.; Olver, A.; Störi, H. Tribological properties of additives for water-based lubricants. *Wear* **2010**, *269*, 804–810. [CrossRef]

204. Wen, P.; Lei, Y.; Li, W.; Fan, M. Synergy between Covalent Organic Frameworks and Surfactants to Promote Water-Based Lubrication and Corrosion-Resistance. *ACS Appl. Nano Mater.* **2020**, *3*, 1400–1411. [CrossRef]

205. Popa, I.; Gillies, G.; Papastavrou, G.; Borkovec, M. Attractive and Repulsive Electrostatic Forces between Positively Charged Latex Particles in the Presence of Anionic Linear Polyelectrolytes. *J. Phys. Chem. B* **2010**, *114*, 3170–3177. [CrossRef] [PubMed]

206. Tadros, T. Chapter 2—Colloid and interface aspects of pharmaceutical science. In *Colloid and Interface Science in Pharmaceutical Research and Development*; Ohshima, H., Makino, K., Eds.; Elsevier: Amsterdam, The Netherlands, 2014; pp. 29–54.

207. Du, Y.; Yuan, X. Coupled hybrid nanoparticles for improved dispersion stability of nanosuspensions: A review. *J. Nanopart. Res.* **2020**, *22*, 261. [CrossRef]

208. Byrd, T.; Walz, J. Interaction Force Profiles between Cryptosporidium parvum Oocysts and Silica Surfaces. *Environ. Sci. Technol.* **2006**, *39*, 9574–9582. [CrossRef] [PubMed]

209. Napper, D.H. Steric stabilization. *J. Colloid Interface Sci.* **1977**, *58*, 390–407. [CrossRef]

210. Zhulina, E.B.; Borisov, O.V.; Priamitsyn, V.A. Theory of steric stabilization of colloid dispersions by grafted polymers. *J. Colloid Interface Sci.* **1990**, *137*, 495–511. [CrossRef]

211. Gerber, P.; Moore, M.A. Comments on the Theory of Steric Stabilization. *Macromolecules* **1977**, *10*, 476–481. [CrossRef]

212. Dutta, N.; Green, D. Impact of Solvent Quality on Nanoparticle Dispersion in Semidilute and Concentrated Polymer Solutions. *Langmuir* **2010**, *26*, 16737–16744. [CrossRef]

213. Munkhbayar, B.; Tanshen, M.R.; Jeoun, J.; Chung, H.; Jeong, H. Surfactant-free dispersion of silver nanoparticles into MWCNT-aqueous nanofluids prepared by one-step technique and their thermal characteristics. *Ceram. Int.* **2013**, *39*, 6415–6425. [CrossRef]

214. Hatami, M.; Ali, F.; Alsabery, A.; Hu, S.; Jing, D.; Hameed, K. mixed convection heat transfer of SiO₂-water and alumina-pao nano-lubricants used in a mechanical ball bearing. *J. Therm. Eng.* **2021**, *7*, 134–161. [CrossRef]

215. Chen, L.; Cheng, M.; Yang, D.; Yang, L. Enhanced Thermal Conductivity of Nanofluid by Synergistic Effect of Multi-Walled Carbon Nanotubes and Fe₂O₃ Nanoparticles. *Appl. Mech. Mater.* **2014**, *548–549*, 118–123. [CrossRef]

216. Botha, S.S.; Ndungu, P.; Bladergroen, B.J. Physicochemical Properties of Oil-Based Nanofluids Containing Hybrid Structures of Silver Nanoparticles Supported on Silica. *Ind. Eng. Chem. Res.* **2011**, *50*, 3071–3077. [CrossRef]

217. Nine, M.J.; Munkhbayar, B.; Rahman, M.S.; Chung, H.; Jeong, H. Highly productive synthesis process of well dispersed Cu₂O and Cu/Cu₂O nanoparticles and its thermal characterization. *Mater. Chem. Phys.* **2013**, *141*, 636–642. [CrossRef]

218. Valdes, M.; Gonzalo, J.; Marín, A.; Rodriguez, M.; Betancur, J. Tribometry: How is friction research quantified? A review. *Int. J. Eng. Res. Technol.* **2020**, *13*, 2596–2610. [CrossRef]

219. Paul, G.; Hirani, H.; Kuila, T.; Murmu, N. Nanolubricants Dispersed with Graphene and its Derivatives: An Assessment and Review of the Tribological Performance. *Nanoscale* **2019**, *11*, 3458–3483. [CrossRef] [PubMed]

220. Sagraloff, N.; Dobler, A.; Tobie, T.; Stahl, K.; Ostrowski, J. Development of an Oil Free Water-Based Lubricant for Gear Applications. *Lubricants* **2019**, *7*, 33. [CrossRef]

221. Zhang, T.; Jiang, F.; Yan, L.; Jiang, Z.; Xu, X. A novel ultrahigh-speed ball-on-disc tribometer. *Tribol. Int.* **2021**, *157*, 106901. [CrossRef]

222. Chen, W.; Amann, T.; Kailer, A.; Rühe, J. Macroscopic Friction Studies of Alkylglucopyranosides as Additives for Water-Based Lubricants. *Lubricants* **2020**, *8*, 11. [CrossRef]

223. Cheng, X.; Jiang, Z.; Wei, D.; Wu, H.; Jiang, L. Adhesion, friction and wear analysis of a chromium oxide scale on a ferritic stainless steel. *Wear* **2019**, *426–427*, 1212–1221. [CrossRef]

224. Pawlak, Z.; Urbaniak, W.; Oloyede, A. The relationship between friction and wettability in aqueous environment. *Wear* **2011**, *271*, 1745–1749. [CrossRef]

225. Liu, Y.; Chen, X.; Li, J.; Luo, J. Enhancement of friction performance enabled by a synergetic effect between graphene oxide and molybdenum disulfide. *Carbon* **2019**, *154*, 266–276. [CrossRef]

226. Wu, H.; Li, Y.; Lu, Y.; Li, Z.; Cheng, X.; Hasan, M.; Zhang, H.; Jiang, Z. Influences of Load and Microstructure on Tribocorrosion Behaviour of High Strength Hull Steel in Saline Solution. *Tribol. Lett.* **2019**, *67*, 124. [CrossRef]

227. Tang, W.; Zhu, X.; Li, Y. Tribological performance of various metal-doped carbon dots as water-based lubricant additives and their potential application as additives of poly(ethylene glycol). *Friction* **2021**, *1*, 1–18. [CrossRef]

228. Kim, H.J.; Shin, D.G.; Kim, D.-E. Frictional behavior between silicon and steel coated with graphene oxide in dry sliding and water lubrication conditions. *Int. J. Precis. Eng. Manuf.-Green Technol.* **2016**, *3*, 91–97. [CrossRef]

229. Rosa, W.O.; Vereda, F.; de Vicente, J. Tribological Behavior of Glycerol/Water-Based Magnetorheological Fluids in PMMA Point Contacts. *Front. Mater.* **2019**, *6*, 32. [CrossRef]

230. Kotia, A.; Borkakoti, S.; Deval, P.; Ghosh, S.K. Review of interfacial layer's effect on thermal conductivity in nanofluid. *Heat Mass Transf.* **2017**, *53*, 2199–2209. [CrossRef]

231. Kotia, A.; Rajkhowa, P.; Rao, G.S.; Ghosh, S.K. Thermophysical and tribological properties of nanolubricants: A review. *Heat Mass Transf.* **2018**, *54*, 3493–3508. [CrossRef]

232. Tang, Z.; Li, S. A review of recent developments of friction modifiers for liquid lubricants (2007–present). *Curr. Opin. Solid State Mater. Sci.* **2014**, *18*, 119–139. [CrossRef]
233. Kong, L.; Sun, J.; Bao, Y. Preparation, characterization and tribological mechanism of nanofluids. *RSC Adv.* **2017**, *7*, 12599–12609. [CrossRef]
234. Liu, L.; Zhou, M.; Jin, L.; Li, L.; Mo, Y.; Su, G.; Li, X.; Zhu, H.; Tian, Y. Recent advances in friction and lubrication of graphene and other 2D materials: Mechanisms and applications. *Friction* **2019**, *7*, 199–216. [CrossRef]
235. Zhao, J.; Huang, Y.; He, Y.; Shi, Y. Nanolubricant additives: A review. *Friction* **2021**, *9*, 891–917. [CrossRef]
236. Sarno, M.; Scarpa, D.; Senatore, A.; Mustafa, W. rGO/GO Nanosheets in Tribology: From the State of the Art to the Future Prospective. *Lubricants* **2020**, *8*, 31. [CrossRef]
237. Laad, M.; Jatti, V.K.S. Titanium oxide nanoparticles as additives in engine oil. *J. King Saud Univ. Eng. Sci.* **2018**, *30*, 116–122. [CrossRef]
238. Khadem, M.; Penkov, O.; Pukha, V.; Maleyev, M.; Kim, D.-E. Ultra-thin carbon-based nanocomposite coatings for superior wear resistance under lubrication with nano-diamond additives. *RSC Adv.* **2016**, *6*, 56918–56929. [CrossRef]
239. Chiñas-Castillo, F.; Spikes, H. Mechanism of Action of Colloidal Solid Dispersions. *J. Tribol.* **2003**, *125*, 552–557. [CrossRef]
240. Rapoport, L.; Leshchinsky, V.; Lvovsky, M.; Nepomnyashchy, O.; Volovik, Y.; Tenne, R. Mechanism of friction of fullerene. *Ind. Lubr. Tribol.* **2002**, *54*, 171–176. [CrossRef]
241. Wu, Y.Y.; Tsui, W.C.; Liu, T.C. Experimental analysis of tribological properties of lubricating oils with nanoparticle additives. *Wear* **2007**, *262*, 819–825. [CrossRef]
242. Aldana, P.U.; Dassenoy, F.; Vacher, B.; Le Mogne, T.; Thiebaut, B. WS$_2$ nanoparticles anti-wear and friction reducing properties on rough surfaces in the presence of ZDDP additive. *Tribol. Int.* **2016**, *102*, 213–221. [CrossRef]
243. Ku, B.-C.; Han, Y.-C.; Lee, J.-E.; Lee, J.-K.; Park, S.-H.; Hwang, Y.-J. Tribological effects of fullerene (C60) nanoparticles added in mineral lubricants according to its viscosity. *Int. J. Precis. Eng. Manuf.* **2010**, *11*, 607–611. [CrossRef]
244. Peng, D.; Kang, Y.; Hwang, R.; Shyr, S.; Chang, Y. Tribological properties of diamond and SiO$_2$ nanoparticles added in paraffin. *Tribol. Int.* **2009**, *42*, 911–917. [CrossRef]
245. Srivyas, P.; Charoo, M.S. A Review on Tribological Characterization of Lubricants with Nano Additives for Automotive Applications. *Tribol. Ind.* **2018**, *40*, 594–623. [CrossRef]
246. Xiao, H.; Liu, S. 2D nanomaterials as lubricant additive: A review. *Mater. Des.* **2017**, *135*, 319–332. [CrossRef]
247. Flores-Castañeda, M.; Camps, E.; Camacho-López, M.; Muhl, S.; García, E.; Figueroa, M. Bismuth nanoparticles synthesized by laser ablation in lubricant oils for tribological tests. *J. Alloy Compd.* **2015**, *643*, S67–S70. [CrossRef]
248. Kato, H.; Komai, K. Tribofilm formation and mild wear by tribo-sintering of nanometer-sized oxide particles on rubbing steel surfaces. *Wear* **2007**, *262*, 36–41. [CrossRef]
249. Song, X.; Zheng, S.; Zhang, J.; Li, W.; Chen, Q.; Cao, B. Synthesis of monodispersed ZnAl$_2$O$_4$ nanoparticles and their tribology properties as lubricant additives. *Mater. Res. Bull.* **2012**, *47*, 4305–4310. [CrossRef]
250. Uflyand, I.E.; Zhinzhilo, V.A.; Burlakova, V.E. Metal-containing nanomaterials as lubricant additives: State-of-the-art and future development. *Friction* **2019**, *7*, 93–116. [CrossRef]
251. Lee, C.-G.; Hwang, Y.-J.; Choi, Y.-M.; Lee, J.-K.; Choi, C.; Oh, J.-M. A study on the tribological characteristics of graphite nano lubricants. *Int. J. Precis. Eng. Manuf.* **2009**, *10*, 85–90. [CrossRef]
252. Lee, K.; Hwang, Y.; Cheong, S.; Choi, Y.; Kwon, L.; Lee, J.; Kim, S.H. Understanding the Role of Nanoparticles in Nano-oil Lubrication. *Tribol. Lett.* **2009**, *35*, 127–131. [CrossRef]
253. Tevet, O.; Von-Huth, P.; Popovitz-Biro, R.; Rosentsveig, R.; Wagner, H.D.; Tenne, R. Friction mechanism of individual multilayered nanoparticles. *Proc. Natl. Acad. Sci. USA* **2011**, *108*, 19901–19906. [CrossRef]
254. Cizaire, L.; Vacher, B.; Le Mogne, T.; Martin, J.M.; Rapoport, L.; Margolin, A.; Tenne, R. Mechanisms of ultra-low friction by hollow inorganic fullerene-like MoS$_2$ nanoparticles. *Surf. Coat. Technol.* **2002**, *160*, 282–287. [CrossRef]
255. Joly-Pottuz, L.; Martin, J.; Dassenoy, F.; Belin, M.; Montagnac, G.; Reynard, B.; Fleischer, N. Pressure-induced exfoliation of inorganic fullerene-like WS2 particles in a Hertzian contact. *J. Appl. Phys.* **2006**, *99*, 023524. [CrossRef]
256. Rapoport, L.; Feldman, Y.; Homyonfer, M.; Cohen, H.; Sloan, J.; Hutchison, J.L.; Tenne, R. Inorganic fullerene-like material as additives to lubricants: Structure–function relationship. *Wear* **1999**, *225–229*, 975–982. [CrossRef]
257. Spikes, H. Friction Modifier Additives. *Tribol. Lett.* **2015**, *60*, 5. [CrossRef]
258. Lin, W.; Klein, J. Control of surface forces through hydrated boundary layers. *Curr. Opin. Colloid Interface Sci.* **2019**, *44*, 94–106. [CrossRef]
259. Klein, J. Hydration lubrication. *Friction* **2013**, *1*, 1–23. [CrossRef]
260. Suresh, S.; Venkitaraj, K.P.; Selvakumar, P.; Chandrasekar, M. Synthesis of Al$_2$O$_3$–Cu/water hybrid nanofluids using two step method and its thermo physical properties. *Colloids Surf. A* **2011**, *388*, 41–48. [CrossRef]
261. Kedzierski, M.A.; Brignoli, R.; Quine, K.T.; Brown, J.S. Viscosity, density, and thermal conductivity of aluminum oxide and zinc oxide nanolubricants. *Int. J. Refrig.* **2017**, *74*, 3–11. [CrossRef]
262. Mehrali, M.; Sadeghinezhad, E.; Latibari, S.T.; Kazi, S.N.; Mehrali, M.; Zubir, M.N.B.M.; Metselaar, H.S.C. Investigation of thermal conductivity and rheological properties of nanofluids containing graphene nanoplatelets. *Nanoscale Res. Lett.* **2014**, *9*, 15. [CrossRef]

263. Kulkarni, D.P.; Das, D.K.; Chukwu, G.A. Temperature dependent rheological property of copper oxide nanoparticles suspension (nanofluid). *J. Nanosci. Nanotechnol.* **2006**, *6*, 1150–1154. [CrossRef]
264. Pawelski, O.; Rasp, W.; Draese, S. Influence of Hydrodynamic Lubricant Entrainment on Friction Effects in Cold-Rolling. *Steel Res.* **1994**, *65*, 488–493. [CrossRef]
265. Qu, J.; Truhan, J.J.; Dai, S.; Luo, H.; Blau, P.J. Ionic liquids with ammonium cations as lubricants or additives. *Tribol. Lett.* **2006**, *22*, 207–214. [CrossRef]
266. Hild, W.; Opitz, A.; Schaefer, J.A.; Scherge, M. The effect of wetting on the microhydrodynamics of surfaces lubricated with water and oil. *Wear* **2003**, *254*, 871–875. [CrossRef]
267. Xia, W.Z.; Zhao, J.W.; Wu, H.; Zhao, X.M.; Zhang, X.M.; Xu, J.Z.; Hee, A.C.; Jiang, Z.Y. Effects of Nano-TiO$_2$ Additive in Oil-in-Water Lubricant on Contact Angle and Antiscratch Behavior. *Tribol. Trans.* **2017**, *60*, 362–372. [CrossRef]
268. Jiang, Z.Y.; Tang, J.; Sun, W.; Tieu, A.K.; Wei, D. Analysis of tribological feature of the oxide scale in hot strip rolling. *Tribol. Int.* **2010**, *43*, 1339–1345. [CrossRef]
269. Hao, L.; Wu, H.; Wei, D.B.; Cheng, X.W.; Zhao, J.W.; Luo, S.Z.; Jiang, L.Z.; Jiang, Z.Y. Wear and friction behaviour of high-speed steel and indefinite chill material for rolling ferritic stainless steels. *Wear* **2017**, *376*, 1580–1585. [CrossRef]
270. Colás, R.; Ramírez, J.; Sandoval, I.; Morales, J.C.; Leduc, L.A. Damage in hot rolling work rolls. *Wear* **1999**, *230*, 56–60. [CrossRef]
271. Najiha, M.S.; Rahman, M.M.; Kadirgama, K. Performance of water-based TiO$_2$ nanofluid during the minimum quantity lubrication machining of aluminium alloy, AA6061-T6. *J. Clean. Prod.* **2016**, *135*, 1623–1636. [CrossRef]
272. Williams, K. Tribology in Metal Working±New Developments. In Proceedings of the Echanical Engineering Conference, London, UK, 7–9 May 1980; pp. 4–6.
273. Zhu, Z.; Sun, J.; Niu, T.; Liu, N. Experimental research on tribological performance of water-based rolling liquid containing nano-TiO$_2$. *J. Nanomater. Nanoeng. Nanosyst.* **2015**, *229*, 104–109. [CrossRef]
274. Jia, T.; Liu, Z.Y.; Hu, H.F.; Wang, G.D. The Optimal Design for the Production of Hot Rolled Strip with "Tight Oxide Scale" by Using Multi-objective Optimization. *ISIJ Int.* **2011**, *51*, 1468–1473. [CrossRef]
275. Jiang, Z.Y.; Tieu, A.K.; Sun, W.H.; Tang, J.N.; Wei, D.B. Characterisation of thin oxide scale and its surface roughness in hot metal rolling. *Mater. Sci. Eng. A* **2006**, *435–436*, 434–438. [CrossRef]
276. Xiong, S.; Liang, D.; Wu, H.; Lin, W.; Chen, J.; Zhang, B. Preparation, characterization, tribological and lubrication performances of Eu doped CaWO$_4$ nanoparticle as anti-wear additive in water-soluble fluid for steel strip during hot rolling. *Appl. Surf. Sci.* **2021**, *539*, 148090. [CrossRef]
277. Cheng, X.; Jiang, Z.; Wei, D.; Hao, L.; Zhao, J.; Jiang, L. Oxide scale characterization of ferritic stainless steel and its deformation and friction in hot rolling. *Tribol. Int.* **2015**, *84*, 61–70. [CrossRef]
278. Yu, X.; Jiang, Z.; Zhao, J.; Wei, D.; Zhou, C.; Huang, Q. Microstructure and microtexture evolutions of deformed oxide layers on a hot-rolled microalloyed steel. *Corros. Sci.* **2015**, *90*, 140–152. [CrossRef]
279. Yu, X.; Jiang, Z.; Zhao, J.; Wei, D.; Zhou, C.; Huang, Q. Effects of grain boundaries in oxide scale on tribological properties of nanoparticles lubrication. *Wear* **2015**, *332–333*, 1286–1292. [CrossRef]
280. Dohda, K.; Boher, C.; Rezai-Aria, F.; Mahayotsanun, N. Tribology in metal forming at elevated temperatures. *Friction* **2015**, *3*, 1–27. [CrossRef]
281. Tominaga, J.; Wakimoto, K.; Mori, T.; Murakami, M.; Yoshimura, T. Manufacture of wire rods with good descaling property. *Trans. ISIJ* **1982**, *22*, 646–656. [CrossRef]
282. Sun, W.; Tieu, A.K.; Jiang, Z.; Zhu, H.; Lu, C. Oxide scales growth of low-carbon steel at high temperatures. *J. Mater. Process. Technol.* **2004**, *155–156*, 1300–1306. [CrossRef]
283. Zhao, J.; Jiang, Z. Thermomechanical processing of advanced high strength steels. *Prog. Mater. Sci.* **2018**, *94*, 174–242. [CrossRef]
284. Verlinden, B.; Driver, J.; Samajdar, I.; Doherty, R.D. *Thermo-Mechanical Processing of Metallic Materials*; Elsevier: Amsterdam, The Netherlands, 2007.
285. Hou, H.; Chen, Q.; Liu, Q.; Dong, H. Grain refinement of a Nb–Ti microalloyed steel through heavy deformation controlled cooling. *J. Mater. Process. Technol.* **2003**, *137*, 173–176. [CrossRef]
286. Cannio, M.; Ponzoni, C.; Gualtieri, M.L.; Lugli, E.; Leonelli, C.; Romagnoli, M. Stabilization and thermal conductivity of aqueous magnetite nanofluid from continuous flows hydrothermal microwave synthesis. *Mater. Lett.* **2016**, *173*, 195–198. [CrossRef]
287. Özerinç, S.; Kakaç, S.; Yazıcıoğlu, A.G. Enhanced thermal conductivity of nanofluids: A state-of-the-art review. *Microfluid. Nanofluid.* **2010**, *8*, 145–170. [CrossRef]
288. Tang, S.; Liu, Z.Y.; Wang, G.D.; Misra, R.D.K. Microstructural evolution and mechanical properties of high strength microalloyed steels: Ultra Fast Cooling (UFC) versus Accelerated Cooling (ACC). *Mater. Sci. Eng. A* **2013**, *580*, 257–265. [CrossRef]
289. Ginzburg, V.B. Steel-rolling technology. *Theory Pract.* **1989**, *328*, 791.
290. Eghbali, B.; Abdollah-zadeh, A. Influence of deformation temperature on the ferrite grain refinement in a low carbon Nb–Ti microalloyed steel. *J. Mater. Process. Technol.* **2006**, *180*, 44–48. [CrossRef]
291. Raulf, M.; Persson, K. Rolling of Steel. In *Encyclopedia of Lubricants and Lubrication*; Mang, T., Ed.; Springer: Berlin/Heidelberg, Germany, 2014; pp. 1663–1680. [CrossRef]
292. Zhang, S.; Ma, T.; Erdemir, A.; Li, Q. Tribology of two-dimensional materials: From mechanisms to modulating strategies. *Mater. Today* **2018**, *26*, 67–86. [CrossRef]

293. Song, J.; She, J.; Chen, D.; Pan, F. Latest research advances on magnesium and magnesium alloys worldwide. *J. Magnes. Alloy* **2020**, *8*, 1–41. [CrossRef]

294. Xie, H.; Jiang, B.; He, J.; Xia, X.; Pan, F. Lubrication performance of MoS_2 and SiO_2 nanoparticles as lubricant additives in magnesium alloy-steel contacts. *Tribol. Int.* **2016**, *93*, 63–70. [CrossRef]

295. Huang, W.; Du, C.; Li, Z.; Liu, M.; Liu, W. Tribological characteristics of magnesium alloy using N-containing compounds as lubricating additives during sliding. *Wear* **2006**, *260*, 140–148. [CrossRef]

296. Huang, W.; Fu, Y.; Wang, J.; Li, Z.; Liu, M. Effect of chemical structure of borates on the tribological characteristics of magnesium alloy during sliding. *Tribol. Int.* **2005**, *38*, 775–780. [CrossRef]

297. Xia, Y.; Jia, Z.; Jia, J. Tribological Behavior of AZ91D Magnesium Alloy against SAE52100 Steel under Ionic Liquid Lubricated Conditions. In *Advanced Tribology*; Springer: Berlin, Germany, 2009; pp. 896–898.

 *lubricants*

Review

Water-Based Lubricants: Development, Properties, and Performances

Md Hafizur Rahman [1], Haley Warneke [1], Haley Webbert [1], Joaquin Rodriguez [1], Ethan Austin [1], Keli Tokunaga [1], Dipen Kumar Rajak [2] and Pradeep L. Menezes [1,*]

[1] Department of Mechanical Engineering, University of Nevada-Reno, Reno, NV 89557, USA; mdhafizurr@unr.edu (M.H.R.); hwarneke@nevada.unr.edu (H.W.); haleywebbert@nevada.unr.edu (H.W.); joaquinrodriguez@nevada.unr.edu (J.R.); ethan.EA98@gmail.com (E.A.); kelit@nevada.unr.edu (K.T.)

[2] Department of Mechanical Engineering, Sandip Institute of Technology & Research Centre, Nashik 422213, India; dipen.pukar@gmail.com

* Correspondence: pmenezes@unr.edu

Abstract: Water-based lubricants (WBLs) have been at the forefront of recent research, due to the abundant availability of water at a low cost. However, in metallic tribo-systems, WBLs often exhibit poor performance compared to petroleum-based lubricants. Research and development indicate that nano-additives improve the lubrication performance of water. Some of these additives could be categorized as solid nanoparticles, ionic liquids, and bio-based oils. These additives improve the tribological properties and help to reduce friction, wear, and corrosion. This review explored different water-based lubricant additives and summarized their properties and performances. Viscosity, density, wettability, and solubility are discussed to determine the viability of using water-based nano-lubricants compared to petroleum-based lubricants for reducing friction and wear in machining. Water-based liquid lubricants also have environmental benefits over petroleum-based lubricants. Further research is needed to understand and optimize water-based lubrication for tribological systems completely.

Keywords: ionic liquids; lubricants; additives; nano-lubricants; tribology

Citation: Rahman, M.H.; Warneke, H.; Webbert, H.; Rodriguez, J.; Austin, E.; Tokunaga, K.; Rajak, D.K.; Menezes, P.L. Water-Based Lubricants: Development, Properties, and Performances. *Lubricants* **2021**, 9, 73. https://doi.org/10.3390/lubricants9080073

Received: 27 June 2021
Accepted: 21 July 2021
Published: 23 July 2021

Publisher's Note: MDPI stays neutral with regard to jurisdictional claims in published maps and institutional affiliations.

1. Introduction

The reduction of friction and wear is a significant challenge for researchers in tribology. In order to minimize friction and wear, a wide variety of lubricants have been developed; including solid (e.g., grease, molybdenum disulfide, polymers, soft metals) [1], gaseous (e.g., air, steam, liquid metal vapor), and liquid (e.g., petroleum oil, bio-derived oils) lubricants [2]. Similarly, a wide variety of lubricant additives have been explored [3].

Liquid lubricants have many advantages over solid or gaseous lubricants for industrial applications. Industrial-scale machines often run under extreme pressure and temperature conditions and may experience tribological failure, due to improper lubrication [4]. Therefore, the use of effective liquid lubricants is essential to minimize energy consumption in such conditions. Liquid lubricants play a significant role in heat removal, corrosion prevention, transfer of wear debris, mechanical noise reduction, and act as a liquid seal between contacts [5]. Therefore, liquid lubricants are typically used in industrial machinery.

The origin of liquid lubricants was probably led by the ancient Egyptians and Sumerians in 3500 B.C.E. [6]. It is believed that to extend the life of wooden axles and wheels, they used a viscous liquid form of petroleum called bitumen. Moreover, animal and vegetable oils and water were utilized as lubricants very often. During this period, water was used as a lubricant by the Chinese as well. However, petroleum-based lubricants always outcompeted water, and therefore, over the centuries, petroleum-based oils had been the most common lubricant in the marketplace.

According to the current reserve to production ratio of fossil-based oil, the world oil reserve might end within the next 50 years [7]. Moreover, fossil fuels generally release toxic materials to the environment, which is a threat to sustainability. Therefore, researchers have realized the need for a cheaper, safer, more eco-friendly, and efficient alternative to petroleum oils for machine lubrication in recent years. Thus, the focus has shifted to water as an alternative lubricant. It is known that water alone is a poor lubricant compared to petroleum oils, firstly due to the low viscosity of water. Moreover, water may accelerate corrosion that is unwanted in metal surfaces. Therefore, researchers have mixed water with different additives to obtain improved tribological performance [8].

Numerous additives could be chosen along with water, based on different applications. In general, the additives could act as antioxidants, anti-foaming agents, corrosion inhibitors, detergents, friction modifiers, wear improvers, metal deactivators, and/or viscosity index improvers [2,9] Also, some additives can support extreme pressure and extreme temperature conditions. Water miscible petroleum oils are generally a common choice to develop WBLs [10]. However, in recent years, some alternative additives received researcher's attention. Some examples of these additives are nanoparticles, such as TiO_2 nano-additive, polyethyleneimine-reduced graphene oxide (PEI-RGO) nanosheets, etc. [11,12]. Moreover, room temperature ionic liquids and bio-derived oils were added to water by researchers to develop WBLs [10]. By using such additives with water, tribological performances were improved. As a result, the overall lifetime of both the lubricant and the tribo-pair could be enhanced [2].

Studies have shown that some WBLs improved the tribological performance compared to petroleum-based lubricants, which resulted in a decrease in the amount of power necessary to run a machine [13]. Their performance was also outstanding as cutting fluids and hydraulic fluids [14]. Moreover, WBLs showed excellent cooling capabilities and environmentally benign attributes [15]. These diversified qualities have made WBLs a promising industrial choice.

Some challenges are evident for the water-based lubricants in their industrial applications. For example, WBLs within gearboxes or metallic machinery could experience an increased risk of corrosion [10,16]. Currently, this issue is being mitigated substantially by incorporating additives. Another potential method of reducing corrosion using WBLs is through galvanic couplings, which provide an electric charge that can further protect against corrosion [16,17]. The other main concerns facing WBLs are their viscosity and the low operating temperature range. This is a downside for applications that experience high friction causing high temperatures, damaging the contact pair. To minimize such problems, water-miscible ionic liquids have played a significant role, as shown in research.

Moreover, recent research shows that petroleum-based oils could be replaced by bio-based oils in many instances [18,19]. Moreover, bio-derived oils have demonstrated superior performance as additives to WBLs [10]. In this review, the state of the art of water-based lubricant additives has been summarized, including solid nanoparticles, ionic liquids, and other oil additives. Besides, the possible challenges of WBLs and their future advancements have been projected and elaborated.

2. Development of Water-Based Lubricants with Different Additives

Water-based liquid lubricants have good thermal conductivity and are environmentally friendly compared to petroleum-based liquid lubricants [20]. Issues, however, revolve around the use of water-based liquid lubricants for metal surfaces, including rusting and corrosion of the metal that the aqueous solution can create within the tribo-system [20]. Therefore, additives are needed to improve the tribological performance of the lubricant in terms of friction, wear, oxidation resistance, corrosion resistance, and anti-foaming, to name a few [21].

There are two general categories of additives: Water-soluble organic compounds and solid particles [20]. Water-soluble organic compounds work well with polar lubricants, such as WBLs [20]. This is because the additive creates a lubrication film on the surfaces involved in the tribo-pair. This film is produced from the organic compound molecules interacting with the polarity or electrostatic attraction of the surface materials [20]. One such example of water-soluble organic compounds includes ionic liquids. Besides, bio-based oils also have potential applications in WBLs. In recent years, solid nanoparticles were also observed as improving the lubrication performance in WBLs. Figure 1 illustrated the overall scenario of water-based lubricant additives and their common advantages.

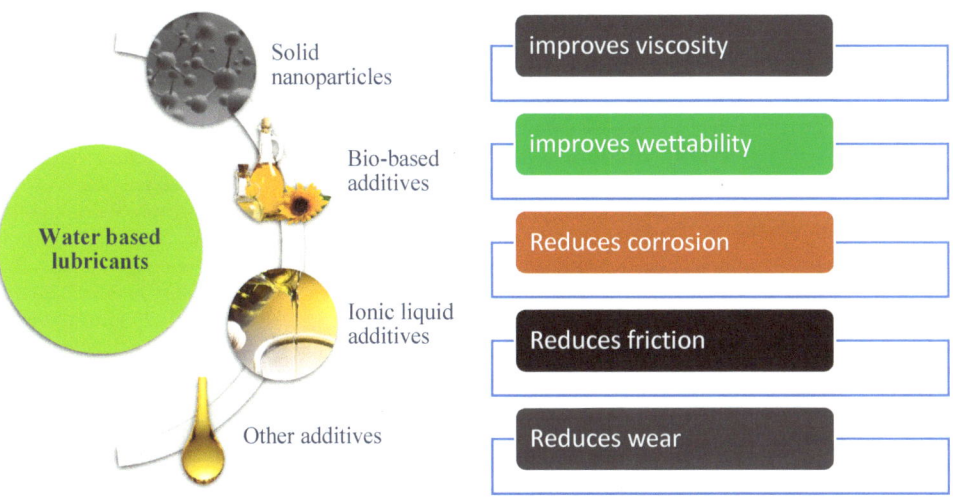

Figure 1. Water-based lubricant additives and their advantages.

2.1. Water-Based Lubricants with Ionic Liquids Additives

Ionic liquids are salts made of cations and anions, with melting points lower than 100 °C. If their melting points are lower than room temperature (25 °C), then they are called room-temperature ionic liquids (RTILs) [22]. Most frequently used cations include imidazolium, pyridinium, ammonium, and phosphonium, while the choice of anions could vary because there are thousands of options available. The most commonly used anions for WBLs were tetrafluoroborate and hexafluorophosphate, as observed in the literature [23]. These anions are often categorized as hydrophobic with weak hydration capacity [23]. However, various combinations of anion and cations are possible to form a particular ionic liquid, and therefore, theoretically, there are 10^{18} different ionic liquids possible to produce through synthesis [24].

Ionic liquid lubricants have gained growing interest among scientists and engineers in the field of tribology within the past two decades [25–27]. This is because the unique properties of ionic liquids can be utilized for creating advanced lubricants for a variety of needs. Moreover, they can improve the lubricant to handle severe conditions, including extreme pressure and extreme temperature, where oils, greases, and solid lubricants fail [22]. The unique properties that make ionic liquids an ideal additive include thermal stability, nonflammability, high conductivity, high polarity, negligible vapor pressure, and good miscibility with water and organic solvents [20–23,28,29]. In short, they can function within a wide temperature range and can form strong and effective adsorption films, due to their highly polar nature [23].

On the other hand, current lubricants are not highly versatile. For instance, lubricants that are used for one type of contact surface material may not be suitable for another (i.e., lubricants for steel-on-steel contact versus lubricants for aluminum-on-ceramics contact) [29]. This is because the material properties of the surface and the lubricant properties are unique to how they behave together under specific applications. Ionic liquids, in comparison, stride ahead, due to versatility. The synthesis of Ionic liquids is also straightforward compared to many petroleum-based lubricant additives [28,30]. This makes the use of ionic liquids potentially more desirable for WBLs.

WBLs could introduce rust on the metal in a harsh environment, due to the corrosive nature of some ionic liquids [20]. This corrosive nature becomes functional within the tribo-system through the hydrolysis between the water and the ionic liquids [23]. The ionic liquid itself can decompose in the presence of water [28]. It was found in the literature that the halogen-containing anions contribute to hydrolysis, and therefore, increases the corrosion rate [23]. Hydrolysis and corrosion could be partially avoided for water-based ionic liquids in a few different ways. The anions in the ionic liquids can be replaced with those that are halogen-free, reducing hydrolysis within the tribo-system [23,31]. In terms of lubricant stability, hydrophobic anions over hydrophilic anions could increase stability and ensure a good lubrication film between the tribo-pair. Lastly, anti-corrosive additives, such as benzotriazole (tetrabutylphosphonium benzotriazole, for example), could help alleviate the corrosive environment [23,32]. Benzotriazole is well miscible with many ionic liquids [23]. The use of benzotriazole is limited, however, because it does not work well for high temperature or low-pressure environments [23]. However, at room temperature application, benzotriazole can play a significant role. More investigations on ionic liquids are needed to enhance high pressure, high-temperature (<100 °C) performance of WBLs.

Researchers have explored the effects of specific ionic liquids as additives for water-based liquid lubricants. Tang et al. [33] discussed the use of carbon dots that were synthesized with ionic liquids to create a carbon dot ionic liquid (CDs-IL) additive for WBLs. In the investigation, ionic liquids, CDs, and CDs-IL were tested as water-based liquid lubricant additives in universal friction and wear tester (four balls, steel-on-steel) to determine the wear and friction reduction of the lubricants. It was determined by comparing ionic liquids, CDs, and CDs-IL as additives that CDs-IL presented the best results in friction and wear performances [33]. The tribological performance of the CDs-IL was much better than that of the ionic liquids and CDs alone, due to the synergistic effect between the CDs and ionic liquid [33].

Aviles et al. [34] experimented with the use of diprotic and triprotic ammonium ionic liquid crystals as additives for water for a sapphire-stainless steel material contact [34]. It was found that, when compared to water, the addition of additives in the lubricant reduced the friction coefficient by as much as 80% [34]. The authors described the additives' advantages as being able to initially reduce the friction coefficient and maintain the reduced friction after the water has evaporated from the solution. The additive, palmitate derivative, presented the best performance by preventing iron oxidation from occurring when in contact with water [34]. Overall, the authors expressed that these findings could help to formulate environmentally friendly water-based liquid lubricants. Zhou et al. [35] discussed ionic liquids as lubricant additives for different types of lubricants. For water-based liquid lubricants, the authors' research observed that 0.25 wt.% [phosphazene][Tf_2N] and 2.0 wt.% [1-Ethyl-3-methylimidazolium][BF_4] used in water-based liquid lubricants significantly reduced the friction of a Si_3N_4-Si_3N_4 material contact [35]. [1-Butyl-3-methylimidazolium][PF_6] also showed reduced friction for both Si_3N_4-Si_3N_4 material contact (2% in water) and steel-steel material contact (2–14.4% in a water solvent) [35]. Imidazolium] [BF_4] was also tested with water for SiO_2 on Si_3N_4, poly-Si on Si_3N_4, and Si_3N_4-Si_3N_4 tribo-pairs [35]. The inclusion of ionic liquids to water-based liquid lubricants does reflect a significant reduction in friction of the rubbing surfaces; more examples are discussed in Section 4.1.

2.2. Water-Based Lubricants with Bio-Derived Additives

Lubricant demand is on the rise from a global perspective. Many lubricants on the market today are petroleum-based. Petroleum-based lubricants can be harmful to the environment and can cause safety hazards for people, animals, and the overall environment. In recent history, government officials have incentivized manufacturing companies to use biodegradable lubricants that are more eco-friendly [36]. One way to reduce the negative impact of lubricants on the ecosystem is to replace petroleum-based lubricants with biodegradable lubricants. Bio-lubricants can reduce environmental impact by providing a renewable source of lubrication that promotes sustainability. Bio-lubricants have evolved, since the push for an eco-friendly alternative to oil-based lubricants. Bio-lubricants were first made with oils found naturally in the environment. These oils have beneficial tribological properties, due to the high lubricity that leads to a reduction in friction and wear [37].

Water-based bio-lubricants could be a sustainable source of lubrication. However, water alone is a poor lubricator, due to the low viscosity of water. Water has also been shown to have corrosive properties. Research has revealed that the addition of certain additives to water could be used to create a lubricant better than water alone [8]. Naturally found oils and fats extracted from biological sources have been used as additives to WBLs to better control tribological properties of WBLs [38]. The transition from traditional petroleum-based lubricants to more modern bio-lubricants involves the integration of chemistry, biology, and engineering.

Nowadays, bio-lubricants are made from materials found naturally in the world, such as biomass [18]. These materials include soybeans, sunflowers, coconuts, plants, to name a few [37]. Table 1 exemplifies a comparison of the factors contributing to the differences in fossil-derived lubricants and biomass-based lubricants [37]. There are many advantages associated with biomass-based lubricants. Bio-lubricants are produced from renewable resources, using the pyrolysis process, and the produced oils contain a significant amount of water. This water content comes from the moisture present in the biomass [39]. For fuel up-gradation, the water needs to be separated [40]; however, the separation step is not essential in water-based lubricants. Ji et al. [41] developed water-based lubricants using biomass-derived levulinic acid (LA) and polyols, such as ethylene glycol and glycerol. The process of how to create these lubricants is shown in Figure 2

Table 1. Comparison of fossil-derived lubricants and biomass-based lubricants.

Fossil-Based Lubricants	Biomass-Based Lubricants
• Well-developed technologies	• Technology in development
• Limited availability	• Unlimited availability, abundant
• Geopolitically sensitive	• Found all over the world
• Water-free	• Water-rich
• Hydrophobic compounds	• Hydrophilic compounds
• Chemistry involves C and H	• Chemistry involves C, H, and O

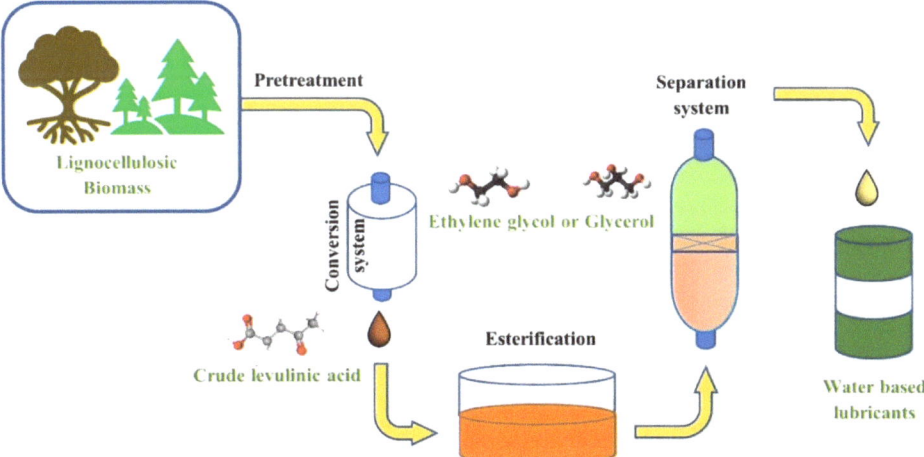

Figure 2. Outline of the processes to create the biomass-derived water-based lubricant. Adapted from [41].

The primary source of water-based bio-lubricants is biomass, which is universally available and cheap. This means the water-based bio-lubricants are widely available and can be implemented effectively at both local and global scales. Moreover, the solutions of bio-lubricants are water-rich, meaning they will produce less harmful pollutants. Since water is renewable, recyclable, and disposable; therefore, water-based bio-lubricants could become a sustainable alternative to petroleum-based lubricants in the future [37]. However, there are some limitations of water-based bio-lubricants as well. The negative impacts include the need for technological advancements, complex logistical planning, and high-level chemistry. These drawbacks require research and development to make the end product better.

There are several potential applications of water-based bio-lubricants. One of such applications could be observed in the automobile industry. The safe and effective use of automobiles requires effective lubrication of all parts to keep the machines operating smoothly. Effective lubrication of all parts that rub together in an automobile decreases energy loss in the entire system. Mineral oils have been used heavily in the past to provide lubrication to automotive machines. However, the world is running out of crude oil, and prices are increasing, making this method of lubrication unsustainable. Therefore, water-based bio-lubricants could be an effective alternative. These additives to water include sunflowers, soybeans, coconut, palm, and other crops. Table 2 exemplifies the oil content presented in plant species found around the world [42].

Table 2. The oil content of plants that can be added to water-based bio-lubricants [43–50]. Reproduced with permission from [42]. Copyright, Elsevier, 2014.

Plant Species	Oil Content (% of Volume)
Jatropha	40–60%
Rapeseed	38–46%
Palm	30–60%
Peanut	45–55%
Olive	45–70%
Coconut	63–65%
Neem	30–50%
Karanja	30–50%
Castor	45–60%
Linseed	35–45%
Moringa	20–36%

Vegetable oils exhibit similar properties to traditional mineral-based oils [51]. Vegetable oils can even sometimes produce better lubricity properties than standard mineral oils. Moreover, vegetable oil's performance was increased by adding boron [52,53] or carbon [54–56], based on solid lubricant additives. An analysis of the properties of vegetable oils and mineral oils was conducted assessing density, viscosity index, shear stability, pour point, clod flow behavior, miscibility with mineral oils, solubility in water, oxidation stability, hydrolytic stability, sludge forming tendency, and seal swelling tendency [42]. From this study, the advantages of vegetable oils over mineral oils as additives to WBLs became evident. Some key advantages of the bio-lubricants were higher lubricity, lower friction losses, decreased automotive emissions, and less toxic environmental effects. On the other hand, one disadvantage of vegetable oil versus standard mineral oils is that they lack adequate oxidative stability for lubrication. Vegetable oils also have low thermal stability, limiting the versatile application of vegetable oils for additives to WBLs [42]. One suggestion for future study is to combine water, vegetable oils, and other additives to make a mixture ideal enough to minimize oxidation effects and make the mixture withstand higher temperatures. Such a mixture could provide an environmental advantage by replacing a significant number of petroleum-based lubricants with bio-based oils and water.

The development of water-based bio-lubricants still has a long way to go to compete with the mineral-based lubricants commonly found today. Current applications are limited, and research and development need to be performed to create new combinations of water-based bio-lubricants that can be used in more processes. With the push for more environmentally friendly manufacturing techniques, water-based bio-lubricants will continue to be developed to promote sustainability.

2.3. Water-Based Lubricants with Solid Nanoparticles

WBLs are low-cost lubricants for metalworking processes. Not only acting as a lubricant, but also as a coolant, WBLs could provide a longer tool life, reduced thermal deformation, and lower friction. However, the lubricant suffers from a low viscosity and corrosive properties that make it unviable for many tribological applications [9]. To counteract the disadvantages of WBLs, the integration of additives is used to improve the properties and performance. In some cases, additives within the lubricant allow for many increased advantages, such as further reduced tool wear and the prevention of corrosion. Furthermore, research into water-based lubricant additives can inhibit fungal or microbial growth [9]. Additives within a water-based lubricant are often used as a surface-active molecule that reduces friction, wear, and corrosion. Some examples of solid nanoparticle additives are graphene, copper, titanium-di-oxide, and many others [9,12,57–60].

One of the most common additives used in oil-based lubricants that are being implemented into WBLs is titanium-based nano-additives. The additive is used to combat the low viscosity and corrosive properties found in WBLs. Specifically, TiO_2 is the most used additive in oil-based lubricants that are beginning to be implemented into WBLs more frequently [12]. The additive TiO_2 has been shown to improve properties overall, such as nontoxicity, low density, friction, and wear [57]. An experiment done by Gu et al. [57] using dual-coated TiO_2 nanoparticles observed friction and wear reduction under an MSR-10D four-ball tribotester at a speed of 1440 rpm at applied loads of 140 N for 10 min. After each test, scanning electron microscope (SEM), atomic force microscope (AFM), and energy dispersive spectrometry (EDS) were used to analyze the wear scars [57]. The team concluded that at low concentrations of TiO_2, the additive improved the anti-wear properties of the lubricant and reduce wear scar diameter. Furthermore, the wear scars became less prominent as the concentrations increased, as well as the coefficient of friction decreased, being able to achieve a minimum value of 0.04 during testing [57].

Another study was done by Wu et al. [12], who tested TiO_2 nano-additive properties using a ball-on-disk tribometer. The experiment consisted of additives at various concentrations during two different methods of testing: One method involving adding the lubricant at set time intervals and the second method involving the lubricant being continuously present throughout the test. The results of each method concluded that the addition of the additive reduced the coefficient of friction and ball wear by 49.5% and 97.8% accordingly [12]. Furthermore, they concluded that the optimal concentration of the TiO_2 additive was 0.8 wt.% at room temperature.

Another water-soluble additive being tested for the lubricant are graphene-based nanoparticles. Testing done on graphene quantum dots (GQDs) additives on steel-to-steel contact proved to have enhanced tribological properties at a concentration of 4 mg/mL. The additive decreased the wear rate by 58.5% and the coefficient of friction by 42.5%. GQDs at lower concentrations were more effective at improving tribological properties than other graphene-based additives [61]. The additive has proven to be one of the most promising advancements within the graphene field.

Nanostructured borates are another additive being created by synthesizing the mixture with magnesium, zinc, aluminum, and titanium particles [62]. The additive could be added to water-based drilling fluids to improve tribological performance. The additive significantly decreased the coefficient of friction by producing a tribo-film on both surfaces [62]. Furthermore, the additive maintained stability during high pressure and temperature conditions.

2.4. Other Water-Based Lubricant Additives

There are many other potential additives for water-based lubricants. One promising additive to be used with WBLs is a water-soluble, rubber seed oil-based sulfonate. Through testing, the rubber seed oil-based sulfonate copolymer noticeably improved the anti-wear, anti-corrosion, and frictional properties of the lubricant. Furthermore, the additive was also able to improve the nonseizure load (P_B value) of the water. At concentrations of 0.5%, the P_B value reached 431 N, which was three times greater than the water-based lubricant without the additive, and such conditions offered a coefficient of friction of 0.085 and a wear scar diameter of 0.78 mm [60]. Rubber seed oil-based sulfonate is becoming a great candidate for additive implementation within WBLs, allowing the lubricant to maintain environmentally friendly attributes and increasing its capabilities for extreme applications.

Furthermore, an ester-based lubricant SMJH-1 was tested for its tribological performances with water-based drilling fluids and was compared to sodium montmorillonite (Na-MMT) base mud. The additive was tested at varying concentrations, while testing the lubricants before and after an aging process. The additive at 1.0% concentration reduced the lubricity coefficient (μ) by 91.4% before aging and 90.7% after the aging process [63]. It was also found that due to the aging process, this would result in increased friction, which they concluded could be the result of increased surface roughness. It was found that SMJH-1 at higher concentrations, the formation of a C=S metal film would be present, which would decrease the average roughness, as well as decrease the lubricity coefficient [63].

3. Properties of Water-Based Lubricants

Properties of the water-based lubricants could be significantly improved by using the above-mentioned additives. Some of the important properties of water-based lubricants are solubility, viscosity, density, atomic structure, wettability. These properties have a crucial effect on the tribological performance of water and are discussed below.

3.1. Viscosity

Viscosity is a measurable physical property that can translate information about the liquid's hydration, solvation, the shape of the infused particles, and the forces acting between the particles [64]. It can also be defined as the internal resistance that a fluid has to flow. Viscosity is an important parameter in Hersey number Equation (1), that is used to define different lubrication regime in the horizontal axis of the Stribeck curve. In the equation, η corresponds to the viscosity, v corresponds to the entrainment speed, and the P is the normal load. Stribeck curve is a popular diagram and could be found in the literature [65,66].

$$Hersey\ number = \frac{\eta.v}{P} \tag{1}$$

Water has a low viscosity which is one quality that makes it a poor lubricant [8]. Additives are usually added to water to improve the viscosity, therefore improving the tribological properties that are needed for lubricants [64,67,68]. For example, the TiO_2 nano-additive increased the overall viscosity of the lubricant and decreased the amount of contact between the metal surfaces [58]. The effect of the viscosity of a lubricant can be seen in:

- Cavity formation
- Decreasing friction
- Film thickness
- Thermal behavior

According to a study performed by Nouri et al. [64], cavity formation was observed using high-resolution images taken throughout a timed interval to view the gradual development of cavities on a single piston-ring assembly. The cavitation was then quantified using a MATLAB program, and it was found that with a decreasing viscosity of the lubricant, the length of the cavities was also decreased. These results were consistent throughout their study, demonstrating viscosity's effect on cavity formation [64].

As per Marx et al. [67], a reduced strain rate curve can be seen when viscosity modifiers are added to lubricants. The viscosity modifiers reduced the amount of friction and power loss at high shaft speeds, due to shear thinning. Their study also addressed film thickness, which is another quality factor that can be altered by viscosity. Film thickness is defined as the layer that is created by a lubricant between the two surfaces in contact [69]. The target thickness can vary depending on the needs of the machine. When a higher-ordered degree of molecule interacts with a surface, a thicker film develops [70]. This aligns with the idea that viscosity is directly linked to the internal properties of a liquid. This is also understandable from the stribeck curve, as increased viscosity increases the film thickness values. Therefore, in the case of water-based lubricants, an increase in the film thickness could be achieved by incorporating suitable additives to increase the viscosity.

Finally, the viscosity of a lubricant used in machines will affect the thermal behavior and heat transferability [71]. Thermal behavior is an important consideration when discussing the type of environment that is encountered by lubricants. A lubricant that heats up with little friction could cause adverse effects to the equipment and the product. The ability to distribute heat to the environment or the product is an additional consideration that must be observed. Depending on the desired requirements of the machines and products being manufactured, the rate at which heat is dissipated by the lubricant will change. The magnitude of change in the viscosity of a lubricant will depend on the additives, the size of the particles, and the percent or fraction of the additives to the solution [71]. Hajmohammadi et al. [72] observed that the viscosity of the Al_2O_3 diathermic oil nanofluids improved with the increase in volume fraction. Moreover, the dynamic viscosity of lubricants increased as the diameter of the nanoparticle decreased [73].

3.2. Density

Density is a property determined by the ratio of the mass of the liquid relative to the volume and can be affected by temperature. Water has a density of 1.00 g cm^{-3}. The density of the liquid additives to water will change the density of the water-based lubricant. While adding solid nanoparticles, density will also be affected by the size and shape of the nanoparticles being used, as well as the concentration of the nanoparticle in the water [74]. For example, the density of a phosphazene-based ionic liquid additive ranges from 1.63–1.65 g cm^{-3} [75]. When added to WBLs, it will have an increased density with the addition of the phosphazene-based ionic liquid. This increase in density demonstrated thicker film formation, making the liquid's suspension of the nanoparticles harder to settle. WBLs containing TiO_2 are found to have a lower density than that of water [58]. Even with a lower density than water, this additive demonstrated friction reduction and a decrease in the amount of required power during the study's drilling process.

3.3. Wettability

Surface wettability is defined by a liquid's ability to maintain contact with a surface through adhesive and cohesive forces [76]. A common method to measure wettability is contact angle. The value of the contact angle determines how the liquid is defined. A contact angle over $150°$ is superhydrophobic. $90°$ to $150°$ is hydrophobic, and $10°$ to $90°$ is hydrophilic [77]. Lastly, super hydrophilic is less than $10°$ [77]. Increased wettability is defined by a smaller contact angle [78], which has been associated with creating a film between surfaces [13]. Water has a reported contact angle of $73.6°$ [13]. The wettability of water can be increased with the addition of nanoparticles [79]. When TiO_2 nanoparticles of various sizes are added to water as a suspension medium, the contact angle of the mixture decreased from 73.6 to 55.22 with an increasing concentration of TiO_2. The contact angle was further decreased by adding sodium dodecylbenzene sulfonate (SDBS) [13]. The dissociation of SDBS in water provides phenyl sulfonic groups. These phenyl sulfonic groups get adsorbed around the nanoparticles and increase the amount of net negative charge of the nanoparticle surface. Therefore, the repulsive forces between TiO_2 nanoparticles increase, and the dispersion stability gets improved [80]. Moreover, added SBDS restricts the agglomeration of nanoparticles which enhances the wettability [13].

Graphene and graphene oxides (GO) demonstrate a reduction in the friction coefficient between magnesium alloy and steel contact, ultimately reducing the wear rate [79]. The GO had a decreased contact angle of 46.5° of water [79]. Compared to graphene, GO is more effective when creating a thin layer used to protect the surfaces in contact [79]. This is further supported by the tribological properties found in the study. The GO lubricant had a more pronounced positive effect on load-carrying capacity and lubrication film endurance [79]. Alkyl polyglucosides (APGs) are another type of additive that can be added to increase the wettability of water [81]. The magnitude of the concentrations will result in either micelles or lyotropic liquid crystals forming. Micelle is formed when molecules are aggregated in a colloidal solution, like that of detergents. On the other hand, lyotropic liquid crystal is formed if an amphiphilic mesogen (a compound that portrays a liquid crystal state) is dissolved in a suitable solvent. There was a decrease in surface tension and contact angle with the addition of APGs [81]. At lower concentrations of APGs, there is a significant decrease in the contact angle [81]. At higher concentrations, there are no significant changes between the measured contact angle values [81]. Another difference was seen between the different types of APGs tested [74]. APGs with 8–10 alkyl groups decrease the contact angle to 41–50°, which is 1.7–1.8 times higher than the contact angle for pure water [81]. The wear is lower in multiple friction couples when APGs are present in the water, acting as a lubricant [81].

1-alkyl-3-methylimidazolium tetrafluoroborate ionic liquid was used for wettability test in literature [82]. With the addition of water, the interfacial structure of ILs was changed [82]. Differing alkyl chain lengths at different concentrations, the wettability was evaluated [82]. There was no significant difference between the different lengths, but there was a decrease in contact angle with an increase in the molar fraction of 1-hexyl-3-methylimidazolium ILs in the liquid [82].

Marine biofouling is an alarming phenomenon that has drawn much attention recently. It is defined as an accumulation of organisms on a submerged surface where they are not wanted. The current solution to this problem is to make: Slippery Liquid Infused Porous Surfaces (SLIPS) [83]. SLIPS are commonly formed using lubricants that are based on silicone or fluorine; neither are well-suited for marine usage [83]. A bio lubricant created using oleic acid, and methyl oleate was suggested to be a substitute for SLIPS being used [83]. Confocal microscopy could be used to observe the wetting state of the gels used to absorb the bio lubricant for testing under ultraviolet (UV) lights [83]. In such conditions, the contact angles before and after UV irradiation for the oleic acid were found to be 57° and 42° accordingly [83]. For the case of methyl oleate, the contact angles were found to be 29° and 7° [83]. Nonionic monoglyceride surfactant and xanthan gum are other additives that have been observed to be substituted for drilling wells [84]. Monoglycerides were observed with varying carbon chain lengths, ranging from 6,8, 10, and 12 [84]. Triglycerides chains were studied at 6, 8, and 10 [84]. The contact angles of each of the aqueous solutions and xanthan gum suspension solutions are displayed in Table 3 [84]. The solution that had the lowest contact angle was the monoglyceride with a carbon chain of 10 added with the xanthan gum [84].

Table 3. The measured contact angle of aqueous solutions and xanthan gum suspensions. Reproduced with permission from [84]. Copyright John Willey and Sons, 2014.

Aqueous Solutions	Contact Angle in Degree	Xanthan Gum Suspensions	Contact Angle in Degree
Water	77.0	Xanthan Gum	69.3
Monoglyceride 6	64.5	Monoglyceride 6 + XG	63.7
Monoglyceride 87	13.7	Monoglyceride 8 + XG	9.6
Monoglyceride 10	12.9	Monoglyceride 10 + XG	7.4

A novel polytetrafluoroethylene with a SiO_2 layer ($PTFE@SiO_2$) is a composite nanoparticle studied in the literature [85]. The successful creation of the nanoparticle was confirmed by a change in the contact angle (using the DSA100 contact angle goniometer) observed. When the nanoparticles were partially developed through the synthesis process, there was a slight increase in the contact angle, going from $35°$ to $50°$ [85]. On the other hand, once the SiO_2 completely enveloped the PTFE, then the wettability got improved, and the contact angle decreased to $32°$ [85].

Coconut diethanol amide (CDEA) and Tween 85 (T-85) are triazine-based covalent organic framework nanomaterial (TriC) that work as additives for WBLs [86]. The wettability test used to observe the contact angle was found by placing droplets on a steel disk that were then recorded using an optical contact angle goniometer [86]. Pure water was found to be $73.3°$, while the CDEA and T-85 were $26.5°$ and $58.6°$, respectively [86]. The addition of the additives decreased the contact angle. The difference between the additives could be explained by the polar groups that allowed hydrogen bonding between the CDEA and the water molecules. The T-85 molecules also have groups that are weakly polar, resulting in a weakening of hydrogen bonding [86].

3.4. Solubility and Other Properties

Water is a low-cost liquid that has many useful properties that make it a promising lubricant. It has a high cooling capacity. In contrast, water also has low viscosity and corrosive properties that make it undesirable as a lubricant in its standard form. To remedy these downsides, additives are mixed with distilled water through various processes to create a water-based lubricant that applies to industry. These lubricants are often used in processes, such as cooling or rolling and even metalworking. WBLs have the distinct advantage of being a coolant and a lubricant. Moreover, the following advantages could be observed: Flushing debris, thermal deformation reduction, surface finish improvements, extended tool life, and lower friction. Nanoparticles and additives are common in creating these effects with WBLs, but their efficacy is often dependent on their solubility in water.

Zhang et al. [87] investigated the feasibility of using metallic nanoparticles as an additive to water. This application is not well studied, due to a few complications with the solubility of the substance with water. The nanoclusters of the compound easily aggregate which hinders their compatibility with water. This issue was resolved using surface-modification techniques for the encapsulations. Such a technique allowed researchers to prevent aggregation and allow for compatibility with lubricants. The nanocomposite material additive that was chosen by Zhang et al. [87] was Cu/SiO_2, and after surface-capping with 3-mercaptopropyl-trimethoxysilane (MPTS), oxidation was entirely prevented, as well as the nanocomposite being compatible with distilled water. The study recommended this water-based lubricant as an effective steel-on-steel solution under moderate loads. Another approach to nanocomposite additives to create WBLs was made by Pei et al. [88]. Their composite of interest was carbon nanotubes or CNT. CNTs are exceptionally strong and possess the abilities of a great WBL additive. However, CNTs are not water-soluble, which poses a problem for researchers. There have been many strategies for this solubility issue, and some will be discussed here.

Polymer chains can be covalently attached to the surfaces of CNTs which gives them improved properties. Linear polymers, copolymers, dendritic, and hyperbranched polymers have been successfully grafted to CNTs through a series of reaction methods [89]. These methods include atom transfer radical polymerization (ATRP), reversible addition and fragmentation chain transfer polymerization (RAFT), and ring-opening polymerization (ROP) [90–95]. Pei et al. [88] ventured to develop a new method of creating water-soluble CNTs. Their approach was to graft water-soluble polymer chains from the surface of MWCNTs. This process was followed by the surface-initiated radical graft polymerization of acrylamide using a redox system consisting of ceric ion and reducing groups (amino) on the surface of multi-walled carbon nanotubes (MWCNTs) as initiators [88,89]. This process is shown in Figure 3.

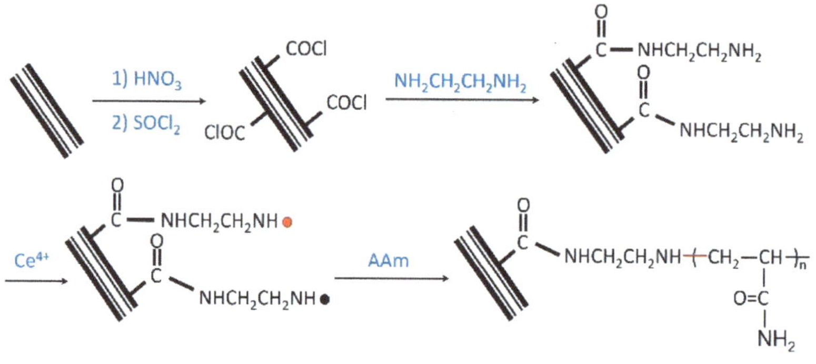

Figure 3. Surface-initiated redox polymerization of acrylamide from MWCNTs. Reproduced with permission from [88]. Copyright Elsevier, 2008.

Pei et al. [88] observed that the load-carrying capacity of the lubricant significantly raised with MWCNTs, while the friction coefficient decreased. They attributed the success of the MWCNTs to their function as nano-ball bearings under moderate load. These results support the further investigation of other related super materials being made compatible with distilled water to be further made into a high-performance water-based lubricant. Solubility in WBLs, as discussed in studies mentioned previously, is an important and challenging task to work around when designing additives to create WBLs.

Sol et al. [14] investigated ionic liquid additives to water as a viable industrial cutting fluid. Their interest in synthesizing an effective water-based lubricant came from their dry machining operations in which high temperatures and contact angles led to excessive tool wear. Ionic Liquids, having high thermal stability, were a great candidate for reducing the tool wear in this case [14]. Sol et al. [14] were very successful in their pursuits of viable water-based cutting fluids. The halogen-free aprotic ionic fluid [THTDP][Deca] was an efficient lubricant for aluminum machining operations. The friction coefficient was reduced by over 70% when compared to water and exhibited minimal aluminum wear [14]. Another successful aspect was the fluid improved the surface finish inside the wear track, as well as reducing adhesion in the pin.

As discussed previously, one of the most interesting current fields of study in tribology is nanoparticles, and their application to WBLs as an additive is no exception. They possess phenomenal wear and friction reduction properties, which makes them a viable candidate for WBLs. One major issue with these nanoparticles is that they are not often water-soluble. Solving this issue has been approached differently by different research groups.

Cui et al. [96] synthesized nanoparticles of different materials and then tested them for their tribological properties as a water-based lubricant additive. Their motivation for this pursuit was that water-lubricated ceramics that could have friction coefficients of less than 0.01 after the running-in process, which is known as superlubricity. The term superlubricity is vague, but it describes a situation in which friction vanishes or nearly vanishes. In addition to these beneficial tribological properties, ceramics are environmentally friendly and economically favorable. Their study used SiO_2, TiO_2, and ZnO nanoparticles and was prepared and tested using a ball-on-plate tribometer [96]. They discovered that the friction reduction capabilities of SiO_2 were much superior to that of the other two additives, although all test subjects showed viability. Additionally, the exceptional compatibility of silica gel and SiO_2 was observed and was shown to be a key factor in its success.

4. Performance of Water-Based Lubricants

One of the most important functions of any lubricant is the reduction of friction between moving components. This reduction also prevents the formation of wear particles in the lubricant. While this effect not only lengthens the life of the parts involved, it also increases the efficiency of the overall system by allowing it to operate at manufactured dimensions for a longer duration. The friction and wear reduction reported in the literature for WBL is summarized in Section 4.1. Moreover, in Table 4, some of the important recent studies have been tabulated, summarizing their key properties, advantages, challenges, and mechanism behind their good performance.

4.1. Friction and Wear Reduction

To absolve the inherent tribological issues of water and to synthesize a high-quality lubricant, high-quality additives are mixed in. Many different groups of additives significantly improve the tribological performance of water. Multiple attempts have been made with countless materials to find the superior lubricant. Oftentimes, the improvement of one aspect of the water-based lubricant happens to the detriment of another property. The best way to utilize additives is to assess the application and choose the best compound for the task. Corrosion inhibitors, for example, prevent corrosion to both the tool and the surface in the system requiring lubrication. The task is done by sometimes creating a protective coating and other times by neutralizing corrosive contaminants in the system depending on the application. Another interesting type of additive is reserve alkalinity additives which regulate the pH of the system which neutralizes any acidic contaminants. Some examples include alkanolamines like monoethanolamine, triethanolamine, and aminomethyl propanol [9]. Emulsifiers are interesting in that they stabilize oil-soluble additives by mitigating tension between incompatible components. The phenomenon that makes this possible is its property to form micelles which are a microscopic aggregation of molecules, as a droplet in a colloidal system [97]. Examples include sodium petroleum sulfonate and alkanol amine salts of fatty acids. Finally, couplers stabilize water dilutable metalworking fluids in the concentrate. This helps stop the separation of components in the system. Some examples are propylene glycol, glycol ethers, and nonionic alkoxylates. Sulfur, phosphorus, and nitrogen are commonly used as additives to water lubricants when applied to ferrous-based equipment, as they were very effective in minimizing friction and wear in these systems [98]. For hydraulic applications, glycols are more commonly used and exhibit successful friction and wear behaviors. As shown by Tomala et al. [98], most additives reduce the coefficient of friction in numerous environments and stresses.

Ethanolamine additives, which are commonly utilized for their corrosion protection and lubrication properties, are often poor at reducing the friction coefficient from that of pure water, and in some cases, even increase it, as can be seen in Figure 4 [9]. Although this can be seen as a disadvantage, ethanolamine additives protect steel surfaces from corrosion effectively. Glycols exhibit some of the best tribological properties regarding wear and friction reduction, which agrees with many published studies [99]. A particularly significant combination of Polyethylene Glycol (PEG) and a water-soluble EP additive to reduce friction and wear was discovered by Yong et al. [99].

Dong et al. [100] tested the viability of proton-type ionic liquids additives for water-based lubricating fluids in several scenarios. Proton-type ionic liquids (PILs) have many advantages over other methods of water-based lubrication. For example, PILs have a very simple synthesis in which they are prepared through a proton transfer reaction between the Bronsted acids and bases, with the advantage of a low-cost, no impurity, and facile approach [101]. Another advantage is their solubility stability. Previous studies that have used WBL additives have had lubricity success, although their synthesis was complicated and costly. Zheng et al. [102] synthesized and experimented with two new water-soluble IL additives in water-glycol and found that these solutions had drastically improved lubricity and extreme pressure properties. These promising results were diminished when it was discovered that the solutions had poor solubility in water and the industrial preparation process was tedious [102].

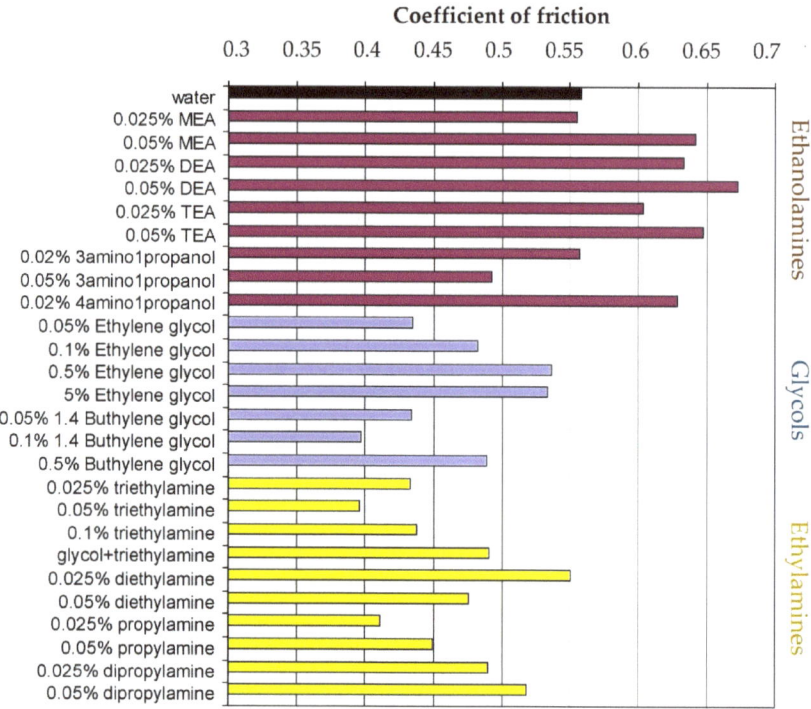

Figure 4. Friction results for chosen additives in different concentrations. Reproduced with permission from [9]. Copyright Elsevier, 2010.

PILs were adsorbed on the friction pair and then formed a double-layer structure on the surface, which together with the formed tribochemical reaction films on the surface effectively buffered the direct contact and collision from micro-convex bodies contributing to the friction reduction and anti-wear properties. Overall, this study further reinforces the capabilities of friction and anti-wear resistance of WBLs. Some additives and techniques create very effective WBLs. As shown in Figure 2, Ji et al. [41] developed a water-based lubricant. Their approach sought to solve common issues with other lubricants. These include properties of nonrenewability and nonbiodegradability, posing a constant threat to ecology and groundwater reserves, with a large proportion of current lubricants being released into the environment. Their results indicated that their most successful lubricant, glycerol ester of levulinic acid (LAGLE), exhibited superior lubricant properties, strong resistance to hydrolytic degradation, and excellent anti-wear performance, implying that the biomass-derived LAGLE was a potential water-based lubricant [41]. Both the LAEGE and another lubricant: Ethylene glycol ester of levulinic acid (LAGLE), showed significant friction and wear resistance abilities. Between LAEGE and LAGLE, LAGLE showed higher resistance to hydrolytic degradation, which would make it a more promising candidate in hydraulic applications.

Another method of improving the tribological performance of water lubricants is utilizing graphene oxide additives. Guo et al. [103] improved the tribological performance of water by adding nanomaterials as a water-based lubricant additive. This synthesis has generated interest in many applications, due to its tribological potential. In their study, they invented a novel nanomaterial named aminated silica-modified graphene oxide (SAG), which combines the layered nanomaterial graphene oxide (GO) and the nanoparticles hydrophilic nano-silica (SiO_2) by 3-aminopropyltriethoxysilane (APTES). Their use of the SAG nanomaterial in DI water dramatically reduced the wear rate of the lubricant in their tests, in addition to producing a very low friction coefficient. The SAG nanomaterial also outperformed the individual GO and nano-silica in improving wear resistance [103].

4.2. Corrosion Reduction

One of the biggest and most prominent drawbacks of WBLs is the increased potential of the lubricant causing corrosion to the machine or machinery. The concern is very prominent, especially for applications that involve gearboxes or bearings [9]. With this problem being one of the main setbacks for WBLs, research and development are ongoing for the best solution to the problem. Research has been carried out to find the optimal percentage of additives and ILs that can provide maximum performance and property improvements.

Additives, such as ionic liquids, have shown significant potential to improve corrosion reduction properties in WBLs. They offer a wide range of variations to test and produce superior water-based lubricants [9,11,81,104–108]. Within this section, ILs and other additives will be analyzed for their anti-corrosion capabilities. An additive with superior corrosion resistance is amino acid ionic liquids (AAILs) by Yang et al. [108]. Corrosion testing was done by using one of the seven types of AAILs that were synthesized by neutralizing acid and alkali. Testing involved the use of low carbon steel pieces in water at 18 mL and 1% $[P_{4444}][Trp]$ (18 mL) within a dry box at 55° C for 24 h. To evaluate the corrosion level properly, the authors used GB6144-85 as the reference standard. They were able to conclude that the inclusion of the additive resulted in weak corrosion with no visible rust that can be observed under normal seeing conditions. Using a scanning electron microscope (SEM), the surface of the piece did not show any significant differences from before the immersion. In comparison, the pieces that were immersed in pure water showed a substantial amount of corrosion, with the surface being covered in black iron oxide. Due to this corrosion, the surface had noticeably many wide and deep grooves on the surface. With additives, the corrosion resulted with grade B, and pure water resulted with grade D; Proving the AAILs increased the corrosion resistance overall [108].

Another additive with corrosion resistance properties is bio-inspired graphene-based coatings done by Chu et al. [107]. Due to the cross-linking effect within the microstructure of the hybrid coating, the additive generated densely stacked lamellar coating, which resulted in increased corrosion resistance. This was concluded through the following testing method–the authors did an electrochemical test using a classical three-electrode cell made up of a platinum plate as a counter electrode, a saturated calomel electrode (SCE) as a reference, and the sample. With the experiment being carried within ambient temperatures within a 3.5 wt.% NaCl electrolyte solution and corrosion parameters being determined using the Tafel extrapolation method. The most significant results were from the addition of polyvinyl alcohol (PVA) within the solution. They determined the additives less than 15 wt.% had weak corrosion resistance and at values greater than 50 wt.% demonstrated very significant corrosion resistance [107]. An additive studied by Liu et al. [11] was PEI-RGO nanosheet that was made into a nano additive for WBLs [11]. The additive was able to double anti-wear and anti-corrosion properties, specifically for WBLs used for steel materials. The results were found from testing 201 stainless steel with 18 wt.% chromium. After stabilizing the austenite phase with manganese at 7 wt.% and decreasing nickel from 8 wt.% to 3 wt.%, it was sanded with sandpaper then washed with water and ethanol. During the use of a water-based solution, the steel was prone to corrosion, due to the oxide layer being constantly deteriorating during the wearing process [11]. Due to this process, iron atoms in the steel were being separated, which would cause iron oxide to continuously form, causing the steel to be under constant corrosion. Though the testing of the additive, PEI-RGO, combatted this process by creating a protective deposited film. The additive was able to stop oxygen molecules from entering the wear scars that would form during the wear process. Demonstrating an additive that so far can be applicable for steel materials that would suffer from corrosion with further testing being planned for nonsteel materials [11].

Table 4. Summary of important water-based lubricant additives with their advantages, mechanisms, and challenges.

WBL Additives	Properties	Advantages	Remarks	Ref.
		Solid nanoparticle additives		
TiO$_2$	• Solid nano additive • Size: 100 nm • High dispersity and stability in pure water	• Power consumption could be lowered during the drilling process using this nanoparticle additive • Improves the load-carrying ability of water • Friction reduction compared to only water • Wear reduction compared to only water	• **Mechanism:** TiO$_2$ forms a dynamic deposition film on the worn surface (observed using SEM, EDS) and separates the contacting surfaces. • **Challenge:** Optimization of concentration is critical.	[57,96]
Titanium Carbide (Ti$_3$C$_2$)	• Solid additive • Nominal and lateral sizes of the Ti$_3$C$_2$ flakes: (0.2–3 μm) and (<1 μm to several μm) accordingly	• Friction and wear decreased with increasing concentration until 5 wt.%. Afterward, friction increased again	• **Mechanism:** Ti$_3$C$_2$ flakes hindered direct contact of the tribo-pair, especially at the edges of the contact interface. • **Challenge:** Optimal concentration is important.	[58]
SiO$_2$	• Solid nano additive • Size: 100 nm	• Silica offered superior friction reduction ability compared to ZnO and TiO$_2$ on ceramic surfaces. • Ultrasmooth surface was observed in ceramic tribopair • Running in period decreased by more than 90% Compared to only water	• **Mechanism:** Silica gel was formed due to the tribo-chemical reaction in the interface. Under SEM photograph, SiO$_2$ nanoparticles were observed as submerged in the silica-gel and were partly visible and partly hidden at the ceramic surface. This surface film was homogeneous and possibly responsible for reducing friction. • **Challenge:** Optimization of concentration is critical.	[96]
Graphene quantum dots (GQD)	• Solid nano additive • GQD are small (2 nm) in size and uniform compared to GO nanosheets (1.5–10 μm)	• A concentration of 4 mg/mL provided friction and wear reduction of 58.5% and 42.5% accordingly, compared to that of only water.	• **Mechanism:** EDS element mapping revealed the higher concentration of C and O in the lubricant film obtained for GQD 4. It means the QOD in water possesses strong polarity and got adsorbed on the steel surface. Moreover, through Raman spectra on the steel surface, the existence of a significant amount of disordered carbon was found, in contrast to FeO, Fe$_2$O$_3$, and Fe$_3$O$_4$ obtained from 0% GQD. • **Challenge:** Appropriate amount of GQD is important.	[61]
ZnO	• Solid nano additive • Size: 100 nm	• ZnO could reduce the friction, but comparatively less than that of SiO$_2$ • Super lubricity was observed for the ZnO additive	• **Mechanism:** Double electric layer, measured through zeta potential, is a key factor for super lubricity. However, the authors suggested that an ultra-smooth surface is required to get the advantage of the double electric layer to reduce friction. In the case of ZnO, the zeta potential was higher (in the negative axis) than that of SiO$_2$. However, since the produced surface was less smooth, the friction reduction was observed less for ZnO compared to that of SiO$_2$. • **Challenge:** Optimization of concentration is critical.	[96]
SiO$_2$/graphene	• Solid nano additive • Decreased contact angle (89° to 60°)	• 0.4 wt.% graphene, mixed with 0.1 wt.% nano-SiO$_2$ reduced the COF by 48.5% and wear volume 79% in comparison of 0.5 wt.% graphene, in AZ31 Mg/AISI 52100 tribopair	• **Mechanism:** (1) SiO$_2$ nanoparticles perhaps acted as a spacer, which potentially restricted the graphene nanosheets from restacking and aggregating. (2) The edge sites and defects of graphene could dissociate water molecules into OH and H. Therefore, dissociative chemisorption could have happened onto the dangling bonds at the edges of graphite. (3) ball bearing effect could have initiated rolling due to the presence of spherical SiO$_2$ nanoparticles rolling, reducing COF. Moreover, the polishing effect could reduce direct contact between tribopairs. • **Challenge:** Optimization of concentration is critical.	[109,110]
Polyethylenimine-reduced graphene oxide (PEI-RGO)	• It is a stable aqueous graphene dispersion • Excellent dispersibility and stability in water-based fluids	• Demonstrates anti-corrosion effect • Reduced the COF (54.6%) • Reduction in wear rate (45.0%)	• **Mechanism:** Lubricating and protective film at contact area was present, as a decrease of iron oxide and an increase of PEI-RGO was observed in SEM image. • **Challenge:** Synthesis of PEI-RGO is complicated, still important to achieving dispersibility.	[11,70]
Multi-walled carbon nanotubes (MWNT)	• MWNTs are water soluble	• The load carrying capacity of the lubricant was increased • Friction was reduced significantly	• **Mechanism:** MWNT anchored with Poly(acrylic acid) on the metal surface. • **Challenge:** MWNT needs further investigations to be used in biological and biomedical applications	[88,89]

Table 4. *Cont.*

WBL Additives	Properties	Advantages	Remarks	Ref.
		Ionic liquid-based additives		
Ionic liquid capped Carbon Dots (CD-IL)	• Low toxicity • The particle size of CD is 4 nm • Uniform morphology • High thermal stability	• 0.015% of CD-IL improved the load-bearing capability of steel-steel pairs from 50N to 80N. • Friction and wear were reduced by 57.5% and 64% accordingly under 40N load, compared to that of only base lubricant.	• **Mechanism:** A synergistic effect was observed between ionic liquid groups and carbon core. Worn surface analyses revealed that CD-IL could form ordered absorption layers quickly on the rubbing surfaces during the process of boundary shear friction by electrostatic interaction between steel surface and ionic liquid groups. • **Challenge:** Obtaining an optimum concentration is important.	[33]
[THTDP] [Deca]	• This is a halogen-free ionic liquid	• Friction was reduced over 70% compared to only water	**Mechanism:** Long alkyl chain helps adsorption film formation and thus, reduces friction	[96]
Ibuprofen-based ionic liquids	• Relatively good hydrolytic stability (ASTMD 2619) • Slight increase in kinematic viscosity • Good solubility (>10%) • Adsorption on the metal surface • Phosphorus-, Sulphur-, and halogen-free	• Corrosion resistance on copper and cast iron strip test • The coefficient of friction and wear volumes reduce with an increase in mass concentration	**Mechanism:** Protective film is formed between the contact pair and reduced the friction and wear. IL offers active polar sites that interact with the tribo-pair.	[104]
Amino acid ionic liquids (AAILs): P_{4444} (Tetrabutylphosphonium)-Histidine, P_{4444}-serine, P_{4444}-tryptophan (Trp), P_{4444}-lysine, P_{4444}-phenylalanine, P_{4444}-cysteine, and P_{4444}-methionine	• Obtained from waste biological proteins • Synthesized using one-step acid-base neutralization • The concentration of P_{4444}-Trp was 0.5% (0.5 mg/mL) • Possesses low toxicity	• All AAILs reduced COF and wear • P_{4444}—Trp exhibited bactericidal properties and low toxicity to plants	• **Mechanism:** P_{4444}-Trp IL forms a physical adsorption film on the surface of the metal, minimizing the direct contact. Moreover, a tribochemical reaction film, formed from the reaction between the active elements on the surface and the adsorption layer, was inferred from the friction-wear experiments, XPS, and TOF-SIMS analyses. • **Challenge:** Optimization of concentration is important	[108]
Benzotriazole: P_{4444}BTA	• Miscibility with many ionic liquids • Kinematic viscosity (373 mm²/sec at 40 °C) is higher than other oils like PAO • The sublimation point of benzotriazole is 100 °C	• Corrosion reduction • Recent research on tetrabutylphosphonium benzotriazole shows it offered lower friction compared to PAO-10 in steel-copper contact at both room temperature and 100 °C temperature	• **Mechanism:** Physical adsorption between the BTA anion and the metal surface was speculated as to the reason for corrosion reduction. The layer was washed away during the ultrasonication before doing the XPS analysis. Moreover, the copper surface could have experienced chemical adsorption from the interaction with BTA⁻, which helped to form a protective film, thus reducing corrosion. • **Challenge:** Low performance at high temperature (above 100 °C)	[23,32]
Ammonium ionic liquids: bis(2-hydroxyethyl) ammonium palmitate (DPA)	• DPA is an ionic liquid crystal with two mesomorphic phase transitions. • At 42.1 °C, it experiences a transition from crystalline solid to liquid crystalline, and at 105.2 °C it melts to isotropic liquid	• Reduced sapphire-steel COF by 80% compared to water • Maintained the friction, even after the evaporation of water from the solution	• **Mechanism:** The lubricant film acted as friction and wear reducer. As per the surface characterization, active groups from the lubricant were observed, indicating the presence of lubricant film. • **Challenge:** As the temperature increased from 75 °C to 110 °C, friction increased for steel-steel tribo-pair.	[34,111]
[Phosphazene][NTf₂]	• Solubility in water (2.5 g/L ± 10%) was less compared to imidazolium salts.	• 0.25% [Phosphazene][NTf₂], and 2% [1-Ethyl-3-methylimidazolium][BF4] reduced the friction in the Si_3N_4-Si_3N_4 contact pair compared to that of only water • Running in the period was also decreased significantly	• **Mechanism:** A potential formation of an electric double layer on the surface could have increased the local viscosity and the load-carrying capacity. • **Challenge:** The solubility was limited in water.	[35,75]
[1-Butyl-3-methylimidazolium][PF₆]	• It is an imidazolium-based ionic liquid	• Reduced friction for Si_3N_4-Si_3N_4 ceramic tribo-pair	• **Mechanism:** The existence of an electric double layer consisting of an equal and oppositely charged region from the ionic liquid could have reduced the friction and wear. • **Challenge:** Halogen containing ionic liquid can experience corrosion in the case of metal contact pairs	[35,112]

Table 4. *Cont.*

WBL Additives	Properties	Advantages	Remarks	Ref.
		Biobased and other WBL additives		
Lithium salt and nonionic surfactant (Li-TW)	• Good solubility in water • Kinematic viscosity of 3.35 (mm²/s) • In-situ forming of ionic liquid without synthesis	• Reduced the corrosion of metal • Friction and wear reduced • Creative of effective protective film	• **Mechanism:** 2% LI-TW additive forms more stable and thicker insulating tribofilm, compared to pure water during the interaction • Challenge: More research needs to be completed	[105]
Water-based oleic acid (OA)/2-acryloylamino-2-methyl-1-propanesulfonic cid (AMPS)	• It is a rubber seed oil-based sulfonate • anti-friction, anti-wear, and anti-corrosion properties	• Additive improved the nonseizure load of water • At 0.5 wt.% concentration improved the nonseizure load of water to 431 N • The lubricant's environmentally friendly attribute is enhanced	• **Mechanism:** (1) Long aliphatic chain of OA-AMPS copolymer could have adsorbed on the metal surface and formed a film. (2) Sulfur elements from OA-AMPS could have reacted with the metal surface under frictional heat under high-pressure interaction and produced ferrous disulfide tribofilms on the metal surface. • Challenge: More research needs to be done since this is a novel copolymer additive.	[60]
SMJH-1	• It is a type of chemical compound from vegetable ester, provided by the Research Institute of Petroleum Engineering, SINOPEC, China	• Decreased the lubricity coefficient	• **Mechanism:** The surfactant was self-assembled at the surface due to the electrostatic repulsive effect. • Challenge: More investigation is needed in regard to water-based cutting fluid.	[63]
Alkyl glucopyranosides (APGs)	• Contact angle 21° at concentration 25 mM • Increased viscosity compared to water	• Lower COF compared to pure water • Coefficient of Friction range from 0.15 to 0.04 depending on chain length and concentration	• **Mechanism:** APG forms an ordered structure of liquid crystalline phase in the gap between friction pairs. Moreover, the interaction between APGs with steel surfaces (100Cr6) forms a wear protective layer. • Challenge: The extent of COF reduction depends on the shearing condition and concentration of the additive.	[8]
Polyethylene glycol (PEG)	• Five different PEGs were used at a 10% concentration	• Friction and wear-reducing ability in ferrous-based equipment	• **Mechanism:** PEG alone could not reduce the friction; rather, the synergy between PEG and water-soluble EP additives can improve the performance. • Challenge: This research found that PEG alone is not a good additive to reduce friction in WBL.	[98,99]
Glycerol ester of levulinic acid (LAGLE)	• Viscosity at 40 °C is 12.28 mm²/s • Flash point: 202 °C • Cloud point −21 °C	• Friction was reduced significantly • Offered strong resistance to hydrolytic degradation	• **Mechanism:** (1) Long chain polar molecules could have enhanced the molecular adsorption on the metallic surface. (2) Synergistic effect of water and polyol ester molecules could have initiated water containing nanofilm, which could have retained shear fluidity in bulk fluid and reduced the friction significantly. • Challenge: More research on lignocellulosic-based WBL is required for future advancements.	[41]

Furthermore, through an in-situ preparation, many multifunctional additives can be created with significant tribological properties and corrosion-resistant capabilities [105,106]. Important additives that were prepared in-situ are benzotriazole tetrabutyl phosphonium [P$_{4444}$][BTA], and benzotriazole tetrabutylammonium [N$_{4444}$][BTA]. Benzotriazole is a corrosion inhibitor for metals that helps increase the corrosion-resistant ability. The additives were able to create adsorption protective film on the iron surface consisting of nitrogen, which would result in no visible corrosion on the surface, outperforming the pieces being exposed to pure water and a Benzotrazole/H$_2$O solution that had visible and substantial corrosion on the surface.

Another important additive that has been created through an in-situ preparation is [Li(nonionic surfactant)]TFSI [105]. Testing determined that the additive could increase the kinematic viscosity of the WBLs better than others at 25 °C. The additive achieved similar anti-corrosion properties as the ones made of benzotriazole, though it was created using a lithium salt and a nonionic surfactant. Another ionic liquid (IL) synthesized and tested by the same research group is [Li(nonionic surfactant)]TFSI; an Ibuprofen-based (L-Ibu) halogen-free ionic liquid [104]. The additive has resulted in greater friction-reducing, extreme pressure, anti-wear properties than commercial additives. The corrosion test consisted of the Ibuprofen IL L-Ibu106, and L-Ibu108 at 2% concentrations within the water. A copper strip and cast-iron strip were soaked independently within each lubricant [104]. After testing, the surfaces of the strips showed no discoloration compared to the strips soaked in water and a water–ethylene glycol solution (W-EG) which showed major discoloration, due to substantial corrosion. Due to this significant anti-corrosion property, the additives were able to achieve a grade A according to GB6144-85, meaning there was no rust, while looking new [104]. Additives and ionic liquids are becoming prime candidates for WBLs to implement to achieve better corrosion-resistant capabilities. Many of the additives and ILs discussed within these sections were able to increase resistance significantly. Often being able to create a protective film along the surface of the metal that would protect any formed wear scars from possible corrosion. In cases where the corrosion testing was evaluated using the GB6144-85 scale, the additives and ILs were able to achieve grade B or A (very little/no rust); Compared to a grade D in base water or another water-based solution/lubricant, which indicated severe/noticeable corrosion.

5. Recent Advancement and Challenges of Water-Based Lubricants

WBLs field has experienced significant development in recent years. Environmental concerns with oil-based lubricants have accelerated the amount of research being performed on WBLs. Petroleum-based lubricants can be toxic and are typically nonbiodegradable. This generates excess waste that can be problematic with the high volume of oil-based lubricants being used. The need for a cleaner, more efficient lubricant to replace traditional oil-based lubricants has driven researchers to develop a more sustainable option. This section will discuss recent advancements in water-based lubrication and the challenges that come along with developing new WBLs. In industrial production, a large percentage of energy loss results from friction and wear [113].

As the world population continues to rise, the need for production also rises. This means that more machines need to be used to satisfy the needs of production. The amount of crude oil the world has available is declining rapidly. In the early 2000s, there were huge efforts to find plant-based lubrication alternatives to reduce the environmental impact of machining. In more recent years, scientific research has been conducted on water-based lubrication options. Water alone cannot be used as a lubricant. Additives must be incorporated with the water to make it suitable to reduce frictional effects caused by machining. This has been one challenge water-based lubrication has presented. Many additives have been tested, but only a few can make a solution that is comparable to an oil-based lubricant. There is a rising need for better lubrication in the metal forming process, due to the high friction created by materials in contact [114]. The excessive amount of friction can lead to several negative effects, including the reduced life cycle of parts, deformation, and heat generation.

Recent research has pointed towards the use of nanoparticles as additives for lubricants. They are small, have no harmful emissions, and are chemically stable [109]. Graphene as a nanoparticle additive has been the focus of researchers recently. The research conducted by the Northwest Normal University indicated that the tribological properties of graphene as an additive decreased wear by nearly four orders of magnitude and friction coefficient by a factor of six [115]. Further, research conducted by the Beijing University of Aeronautics and Astronautics found that graphene as a water additive for steel-on-steel contact offers 81.3% friction reduction [116]. The use of graphene has been highly efficient and is continually being implemented for more machining processes. Another nanoparticle that has been widely accepted as an efficient additive to WBLs is SiO_2. SiO_2 is advantageous, due to the low-cost nature of adding this nanoparticle to water. SiO_2 creates a rolling action between surfaces in contact, which leads to a reduction in friction [109]. Nanoparticles can also be combined to make new additives in the ideal combinations. Recent research suggests that the lubricity of WBLs has been enhanced with the introduction of nanoparticles to the water. Water-based nanolubricants are one of the ideal formulations of WBLs, due to their eco-friendliness and overall simplicity. Another added benefit is the low-cost nature of producing this type of formulation. This advancement on the lubrication front has been applied to the industrial scale hot steel rolling process. Water-based nanolubricants that contain TiO_2 have reduced the rolling force that is present in the hot rolling process. The nanoparticles took the places between the oxide layers on the workpiece surface, while the rolling process was carried out (Figure 5) [117]. Therefore, a stronger film was formed. Moreover, the friction reduction was facilitated by the wettability of the TiO$_2$ [117]. However, more studies are needed to completely understand the mechanism of such nanolubricants.

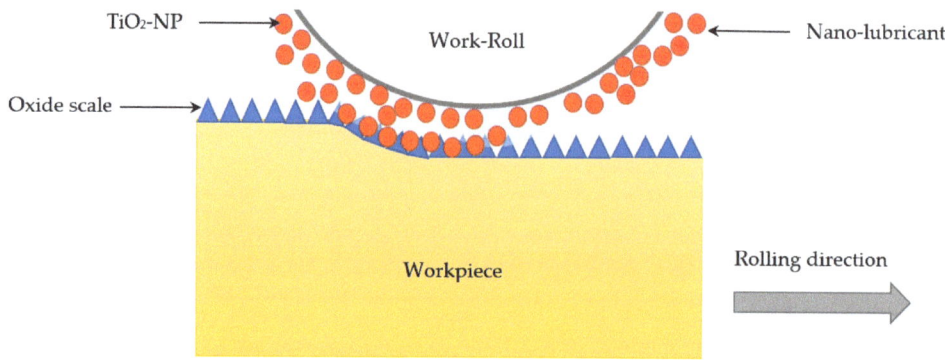

Figure 5. Schematic illustration of lubrication mechanisms using water-based nanolubricants containing TiO$_2$ nanoparticles during hot steel rolling. Adopted from [117].

At present, several studies have been focused on using WBLs to achieve superlubricity or very low friction. To achieve superlubricity, the frictional coefficient needs to be lower than 0.01 at the nanoscale [118]. Experimentation supported by the National Science Foundation of China (NSFC), revealed that using the running-in process on WBLs with sulfuric acid can generate super low friction. In the process, a silica layer forms from a tribochemical process and allows the surfaces to glide over one another more easily [118]. This process, as a result, reduces the coefficient of friction between surfaces.

Overall, water-based lubrication is at the forefront of technological advancement to meet the world's desire for methods that promote sustainable manufacturing. The new type of lubricant has presented many challenges; however, research has overcome some of the challenges, and scientists have been able to produce several WBLs in large quantities. The governmental push for more environmentally friendly lubricants has quickly accelerated the advancements of water-based lubricant technologies around the world. As time goes on, manufacturing demand and capabilities will continue to rise. This will result in a greater demand for eco-friendly lubricants. Water-based lubrication is one method of machine lubrication that promotes environmental awareness and sustainable growth. Recently, the use of additives in combination with water has provided the world with a positive outlook on the lubrication front. Nanoparticles are generally cheap and easily accessible, making them an ideal additive for lubrication.

6. Conclusions

Water-based lubricants could be an environmentally friendly alternative to petroleum-based lubricants. However, the issue with WBLs is that they have poor tribological properties: Poor friction reduction, high corrosiveness, and low viscosity. Additives have shown great potential for improving the nature of WBLs. Utilizing additives with WBLs is gaining interest among scientists and engineers, not only because of the potential for being environmentally friendly, but also because there is a potential for a reusable liquid lubricant. Further exploring the use and development of additives also creates lubricants that can be used for multiple purposes. Additives can change the properties and behavior of the lubricant, resulting in high-performance liquid lubricants designed for specific tribosystems. Properties include a higher viscosity and density to improve the thermal behavior and create thicker fluid films within the lubricant. There are also improved friction, wear,

Lubricants **2021**, 9, 73

and corrosion performance for water-based liquid lubricants. The atomic structure of the lubricant changes based on the type of additive that is used. Ionic liquids, bio-additives, titanium-based nano-additive, water-soluble rubber seed oil-based sulfonate, graphene-based nanoparticles, and nanostructured borates are all potential additives that can be used with water-based liquid lubricants to improve the behavior and properties of the lubricant. A potential challenge for WBLs is that the manufacturing demand and capabilities will increase over time as additives used for water-based liquid lubricants become a viable option. Current research is focusing on the advancement and development of additives. Therefore, continuing the research and development for additives used in combination with water-based liquid lubricants is important. This will further expand the usage of water-based lubricants over petroleum-based lubricants as an eco-friendly alternative.

Author Contributions: Conceptualization, M.H.R. and P.L.M.; Methodology, M.H.R., H.W. (Haley Warneke), H.W. (Haley Webbert), J.R., E.A., K.T.; Writing-original draft, H.W. (Haley Warneke), H.W. (Haley Webbert), J.R., E.A., K.T., M.H.R.; writing—review and editing, M.H.R., D.K.R., P.L.M.; visualization, M.H.R.; supervision, M.H.R.; project administration, P.L.M. All authors have read and agreed to the published version of the manuscript.

Funding: We acknowledge the support and facilities from the Department of Mechanical Engineering at the University of Nevada at Reno.

Acknowledgments: We acknowledge the technical help of Ashish Kasar during this review.

Conflicts of Interest: The authors declare no conflict of interest. The funders had no role in the design of the study; in the collection, analyses, or interpretation of data; in the writing of the manuscript, or in the decision to publish the results.

References

1. Scharf, T.; Prasad, S. Solid lubricants: A review. *J. Mater. Sci.* **2013**, *48*, 511–531. [CrossRef]
2. Menezes, P.L.; Ingole, S.P.; Nosonovsky, M.; Kailas, S.V.; Lovell, M.R. *Tribology for Scientists and Engineers*; Springer: New York, NY, USA, 2013.
3. Bart, J.; Gucciardi, E.; Cavallaro, S. Advanced lubricant fluids. In *Biolubricants: Science and Technology*; Woodhead Publishing Ltd.: Cambridge, UK, 2013; pp. 824–846.
4. Panja, S.K. Tribological Properties of Ionic Liquids. In *Tribology*; IntechOpen: London, UK, 2020.
5. Cai, M.; Yu, Q.; Liu, W.; Zhou, F. Ionic liquid lubricants: When chemistry meets tribology. *Chem. Soc. Rev.* **2020**. [CrossRef]
6. Fitch, J. The History of Lubrication. Available online: https://www.machinerylubrication.com/Read/579/lubrication-genesis (accessed on 7 April 2021).
7. Ritchie, H. How Long before We Run Out of Fossil Fuels? Available online: https://ourworldindata.org/how-long-before-we-run-out-of-fossil-fuels (accessed on 7 April 2021).
8. Wei, C.; Tobias, A.; Andreas, K.; Jurgen, R. Macroscopic Friction Studies of Alkylglucopyranosides as Additives for Water-Based Lubricants. *Lubricants* **2020**, *8*, 11. [CrossRef]
9. Tomala, A.; Karpinska, A.; Werner, W.S.M.; Olver, A.; Stori, H. Tribological Properties of Additives for Water-Based Lubricants. *Wear* **2010**, *296*, 804–810. [CrossRef]
10. Sagraloff, N.; Dobler, A.; Tobie, T.; Stahl, K.; Ostrowski, J. Development of an Oil Free Water-Based Lubricant for Gear Applications. *Lubricants* **2019**, *7*, 33. [CrossRef]
11. Liu, C.; Guo, Y.; Wang, D. PEI-RGO Nanosheets as a Nano-Additive for Enhancing the Tribological Properties of Water-Based Lubricants. *Tribol. Int.* **2019**, *140*. [CrossRef]
12. Wu, H.; Zhao, J.; Xia, W.; Cheng, X.; He, A.; Yun, J.H.; Wang, L.; Huang, H.; Jiao, S.; Huang, L.; et al. A study of the tribological behaviour of TiO2 nano-additive water-based lubricants. *Tribol. Int.* **2017**, *109*, 398–408. [CrossRef]
13. Wu, H.; Jia, F.; Li, Z.; Lin, F.; Huo, M.; Huang, S.; Sayyar, S.; Jiao, S.; Huang, H.; Jiang, Z. Novel water-based nanolubricant with superior tribological performance in hot steel rolling. *Int. J. Extrem. Manuf.* **2020**, *2*. [CrossRef]
14. Del Sol, I.; Gámez, A.; Rivero, A.; Iglesias, P. Tribological performance of ionic liquids as additives of water-based cutting fluids. *Wear* **2019**, *426*, 845–852. [CrossRef]
15. Wang, H.; Liu, Y.; Liu, W.; Liu, Y.; Wang, K.; Li, J.; Ma, T.; Eryilmaz, O.L.; Shi, Y.; Erdemir, A. Superlubricity of polyalkylene glycol aqueous solutions enabled by ultrathin layered double hydroxide nanosheets. *ACS Appl. Mater. Interfaces* **2019**, *11*, 20249–20256. [CrossRef]
16. Kailer, A.; Amann, T. Development of Water-Based Lubricants. Available online: https://www.iwm.fraunhofer.de/en/services/tribology/wear-protection-advanced-ceramics/development-of-water-based-lubricants.html (accessed on 2 February 2021).

17. Supekar, S.D.; Graziano, D.J.; Skerlos, S.J.; Cresko, J. Comparing energy and water use of aqueous and gas-based metalworking fluids. *J. Ind. Ecol.* **2020**, *24*, 1158–1170. [CrossRef]
18. Rahman, M.H.; Bhoi, P.R.; Saha, A.; Patil, V.; Adhikari, S. Thermo-catalytic co-pyrolysis of biomass and high-density polyethylene for improving the yield and quality of pyrolysis liquid. *Energy* **2021**, *225*, 120231. [CrossRef]
19. Rahman, M.H. *Catalytic Co-pyrolysis of Pinewood and Waste Plastics for Improving the Selectivity of Hydrocarbons and the Quality of Pyrolysis Oil*; Georgia Southern University: Statesboro, GA, USA, 2020.
20. Wang, B.; Tang, W.; Liu, X.; Huang, Z. Synthesis of ionic liquid decorated muti-walled carbon nanotubes as the favorable water-based lubricant additives. *Appl. Phys. A* **2017**, *123*, 1–11. [CrossRef]
21. Khanmohammadi, H.; Wijanarko, W.; Espallargas, N. Ionic Liquids as Additives in Water-based Lubricants: From Surface Adsorption to Tribofilm Formation. *Tribol. Lett.* **2020**, *68*. [CrossRef]
22. Carrion, F.-J.; Sanes, J.; Bermudez, M.-D.; Jimenez, A.-E. Ionic Liquids as Advanced Lubricant Fluids. *Molecules* **2009**, *14*, 2888–2908. [CrossRef]
23. Zhou, F.; Liang, Y.; Liu, W. Ionic Liquid Lubricant: Designed Chemistry for Engineering Applications. *Chem. Soc. Rev.* **2009**. [CrossRef]
24. Plechkova, N.V.; Seddon, K.R. Applications of ionic liquids in the chemical industry. *Chem. Soc. Rev.* **2008**, *37*, 123–150. [CrossRef]
25. Reeves, C.J.; Siddaiah, A.; Menezes, P.L. Tribological study of imidazolium and phosphonium ionic liquid-based lubricants as additives in carboxylic acid-based natural oil: Advancements in environmentally friendly lubricants. *J. Clean Prod.* **2018**, *176*, 241–250. [CrossRef]
26. Reeves, C.J.; Siddaiah, A.; Menezes, P.L. Friction and Wear Behavior of Environmentally Friendly Ionic Liquids for Sustainability of Biolubricants. *J. Tribol.* **2019**, *141*. [CrossRef]
27. Reeves, C.J.; Kasar, A.K.; Menezes, P.L. Tribological performance of environmental friendly ionic liquids for high-temperature applications. *J. Clean Prod.* **2021**, *279*, 123666. [CrossRef]
28. Van Rensselar, J. Unleashing the potential of ionic liquids. *Tribol. Lubr. Technol.* **2010**, *66*, 24–31.
29. Ye, C.; Liu, W.; Chen, Y.; Yu, L. Room-temperature ionic liquids: A novel versatile lubricant. *Chem. Commun.* **2001**, 2244–2245. [CrossRef]
30. Wasserscheid, P.; Welton, T. *Ionic Liquids in Synthesis*; John Wiley & Sons: Hoboken, NJ, USA, 2008.
31. Westerholt, A.; Weschta, M.; Bosmann, A.; Tremmel, S.; Korth, Y.; Wolf, M.; Schlücker, E.; Wehrum, N.; Lennert, A.; Uerdingen, M. Halide-free synthesis and tribological performance of oil-miscible ammonium and phosphonium-based ionic liquids. *ACS Sustain. Chem. Eng.* **2015**, *3*, 797–808. [CrossRef]
32. Zhang, S.; Ma, L.; Dong, R.; Zhang, C.; Sun, W.; Fan, M.; Yang, D.; Zhou, F.; Liu, W. Study on the synthesis and tribological properties of anti-corrosion benzotriazole ionic liquid. *RSC Adv.* **2017**, *7*, 11030–11040. [CrossRef]
33. Tang, W.; Wang, B.; Li, J.; Li, Y.; Zhang, Y.; Quan, H.; Huang, Z. Facile Pyrolysis Synthesis of Ionic Liquid Capped Carbon Dots and Subsequent Application as the water-Based Lubricant Additives. *J. Mater. Sci.* **2019**, *54*, 1171–1183. [CrossRef]
34. Meng, Y.; Xu, J.; Jin, Z.; Prakash, B.; Hu, Y. A review of recent advances in tribology. *Friction* **2020**, *8*. [CrossRef]
35. Zhou, Y.; Qu, J. Ionic Liquids as Lubricant Additives: A Review. *Appl. Mater. Interfaces* **2017**, *9*, 3209–3222. [CrossRef]
36. Bremmer, B.; Plonsker, L.; Martin, J. *Bio-Based Lubricants*; United Soybean Board: Chesterfield, MO, USA, 2013.
37. Reeves, C.; Siddaiah, A.; Menezes, P. Ionic Liquids: A Plausible Future of Bio-lubricants. *J. Bio Tribo Corros.* **2017**. [CrossRef]
38. Gucciardi, E.; Bart, J.; Cavallaro, S. *Biolubricants: Science and Technology*; Woodhead Publishing Limites: Cambridge, UK, 2013.
39. Bhoi, P.; Ouedraogo, A.; Soloiu, V.; Quirino, R. Recent advances on catalysts for improving hydrocarbon compounds in bio-oil of biomass catalytic pyrolysis. *Renew. Sustain. Energy Rev.* **2020**, *121*, 109676. [CrossRef]
40. Lindfors, C.; Kuoppala, E.; Oasmaa, A.; Solantausta, Y.; Arpiainen, V. Fractionation of bio-oil. *Energy Fuels* **2014**, *28*, 5785–5791. [CrossRef]
41. Ji, H.; Zhang, X.; Tan, T. Preparation of a Water-Based Lubricant from Lignocellulosic Biomass and Its Tribological Properties. *Ind. Eng. Chem. Res.* **1999**. [CrossRef]
42. Mobarak, H.; Mohamad, N.; Masjuki, H.; Kalam, M.; Mahmud, K.; Habibullah, M.; Ashraful, A. The prospects of biolubricants as alternatives in automotive applications. *Renew. Sustain. Energy Rev.* **2014**, *33*, 34–43. [CrossRef]
43. Sharma, Y.; Singh, B. An ideal feedstock, kusum (*Schleichera triguga*) for preparation of biodiesel: Optimization of parameters. *Fuel* **2010**, *89*, 1470–1474. [CrossRef]
44. Karaosmanoglu, F.; Tuter, M.; Gollu, E.; Yanmaz, S.; Altintig, E. Fuel properties of cottonseed oil. *Energy Sources* **1999**, *21*, 821–828. [CrossRef]
45. Usta, N.; Aydoğan, B.; Çon, A.; Uğuzdoğan, E.; Özkal, S. Properties and quality verification of biodiesel produced from tobacco seed oil. *Energy Convers. Manag.* **2011**, *52*, 2031–2039. [CrossRef]
46. Singh, D.; Singh, S. Low cost production of ester from non edible oil of Argemone mexicana. *Biomass Bioenergy* **2010**, *34*, 545–549. [CrossRef]
47. Wang, R.; Hanna, M.A.; Zhou, W.-W.; Bhadury, P.S.; Chen, Q.; Song, B.-A.; Yang, S. Production and selected fuel properties of biodiesel from promising non-edible oils: *Euphorbia lathyris* L., *Sapium sebiferum* L. and *Jatropha curcas* L. Bioresour. Technol. **2011**, *102*, 1194–1199. [CrossRef]
48. Li, X.; He, X.-Y.; Li, Z.-L.; Wang, Y.-D.; Wang, C.-Y.; Shi, H.; Wang, F. Enzymatic production of biodiesel from Pistacia chinensis bge seed oil using immobilized lipase. *Fuel* **2012**, *92*, 89–93. [CrossRef]

49. Wang, R.; Zhou, W.-W.; Hanna, M.A.; Zhang, Y.-P.; Bhadury, P.S.; Wang, Y.; Song, B.-A.; Yang, S. Biodiesel preparation, optimization, and fuel properties from non-edible feedstock, *Datura stramonium* L. *Fuel* **2012**, *91*, 182–186. [CrossRef]
50. Mofijur, M.; Masjuki, H.; Kalam, M.; Hazrat, M.; Liaquat, A.; Shahabuddin, M.; Varman, M. Prospects of biodiesel from Jatropha in Malaysia. *Renew. Sustain. Energy Rev.* **2012**, *16*, 5007–5020. [CrossRef]
51. Reeves, C.J.; Menezes, P.L.; Jen, T.-C.; Lovell, M.R. The influence of fatty acids on tribological and thermal properties of natural oils as sustainable biolubricants. *Tribol. Int.* **2015**, *90*, 123–134. [CrossRef]
52. Lovell, M.R.; Kabir, M.; Menezes, P.L.; Higgs III, C.F. Influence of boric acid additive size on green lubricant performance. *Philos. Trans. R. Soc. A Math. Phys. Eng. Sci.* **2010**, *368*, 4851–4868. [CrossRef] [PubMed]
53. Reeves, C.J.; Menezes, P.L.; Lovell, M.R.; Jen, T.-C. The size effect of boron nitride particles on the tribological performance of biolubricants for energy conservation and sustainability. *Tribol. Lett.* **2013**, *51*, 437–452. [CrossRef]
54. Omrani, E.; Siddaiah, A.; Moghadam, A.D.; Garg, U.; Rohatgi, P.; Menezes, P.L. Ball Milled Graphene Nano Additives for Enhancing Sliding Contact in Vegetable Oil. *Nanomaterials* **2021**, *11*, 610. [CrossRef] [PubMed]
55. Omrani, E.; Menezes, P.L.; Rohatgi, P.K. Effect of micro-and nano-sized carbonous solid lubricants as oil additives in nanofluid on tribological properties. *Lubricants* **2019**, *7*, 25. [CrossRef]
56. Siddaiah, A.; Kasar, A.K.; Manoj, A.; Menezes, P.L. Influence of environmental friendly multiphase lubricants on the friction and transfer layer formation during sliding against textured surfaces. *J. Clean Prod.* **2019**, *209*, 1245–1251. [CrossRef]
57. Gu, Y.; Zhao, X.; Liu, Y.; Lv, Y. Preparation and Tribological Properties of Dual-Coated TiO_2 Nanoparticles as Water-Based Lubricant Additives. *J. Nanomater.* **2014**, *2014*, 785680. [CrossRef]
58. Nguyen, H.T.C.K.-H. Assessment of Tribological Properties of Ti_3C_2 as a Water-Based Lubricant Additive. *Materials* **2020**, *13*, 5545. [CrossRef]
59. Liu, T.; Zhou, C.; Gao, C.; Zhang, Y.; Yang, G.; Zhang, P.; Zhang, S. Preparation of $Cu@SiO_2$ composite nanoparticle and its tribological properties as water-based lubricant additive. *Lubr. Sci.* **2020**, *32*, 69–79. [CrossRef]
60. Ding, H.; Wang, M.; Li, M.; Huang, K.; Li, S.; Xu, L.; Yang, X.; Xia, J. Synthesis of a water-soluble, rubber seed oil–based sulfonate and its tribological properties as a water-based lubricant additive. *J. Appl. Polym. Sci.* **2018**, *135*, 46119. [CrossRef]
61. Qiang, R.; Hu, L.; Hou, K.; Wang, J.; Yang, S. Water-Soluble Graphene Quantum Dots as High-Performance Water-Based Lubricant Additive for Steel/Steel Contact. *Tribol. Lett.* **2019**, *67*, 1–9. [CrossRef]
62. Saffari, H.R.M.; Soltani, R.; Alaei, M.; Soleymani, M. Tribological properties of water-based drilling fluids with borate nanoparticles as lubricant additives. *J. Pet. Sci. Eng.* **2018**, *171*, 253–259. [CrossRef]
63. Dong, X.; Wang, L.; Yang, X.; Lin, Y.; Xue, Y. Effect of ester based lubricant SMJH-1 on the lubricity properties of water based drilling fluid. *J. Pet. Sci. Eng.* **2015**, *135*, 161–167. [CrossRef]
64. Nouri, J.M.; Vasilakos, I.; Yan, Y.; Reyes-Aldasoro, C.-C. Effect of Viscosity and Speed on Oil Cavitation Development in a Single Piston-Ring Lubricant Assembly. *Lubricants* **2019**, *7*, 88. [CrossRef]
65. Hamrock, B.J.; Dowson, D. *Ball Bearing Lubrication: The Elastohydrodynamics of Elliptical Contacts*; ASME: New York, NY, USA, 1981.
66. Menezes, P.L.; Kailas, S.V. Effect of roughness parameter and grinding angle on coefficient of friction when sliding of Al–Mg alloy over EN8 steel. *J. Tribol.* **2006**, *128*, 697–704. [CrossRef]
67. Marx, N.; Fernández, L.; Barceló, F.; Spikes, H. Shear Thinning and Hydrodynamic Friction of Viscosity Modifier-Containing Oils. Part II: Impact of Shear Thinning on Journal Bearing Friction. *Tribol. Lett.* **2018**, *66*, 91. [CrossRef]
68. MacConochie, I.O.; Newman, W.H. The effect of lubricant viscosity on the lubrication of gear teeth. *Wear* **1961**, *4*, 10–21. [CrossRef]
69. Larsson, R. EHL Film Thickness Behavior. In *Encyclopedia of Tribology*; Wang, Q.J., Chung, Y.-W., Eds.; Springer: Boston, MA, USA, 2013; pp. 817–827.
70. Shen, M.; Luo, J.; Wen, S.; Yao, J. Nano-tribological properties and mechanisms of the liquid crystal as an additive. *Chin. Sci. Bull.* **2001**, *46*, 1227–1232. [CrossRef]
71. Ramezanizadeh, M.; Ahmadi, M.H.; Nazari, M.A.; Sadeghzadeh, M.; Chen, L. A review on the utilized machine learning approaches for modeling the dynamic viscosity of nanofluids. *Renew. Sustain. Energy Rev.* **2019**, *114*, 109345. [CrossRef]
72. Hajmohammadi, M.R. Assessment of a lubricant based nanofluid application in a rotary system. *Energy Convers. Manag.* **2017**, *146*, 78–86. [CrossRef]
73. Corcione, M. Empirical correlating equations for predicting the effective thermal conductivity and dynamic viscosity of nanofluids. *Energy Convers. Manag.* **2011**, *52*, 789–793. [CrossRef]
74. Najiha, M.S.; Rahman, M.M.; Kadirgama, K. Performance of water-based TiO_2 nanofluid during the minimum quantity lubrication machining of aluminium alloy, AA6061-T6. *J. Clean Prod.* **2016**, *135*, 1623–1636. [CrossRef]
75. Omotowa, B.A.; Phillips, B.S.; Zabinski, J.S.; Shreeve, J.n.M. Phosphazene-Based Ionic Liquids: Synthesis, Temperature-Dependent Viscosity, and Effect as Additives in Water Lubrication of Silicon Nitride Ceramics. *Inorg. Chem.* **2004**, *43*, 5466–5471. [CrossRef] [PubMed]
76. Moldoveanu, S.C.; David, V. (Eds.) Chapter 7—RP-HPLC Analytical Columns. In *Selection of the HPLC Method in Chemical Analysis*; Elsevier: Boston, MA, USA, 2017; pp. 279–328.
77. Cabezudo, N.; Sun, J.; Andi, B.; Ding, F.; Wang, D.; Chang, W.; Luo, X.; Xu, B.B. Enhancement of surface wettability via micro- and nanostructures by single point diamond turning. *Nanotechnol. Precis. Eng.* **2019**, *2*, 8–14. [CrossRef]
78. Pawlak, Z.; Urbaniak, W.; Oloyede, A. The relationship between friction and wettability in aqueous environment. *Wear* **2011**, *271*, 1745–1749. [CrossRef]

79. Hu, Y.; Wang, Y.; Zeng, Z.; Zhao, H.; Ge, X.; Wang, K.; Wang, L.; Xue, Q. PEGlated graphene as nanoadditive for enhancing the tribological properties of water-based lubricants. *Carbon* **2018**, *137*, 41–48. [CrossRef]
80. Wang, X.-J.; Zhu, D.-S. Investigation of pH and SDBS on enhancement of thermal conductivity in nanofluids. *Chem. Phys. Lett.* **2009**, *470*, 107–111. [CrossRef]
81. Sułek, M.W.; Ogorzałek, M.; Wasilewski, T.; Klimaszewska, E. Alkyl Polyglucosides as Components of Water Based Lubricants. *J. Surfactants Deterg.* **2013**, *16*, 369–375. [CrossRef] [PubMed]
82. Bhattacharjee, S.; Chakraborty, D.; Khan, S. Wetting behavior of aqueous 1-alkyl-3-methylimidazolium tetrafluoroborate {[Cn MIM][BF4] (*n* = 2, 4, 6)} on graphite surface. *Chem. Eng. Sci.* **2021**, *229*, 116078. [CrossRef]
83. Basu, S.; Hanh, B.M.; Isaiah Chua, J.Q.; Daniel, D.; Ismail, M.H.; Marchioro, M.; Amini, S.; Rice, S.A.; Miserez, A. Green biolubricant infused slippery surfaces to combat marine biofouling. *J. Colloid Interface Sci.* **2020**, *568*, 185–197. [CrossRef]
84. Nunes, D.G.; da Silva, A.d.P.M.; Cajaiba, J.; Pérez-Gramatges, A.; Lachter, E.R.; Nascimento, R.S.V. Influence of glycerides–xanthan gum synergy on their performance as lubricants for water-based drilling fluids. *J. Appl. Polym. Sci.* **2014**, *131*. [CrossRef]
85. Lakshmi, R.V.; Bera, P.; Anandan, C.; Basu, B.J. Effect of the size of silica nanoparticles on wettability and surface chemistry of sol–gel superhydrophobic and oleophobic nanocomposite coatings. *Appl. Surf. Sci.* **2014**, *320*, 780–786. [CrossRef]
86. Wen, P.; Lei, Y.; Li, W.; Fan, M. Synergy between Covalent Organic Frameworks and Surfactants to Promote Water-Based Lubrication and Corrosion Resistance. *ACS Appl. Nano Mater.* **2020**, *3*, 1400–1411. [CrossRef]
87. Zhang, C. Preparation and Tribological Properties of Water-Soluble Copper/Silica Nanocomposite as a Water-Based Lubricant Additive. *Appl. Surf. Sci.* **2012**. [CrossRef]
88. Pei, X.; Hu, L.; Liu, W.; Hao, J. Synthesis of Water-Soluble Carbon Nanotubes via Surface Initiated Redox Polymerization and Their Tribological Properties as Water-Based Lubricant Additive. *Eur. Polym. J.* **2008**. [CrossRef]
89. Wang, Z.; Liu, Q.; Zhu, H.; Liu, H.; Chen, Y.; Yang, M. Dispersing multi-walled carbon nanotubes with water soluble block copolymers and their use as supports for metal nanoparticles. *Carbon* **2007**, *45*, 285–292. [CrossRef]
90. Yao, Z.; Braidy, N.; Botton, G.A.; Adronov, A. Polymerization from the surface of single-walled carbon nanotubes-preparation and characterization of nanocomposites. *J. Am. Chem. Soc.* **2003**. [CrossRef]
91. Gao, M.; Huang, S.; Dai, L.; Wallace, G.; Gao, R.; Wang, Z. Aligned coaxial nanowires of carbon nanotubes sheathed with conducting polymers. *Zuschriften* **2000**. [CrossRef]
92. Kong, H.; Gao, C.; Yan, D. Functionalization of multiwalled carbon nanotubes by atom transfer radical polymerization and defunctionalization of the products. *Macromolecules* **2004**. [CrossRef]
93. Wu, H.X.; Tong, R.; Qiu, X.Q.; Yang, H.F.; Lin, Y.H.; Cai, R.F.; Qian, S.X. Functionalization of multiwalled carbon nanotubes with polystyrene under atom transfer radical polymerization conditions. *Carbon* **2007**, *45*, 152–159. [CrossRef]
94. Cui, J.; Wang, W.; You, Y.; Liu, C.; Wang, P. Functionalization of multiwalled carbon nanotubes by reversible addition fragmentation chain-transfer polymerization. *Polymer* **2004**, *45*, 8717–8721. [CrossRef]
95. Rahman, M.H.; Bhoi, P.R. An Overview of Non-biodegradable Bioplastics. *J. Clean Prod.* **2021**, 126218. [CrossRef]
96. Cui, Y.; Ding, M.; Sui, T.; Zheng, W.; Qiao, G.; Yan, S.; Liu, X. Role of Nanoparticle Materials as Water-Based Lubricant Additives for Ceramics. *Tribol. Int.* **2019**. [CrossRef]
97. Gresham, R.M. The Mysterious World of MWF Additives. Available online: https://www.stle.org/images/pdf/STLE_ORG/BOK/OM_OA/Friction_Tribology/The%20Mysterious%20World%20of%20MWF%20Additives_tlt%20article_Sept06.pdf (accessed on 7 April 2021).
98. Ma, H.; Li, J.; Chen, H.; Zuo, G.; Yu, Y.; Ren, T.; Zhao, Y. XPS and XANES characteristics of tribofilms and thermal films generated by two P- and/or S-containing additives in water-based lubricant. *Tribol. Int.* **2009**. [CrossRef]
99. Yong, W.; Qunji, X.; Lili, C. Tribological properties of some water-based lubricants containing polyethylene glycol under boundary lubrication conditions. *J. Synth. Lubr.* **1997**. [CrossRef]
100. Dong, R.; Yu, Q.; Bai, Y.; Ma, Z.; Zhang, J.; Zhang, C.; Yu, B.; Zhou, F.; Liu, W.; Cai, M. Towards Superior Lubricity and Anticorrosion Performances of Proton-Type Ionic Liquids Additives for Water-Based Lubricating Fluids. *Chem. Eng. J.* **2019**. [CrossRef]
101. Espinosa, T.; Sanes, J.; Jiménez, A.E.; Bermúdez, M.D. Bermúdez. Protic ammonium carboxylate ionic liquid lubricants of OFHC copper Wear. *Wear* **2013**, *303*, 495–509. [CrossRef]
102. Zheng, G.; Zhang, G.; Ding, T.; Xiang, X.; Li, F.; Ren, T.; Liu, S.; Zheng, L. Tribological properties and surface interaction of novel water-soluble ionic liquid in water-glycol. *Tribol. Int.* **2017**. [CrossRef]
103. Guo, M.L.T.; Tsao, C.-Y.A. Tribological Behavior of Self-Lubricating Aluminium/SiC/Graphite Hybrid Composites Synthesized by the Semi-Solid Powder-Densification Method. *Compos. Sci. Technol.* **1999**. [CrossRef]
104. Wang, Y.; Yu, Q.; Cai, M.; Shi, L.; Zhou, F.; Liu, W. Ibuprofen-Based Ionic Liquids as Additives for Enhancing the Lubricity and Antiwear of Water–Ethylene Glycol Liquid. *Tribol. Lett.* **2017**, *65*, 1–13. [CrossRef]
105. Wang, Y.; Yu, Q.; Cai, M.; Shi, L.; Zhou, F.; Liu, W. Synergy of lithium salt and non-ionic surfactant for significantly improved tribological properties of water-based fluids. *Tribol. Int.* **2017**, *113*, 58–64. [CrossRef]
106. Fan, M.; Du, X.; Ma, L.; Wen, P.; Zhang, S.; Dong, R.; Sun, W.; Yang, D.; Zhou, F.; Liu, W. In situ preparation of multifunctional additives in water. *Tribol. Int.* **2019**, *130*, 317–323. [CrossRef]
107. Chu, J.H.; Tong, L.B.; Zhang, J.B.; Kamado, S.; Jiang, Z.H.; Zhang, H.J.; Sun, G.X. Bio-inspired graphene-based coatings on Mg alloy surfaces and their integrations of anti-corrosive/wearable performances. *Carbon* **2019**, *141*, 154–168. [CrossRef]

108. Yang, Z.; Sun, C.; Zhang, C.; Zhao, S.; Cai, M.; Liu, Z.; Yu, Q. Amino acid ionic liquids as anticorrosive and lubricating additives for water and their environmental impact. *Tribol. Int.* **2021**, *153*, 106663. [CrossRef]
109. Xie, H.; Dang, S.; Xiang, L.; Jiang, B.; Zhou, S.; Sheng, H.; Yang, T.; Pan, F. Tribological performances of SiO$_2$/graphene combinations as water-based lubricant additives for magnesium alloy rolling. *Appl. Surf. Sci.* **2019**, *475*, Fusheng. [CrossRef]
110. Rietsch, J.-C.; Brender, P.; Dentzer, J.; Gadiou, R.; Vidal, L.; Vix-Guterl, C. Evidence of water chemisorption during graphite friction under moist conditions. *Carbon* **2013**, *55*, 90–97. [CrossRef]
111. Avilés, M.; Carrión, F.; Sanes, J.; Bermúdez, M. Bio-based ionic liquid crystal for stainless steel-sapphire high temperature ultralow friction. *Wear* **2021**, 204020. [CrossRef]
112. Phillips, B.; Zabinski, J. Ionic liquid lubrication effects on ceramics in a water environment. *Tribol. Lett.* **2004**, *17*, 533–541. [CrossRef]
113. Perry, S.S.; Tysoe, W.T. Frontiers of fundamental tribological research. *Tribol. Lett.* **2005**, *19*. [CrossRef]
114. Selvam, B.; Marimuthu, P.; Narayanasamy, R.; Senthilkumar, V.; Tun, K.S.; Gupta, M. Effect of temperature and strain rate on compressive response of extruded magnesium nano-composite. *J. Magnes. Alloy.* **2015**, *3*. [CrossRef]
115. Berman, D.; Erdemir, A.; Sumant, A. Few layer graphene to reduce wear and friction on sliding steel surfaces. *Carbon* **2013**, *54*, 454–459. [CrossRef]
116. Liang, S.; Yi, M.; Shen, Z.; Liu, L.; Ma, S.; Zhang, X. In-situ exfoliated graphene for high-performance water-based lubricants. *Carbon* **2016**, *96*, 1181–1190. [CrossRef]
117. Wu, H.; Kamila, H.; Huo, M.; Lin, F.; Huang, S.; Huang, H.; Jiao, S.; Xing, Z.; JIang, Z. Eco-friendl water-based nanolubricants for industrial-scale hot steel rolling. *Lubricants* **2020**, *8*, 96. [CrossRef]
118. Deng, M.; Li, J.; Zhang, C.; Ren, J.; Zhou, N.; Luo, J. Investigation of running-in process in water-based lubrication aimed at achieving super-low friction. *Tribol. Int.* **2016**, *102*, 257–264. [CrossRef]

 *lubricants*

Article

Friction and Wear Characteristics of Aqueous ZrO₂/GO Hybrid Nanolubricants

Shuiquan Huang [1,2], Zhen Wang [1], Longhua Xu [1] and Chuanzhen Huang [1,*]

[1]   School of Mechanical Engineering, Yanshan University, Qinhuangdao 066004, China;
    shuiquan.huang@ysu.edu.cn (S.H.); wangzhen@ysu.edu.cn (Z.W.); longhua1357@ysu.edu.cn (L.X.)
[2]   School of Mechanical and Mining Engineering, The University of Queensland, Brisbane 4072, Australia
*   Correspondence: huangchuanzhen@ysu.edu.cn

Abstract: Aqueous nanolubricants containing ZrO₂ nanoparticles, graphene oxide (GO) nanosheets, or hybrid nanoparticles of ZrO₂ and GO were formulated using a cost-effective ultrasonication de-agglomeration method. The friction and wear characteristics of these water-based nanolubricants were systematically investigated using a block-on-ring testing configuration with a stainless- and alloy steel contact pair. The concentrations and mass ratios of nanoadditives were varied from 0.02 to 0.10 wt.% and 1:5 to 5:1, respectively, to obtain optimal lubrication performance. The application of a 0.06 wt.% 1:1 ZrO₂/GO hybrid nanolubricant resulted in a 57% reduction in COF and a 77% decrease in wear volume compared to water. The optimised ZrO₂/GO hybrid nanolubricant was found to perform better than pure ZrO₂ and GO nanolubricant in terms of tribological performance due to its synergistic lubrication effect, which showed up to 54% and 41% reductions in friction as well as 42% and 20% decreases in wear compared with 0.06 wt.% ZrO₂ and 0.06 wt.% GO nanolubricants. The analysis of wear scars revealed that using such a ZrO₂/GO hybrid nanolubricant yielded a smooth worn surface, with 87%, 45%, and 33% reductions in S_a compared to water and 0.06 wt.% ZrO₂ and 0.06 wt.% GO nanolubricants. The superior tribological performance can be ascribed to the combination of the rolling effect of ZrO₂ nanoparticles and the slipping effect of GO nanosheets.

Keywords: water-based nanolubricant; zirconia; graphene oxide; hybrid nanoparticles; friction; wear

Citation: Huang, S.; Wang, Z.; Xu, L.; Huang, C. Friction and Wear Characteristics of Aqueous ZrO₂/GO Hybrid Nanolubricants. *Lubricants* **2022**, *10*, 109. https://doi.org/ 10.3390/lubricants10060109

Received: 12 May 2022
Accepted: 31 May 2022
Published: 1 June 2022

Publisher's Note: MDPI stays neutral with regard to jurisdictional claims in published maps and institutional affiliations.

1. Introduction

Mineral-oil-based lubricants are commonly used for industrial lubrication; however, their usage is frequently seen to raise environmental, health, and economic concerns due to their non-biodegradable nature and inherent toxicity [1–5]. In this context, water-based lubricants have recently attracted increasing academic and industrial interest, because they are clean and environmentally friendly [6–11]. However, water alone cannot offer satisfactory lubrication because of its low lubricity and weak load-carrying capability. High-performance additives are therefore introduced into water-based lubricants for enhancing their friction-reduction and anti-wear capacities.

Innovative nanoparticles with a near-spherical/rounded shape and a high load-carrying capacity including SiO₂, Al₂O₃, and ZrO₂ [12–16], carbon-based nanosheets of low shear resistance such as graphene and graphene oxide (GO) [17–20], and clay nanomaterials of low cost such as montmorillonite and zeolite [21–23] have recently been applied to water-based lubrication for achieving improved tribological and machining performance [24–27]. For example, zero-dimensional (0D) Al₂O₃ nanoparticles of 30 nm were added to water to prepare an Al₂O₃ nanolubricant for alloy steel on stainless steel contact [15]. The nanolubricant formed a dynamically balanced tribo-layer during sliding, improving surface asperity contact, and thus lowered friction and wear by 27% and 22% compared with a water lubricant. The friction and wear of an alloy steel ball sliding against a low-carbon microalloyed steel disk was considerably reduced when using a water-based TiO₂ nanolubricant of 20 nm, because of its mending and rolling effects [28]. Two-dimensional graphene nanosheets were dispersed in water for chromium steel on carbon steel contact, which resulted in respective 53% and 91% reductions in friction and wear rate compared with water attributed to graphene's slipping effect [29]. An aqueous nanolubricant containing GO nanosheets of 2 nm thickness was found to produce a self-lubricating film to mitigate asperity ploughing during sliding, thus lowering the contact friction of an alloy steel ball on stainless-steel plate by around 45% in comparison to deionised water [30]. Due to their satisfactory tribological performance, water-based nanolubricants have recently been applied to different processes, such as rolling [31], turning [32], milling [21], and grinding [25], for achieving high-performance and green manufacturing. For example, a 4.0 wt.% TiO_2 nanolubricant was innovatively used to hot roll low-carbon microalloyed steels, which performed comparably to a conventional 1.0 vol.% oil-in-water lubricant in terms of rolled strip surface quality and rolling forces at rolling temperatures of 850 and 950 ∘C [31]. Water-based nanolubricants containing CuO, ZnO, Fe_2O_3, or Al_2O_3 were also developed for turning AISI 4340 steels [32]. The CuO nanolubricant presented a great heat-carrying capacity, and it produced a protective film that resulted in a decreased cutting force and reduced tool wear. Water-based GO nanolubricants were employed for grinding GaAs wafers [24] and GGG laser crystals [25], which significantly reduced friction at the abrasives–workpiece interface, resulting in a much lower grinding force and improved surface quality compared with an emulsion coolant. In recent years, due to their excellent tribological performance and low-cost and environment-friendly nature, a water-based lubricant using montmorillonite clay nanoparticles as additives were employed to mill AISI 1018 steel plates [22] and AISI 4340 steel bars [33]. The nanolubricant performed better than a conventional mineral-oil-based emulsion in terms of machined surface quality and tool life. Although much work has shown promising tribological and machining results when using 0D or 2D nanomaterials as lubricant additives, these nanoadditives in water lubricants have a tendency to agglomerate or restack due to their high surface energy and activity. This causes sedimentation, increasing particle size, and limits formation of self-lubricating films, thus resulting in failure of lubrication [34].

In recent years, the use of hybrid nanoparticles as lubricant additives, such as Al_2O_3/ MoS_2 [35], TiO_2/GO [16], and TiO_2/montmorillonite clay [33], has been reported to be an effective approach to achieve improved lubrication performance. For instance, Huang et al. [34] synthesised an aqueous hybrid nanolubricant through dispersing Al_2O_3 nanoparticles in a water-based GO nanosheet solution, and found that such dimensional integration could relieve Al_2O_3 agglomeration and restrain GO restacking in water. The Al_2O_3/GO nanolubricant thus produced a significant decrease in friction and wear, as well as a considerable improvement in the worn surface quality compared to the nanolubricant only containing Al_2O_3 or GO nanoadditives. In addition to their environment-friendly and low-cost nature, ZrO_2 nanoparticles have demonstrated excellent desirable properties for water-based lubrication, such as high load-carrying capacity and excellent durability [16]. Hydrophilic GO nanosheets can form a stable suspension in water [19,30], which makes them attractive additives for water-based lubricants. However, until now, few efforts have been made to explore the tribological characteristics and mechanisms of water-based lubricants with ZrO_2/GO hybrid nanoparticles as additives [16].

In this work, aqueous hybrid nanolubricants were formulated by ultrasonically mixing ZrO_2 nanoparticles in GO suspensions. The tribological properties of the developed nanolubricants were systematically investigated utilizing a block-on-ring testing approach with stainless steel on alloy steel contact, with water serving as a benchmark lubricant. The concentrations and mass ratios of nanoadditives were changed to gain an optimal lubrication performance. The roles of nanoadditives in water-based lubrication were investigated and discussed.

2. Materials and Methods

2.1. Formulation of Nanolubricants

Zero-dimensional-ZrO_2 nanoparticles of ~30 nm in diameter (purchased from XF-NANO Co. Ltd., Nanjing, China) and 2D-GO nanosheets (supplied by Hengqiu Graphene Technology Co. Ltd., Suzhou, China) were used as nanoadditives. The GO nanosheets had a lateral diameter of ~5 μm and a thickness of 1~2 nm. Distilled water was employed as a base solvent. ZrO_2/GO hybrid nanolubricants were synthesised as follows. ZrO_2 nanoparticles and GO nanosheets were mechanically mixed in water at mass ratios of ZrO_2 to GO of 1:5, 2:4, 3:3, 4:2, and 5:1 with a fixed additive content of 0.06 wt.% [30]. The ZrO_2 and GO mixture suspensions were then processed using a 450 W high-intensity ultrasonic probe for 30 min to dimensionally integrate ZrO_2 with GO. The ZrO_2/GO nanolubricants with different contents of 0.02, 0.04, 0.06, 0.08, and 0.10 wt.% at a constant ZrO_2 to GO mass ratio of 1:1 were also formulated to optimise lubrication performance [34]. A ZrO_2 or GO nanolubricant with a 0.06 wt.% additive content was prepared for comparison.

The morphologies of ZrO_2 nanoparticles, GO nanosheets, and ZrO_2/GO hybrid nanoparticles dispersing in water were examined using transmission electron microscopy (TEM, JEM-2100, JEOL, Japan). In Figure 1a, ZrO_2 nanoparticles have a uniform size distribution with near-spherical/rounded shapes, but they are agglomerated in water, forming network clusters. The TEM image in Figure 1b shows that GO nanosheets are super-thin with micro-scale wide coverage, and are highly transparent with wrinkles on their surfaces. Figure 1c demonstrates that ZrO_2 nanoparticles are sparsely deposited onto GO nanosheets. The agglomeration sizes of ZrO_2 nanoparticles are much smaller than those shown in Figure 1a, indicating a better dispersion performance with dimensional hybridization of 0D nanoparticles and 2D nanosheets.

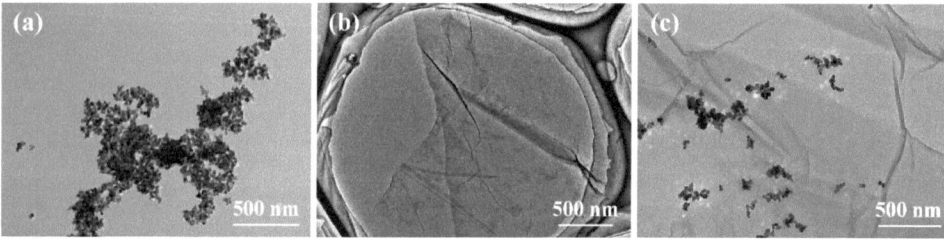

Figure 1. TEM images showing (**a**) GO nanosheets, (**b**) ZrO_2 nanoparticles, and (**c**) ZrO_2/GO hybrid nanoparticles dispersed in water.

2.2. Tribological Tests

The friction and wear performance of the formulated nanolubricants was assessed using a block-on-ring contact pair on a multi-specimen tribometer (UMT-3, Bruker, USA) [16]. Water was utilised as a benchmark lubricant for comparison. The block used was made of AISI 304 stainless steel and was ground with SiC abrasive paper of 15.3 μm before each test. The ring employed was made of AISI 52100 Cr alloy steel and was processed using a 15.3 μm SiC abrasive paper to maintain its surface texture. Considering the potential application in metal forming processes, the testing was performed at normal loads of 15, 20, 25, and 30 N, and sliding speeds of 200, 300, 400, and 500 mm/s with a fixed sliding distance of 12 m [5,31]. Each test was repeated three times, and average values were reported.

The steel blocks were ultrasonically cleaned in an acetone bath for 5 min after friction testing. The wear scar widths of the blocks were measured using a laser confocal microscope (LEXT OLS4100, Olympus, Tokyo, Japan). The wear volume (Vw, in mm^3) was then calculated using [23]:

$$V_{\mathrm{w}} = \frac{D^2 t}{8}\left[2\sin^{-1}\frac{\mathrm{b}}{D} - \sin\left(2\sin^{-1}\frac{\mathrm{b}}{D}\right)\right] \tag{1}$$

where D is the diameter of the ring (35 mm), t is the width of the block (6 mm), and b is the average width of the wear scar. A scanning electron microscope (SEM, JEOL, JSM-6610, Tokyo, Japan) equipped with an energy-dispersive X-ray spectroscope (EDS) module and an inVia confocal Raman microscope (Renishaw plc., London, UK) was used to analyse the worn surfaces of the blocks.

3. Results and Discussion

3.1. Friction and Wear Properties of Nanolubricants

Figure 2 displays the coefficient of friction (COF) curves achieved from the tribological tests using water, 0.06 wt.% ZrO_2, 0.06 wt.% GO, and 0.03–0.03 wt.% ZrO_2/GO nanolubricants with a normal force of 20 N and a sliding speed of 300 mm/s. For all four tested lubricants, the COF reaches a steady state after running in for approximately 3 m. Water has a high COF of over 0.5 in the steady wear state. The addition of nanoadditives in water yields smaller COF values; particularly, the ZrO_2/GO nanolubricant produces the lowest COF value of approximately 0.23 among all the three tested nanolubricants, as shown in Figure 2. This result clearly indicates that the application of the hybrid nanolubricant produced the best interfacial lubrication.

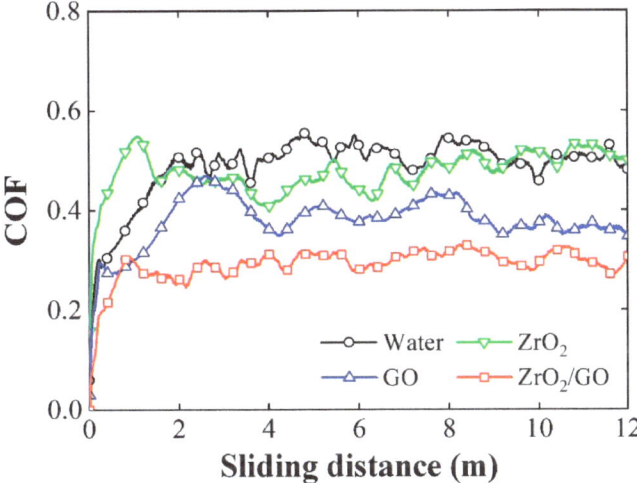

Figure 2. COF time histories generated using water, 0.06 wt.% ZrO_2, 0.06 wt.% GO, and 0.03–0.03 wt.% ZrO_2/GO nanolubricants. The friction tests were conducted at a normal force of 20 N and a sliding speed of 300 mm/s.

The effects of the mass ratio of ZrO_2 to GO and nanoadditive content on the averaged COF using ZrO_2/GO hybrid nanolubricants are shown in Figure 3. Note that the ZrO_2/GO content was fixed at 0.06 wt.%, and the COF values of water, the 0.06 wt.% ZrO_2, and 0.06 wt.% GO nanolubricants were plotted in the figure for comparison. It can be seen in Figure 3a that the addition of ZrO_2 or GO in water generates a lower COF value, indicating an improved interfacial lubrication performance. When ZrO_2/GO hybrid nanolubricants were employed, the COF decreased slightly and then increased significantly as the ZrO_2 to GO mass ratio changed from 1:5 to 5:1. The hybrid nanolubricant containing 0.03 wt.% ZrO_2 and 0.03 wt.% GO yielded the lowest COF, with its values being 57%, 54%, and 41% lower than those of the baseline water and the ZrO_2 and GO nanolubricants, respectively. The results suggest that such dimensional integration indeed helped improve the anti-friction performance of each nanoadditive. In other words, a synergistic lubricating action is in effect when using ZrO_2/GO nanocomposites as an additive in water [33,34].

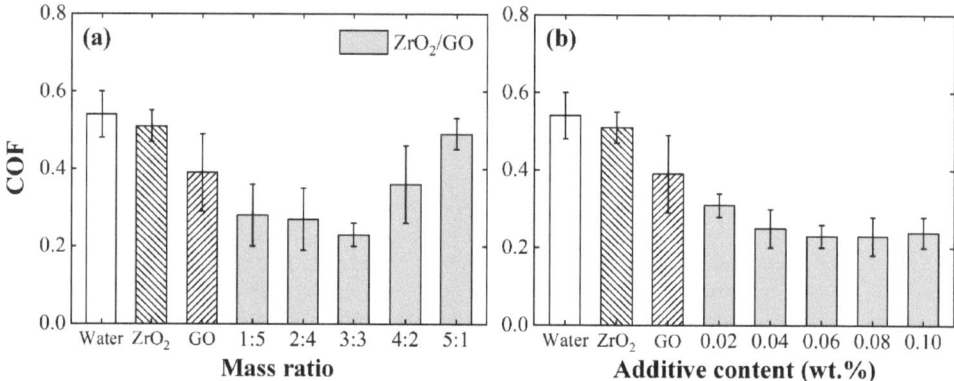

Figure 3. Effects of (**a**) ZrO_2 to GO mass ratio and (**b**) nanoadditive content on averaged COF with ZrO_2/GO nanolubricants at a testing load of 20 N and a sliding speed of 300 mm/s. The COF values of water and 0.06 wt.% ZrO_2 and 0.06 wt.% GO nanolubricants were used for comparison.

Figure 3b shows that the nanolubricants generate lower COF values than water, clearly confirming their good friction-reduction performance. When ZrO_2/GO nanolubricants were used, the increase in additive content from 0.02 to 0.06 wt.% resulted in a substantial decrease in COF, and a further increase in content had an insignificant effect on the COF. This result indicates that nanoadditive content indeed has an optimal value regarding lubrication performance, as reported in He et al. [15]. Again, the ZrO_2/GO hybrid nanolubricant produces lower interfacial friction compared to the ZrO_2 and GO nanolubricants for all the tested contents, as illustrated in Figure 3b, which was most likely due to its synergistic lubrication effect.

Figure 4 shows the effects of nanoadditive mass ratio and content on the averaged wear volume (WV) produced using the ZrO_2/GO hybrid nanolubricants at a normal load of 20 N and a sliding speed of 300 mm/s. Similarly, the WV values of water and the 0.06 wt.% ZrO_2 and 0.06 wt.% GO nanolubricants were included in the figure for comparison. The ZrO_2 and GO nanolubricants generated significantly less wear than water, with a higher WV value for the ZrO_2 nanolubricant due to the polishing effect of the ZrO_2 nanoparticles [7]. As shown in Figure 4a, for the ZrO_2/GO nanolubricants the effect of mass ratio on the wear loss is inconsistent. The hybrid nanolubricant with a ZrO_2-GO mass ratio of 3 to 3 generates the smallest wear volume, showing up to 42% and 20% reductions compared with the ZrO_2 and GO nanolubricants. A further increase in the mass percentage of ZrO_2 nanoparticles causes a substantial increase in wear attributed to the increased polishing effect as more ZrO_2 nanoparticles are involved in sliding.

Figure 4b depicts that the GO-related nanolubricants produce smaller wear volume values than the baseline water and the ZrO_2 nanolubricant. This is likely because 2D-GO nanosheets have better lubricity than 0D nanoparticles in water-based lubrication [36]. The wear loss of the ZrO_2/GO hybrid nanolubricants is significantly less than that of water at all the tested contents, as displayed in Figure 4b, indicating improved wear resistance. Again, the hybrid nanolubricant containing 0.03 wt.% ZrO_2 and 0.03 wt.% GO yielded the least wear.

The friction and wear results clearly indicate that aqueous nanolubricants with ZrO_2/GO hybrid nanoadditives could produce a synergistic lubrication effect in friction and wear reduction. The nanoadditive mass ratio and content indeed had optimal values in terms of friction-reduction and anti-wear performance. The ZrO_2/GO nanolubricant of a ZrO_2 to GO mass ratio of 1:1 with a content of 0.06 wt.% was thus selected for further tests.

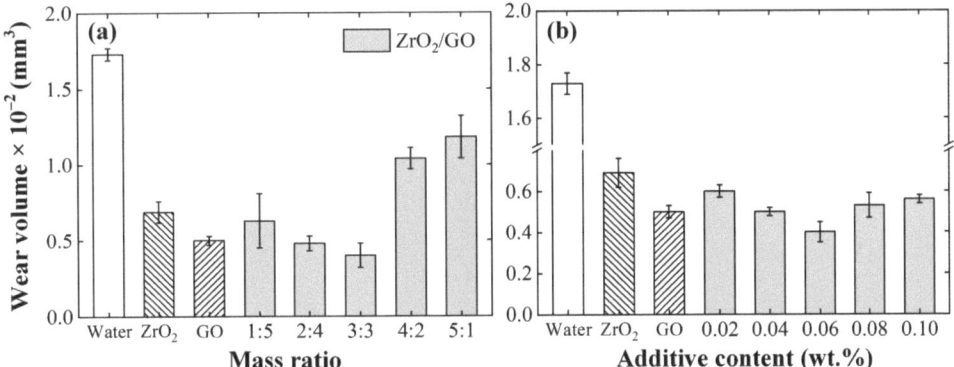

Figure 4. Effects of (**a**) ZrO_2 to GO mass ratio and (**b**) nanoadditive content on averaged wear volume using ZrO_2/GO nanolubricants. The wear volume values of water and 0.06 wt.% ZrO_2 and 0.06 wt.% GO nanolubricants were used for comparison. Applied load = 20 N and sliding speed = 300 mm/s.

3.2. Effect of Testing Conditions on Friction-Reduction Performance

The COF values of the 0.06 wt.% ZrO_2, 0.06 wt.% GO, and 0.03–0.03 wt.% ZrO_2/GO nanolubricants are plotted in Figure 5, as a function of applied load or sliding speed. The results achieved with water were used as benchmarks. In Figure 5a, a higher load yields a higher COF for all four tested lubricants as a result of the increased pressure of contact. However, the ZrO_2/GO hybrid nanolubricant produced the smallest COF compared to those obtained using water and the ZrO_2 and GO nanolubricants, clearly demonstrating its better performance in reducing friction. It can be seen from Figure 5b that the COF decreases with the increase in sliding speed for all the tested lubricants because of the reduced ploughing intensity of surface asperities. The ZrO_2 nanolubricant generated slightly smaller COF in comparison to water at all the tested speeds. The ZrO_2/GO hybrid nanolubricant, however, resulted in the lowest COF, followed by the GO nanolubricant.

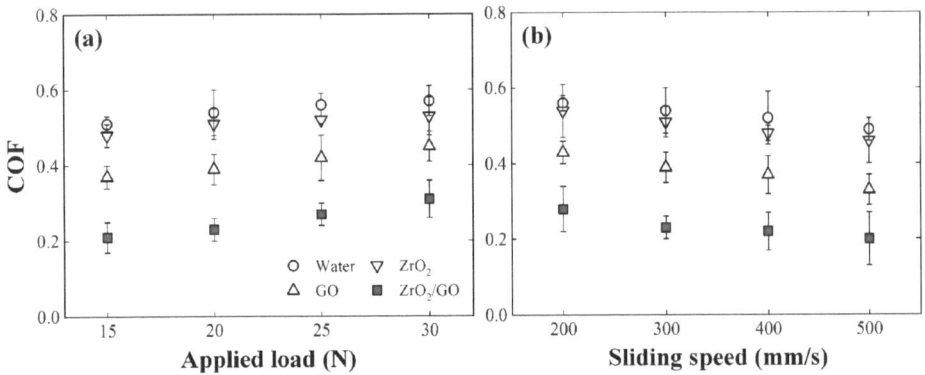

Figure 5. Effects of (**a**) applied load and (**b**) sliding speed on COF produced using water and 0.06 wt.% ZrO_2, 0.06 wt.% GO, and 0.03–0.03 wt.% ZrO_2/GO nanolubricants.

3.3. Effect of Testing Conditions on Anti-Wear Performance

Figure 6 displays the effects of testing load and speed on the wear volume produced using water and the ZrO_2, GO, and ZrO_2/GO nanolubricants. For all four tested lubricants, the wear volume increases substantially with the increase in applied load from 15 to 30 N. This is because a higher load could increase the ploughing intensity of surface asperities of the ring, thus resulting in the higher material removal of the block. The ZrO_2/GO hybrid nanolubricant yields the lowest wear volume, as shown in Figure 6a, indicating the significantly improved asperity contact with the hybrid nanoadditives. In Figure 6b, a faster sliding speed generates less wear for all the tested lubricants, with the smallest wear volume achieved by the ZrO_2/GO hybrid nanolubricant. The results again demonstrate the effectiveness of the hybrid nanoparticles in improving the wear resistance of water.

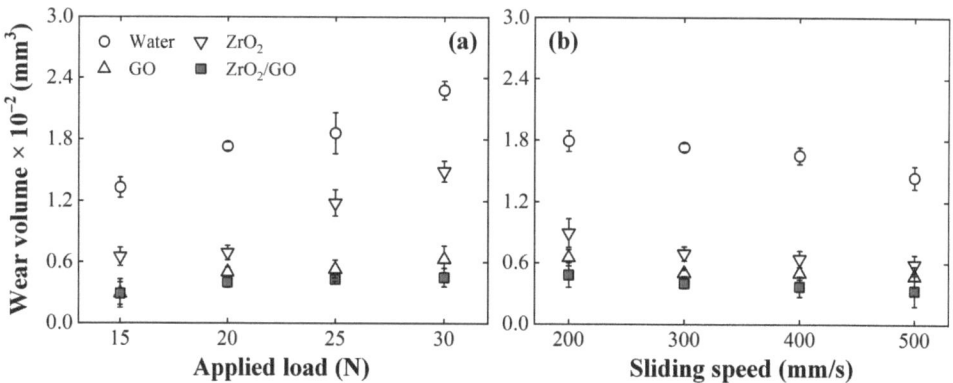

Figure 6. Effects of (**a**) applied load and (**b**) sliding speed on wear volume produced using water and 0.06 wt.% ZrO_2, 0.06 wt.% GO, and 0.03–0.03 wt.% ZrO_2/GO nanolubricants.

3.4. Worn Surface Characteristics

Figure 7 presents the SEM micrographs of the worn surfaces lubricated with water and the 0.06 wt.% ZrO_2, 0.06 wt.% GO, and 0.03–0.03 wt.% ZrO_2/GO nanolubricants at a normal force of 20 N and a sliding speed of 300 mm/s. In Figure 7a, delamination with slight grooves is clearly observed on the worn surface produced with water. During sliding, water had a weak capacity to relieve the "cold welding" of surface asperities due to its poor lubricity, thus resulting in the occurrence of material adhesion and peeling (see Figure 7b). The addition of ZrO_2 or GO improved asperity contact, thus producing a smoother surface with more visible ploughing grooves in comparison with water, as displayed in Figure 7b,c. The corresponding enlarged images shown in Figure 7f,g demonstrate that the sliding with the ZrO_2 and GO nanolubricants is mainly in the abrasive wear regime. It can be seen in Figure 7d,h that the ZrO_2/GO nanolubricant yields the smoothest worn surface and has no delamination in the sliding area due to its excellent synergistic lubrication performance. The performance of the ZrO_2/GO nanolubricant in surface quality improvement was also quantitatively compared with water and the ZrO_2 and GO lubricants. As depicted in Figure 8, the ZrO_2/GO nanolubricant produces the lowest surface roughness, with 87%, 45%, and 33% reductions in S_a compared to water and the ZrO_2 and GO nanolubricants, respectively.

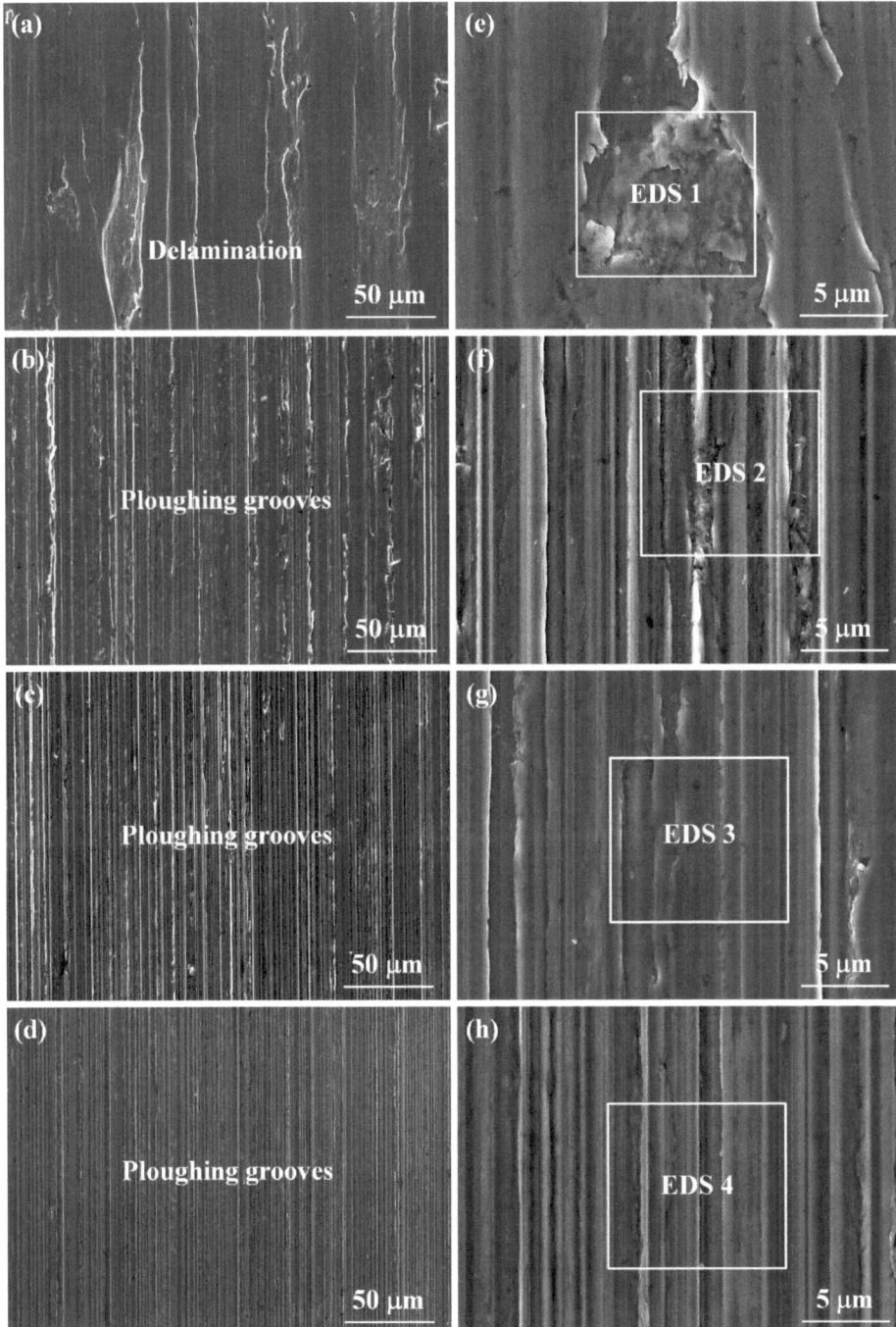

Figure 7. Worn surfaces generated using (**a**) water and (**b**) 0.06 wt.% ZrO$_2$, (**c**) 0.06 wt.% GO, and (**d**) 0.03–0.03 wt.% GO/ZrO$_2$ nanolubricants; (**e**–**h**) are their corresponding enlarged SEM images. Testing load = 20 N and sliding speed = 300 mm/s.

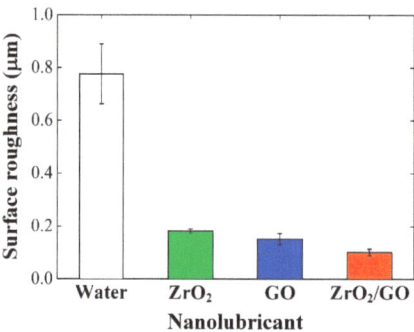

Figure 8. Averaged surface roughness (S_a) values achieved using water and 0.06 wt.% ZrO_2, 0.06 wt.% GO, and 0.03–0.03 wt.% ZrO_2/GO nanolubricants with an applied load of 20 N and a sliding speed of 300 mm/s.

3.5. Lubrication Mechanism Analysis

Understanding the roles of nanoadditives in water is crucial to optimizing the aqueous nanolubricant formulation for any specific mechanical contact system. To date, a number of lubrication mechanisms have been proposed to explain the improvement in lubrication with the application of nanoparticles as additives in water lubricants [7,28], including the ball bearing/rolling effect, polishing/smoothing effect, protective/lubricating film effect, and synergistic lubrication effect, as schematically illustrated in Figure 9. Nanoparticles with high hardness and spherical/rounded shapes can serve not only as tiny balls to generate a hybrid friction process of rolling and sliding to achieve decreased friction (see Figure 9a), but also polishing media to help smoothen surface asperities during sliding, thereby producing improved surface quality (see Figure 9b). Nanoparticles that possess moderate hardness and high chemical inertness likely form a protective film at the sliding area through a tribo-physical reaction, as displayed in Figure 9c. The film functions as a tribo-layer to lower friction and wear through improving asperity contact. Nanoadditives that have low hardness and lamellar structures, however, can deposit onto contact surfaces during sliding to generate a lubricating film through tribo-chemical or physical reactions, as shown in Figure 9d. The film can considerably lower friction and wear through relieving asperity ploughing. As illustrated in Figure 9e, hybrid nanoadditives consisting of two different dimensional nanomaterials can simultaneously obtain friction reduction and surface quality improvement through a synergistic lubrication effect produced by tribo-layer formation and asperity smoothing.

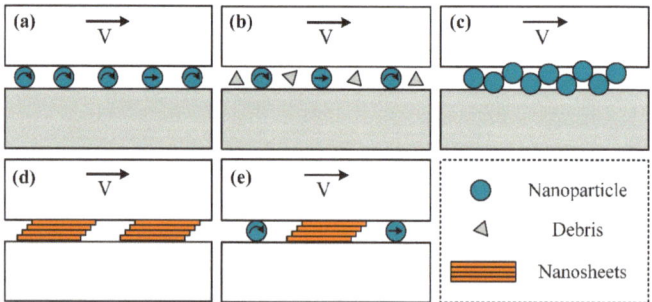

Figure 9. Lubrication mechanisms of nanoadditives suspended in water, including (**a**) rolling effect, (**b**) polishing effect, (**c**) protective film effect, (**d**) lubricating film effect, and (**e**) synergistic lubrication effect [7]. Reproduced with permission from [7]. Copyright Elsevier, 2022.

To understand the lubrication mechanism of ZrO_2, GO, and ZrO_2/GO nanolubricants, the corresponding worn surfaces were analysed using EDS and Raman microscopy. Figure 10a–d display the EDS spectra conducted on the squared areas marked on the worn block surfaces shown in Figure 7e–h, respectively, and the typical elements and their corresponding weight percentages are listed in Table 1. Zr was clearly detected on the worn surface lubricated with the ZrO_2 nanolubricant, as can be seen in Figure 10b, suggesting that during sliding the ZrO_2 nanoparticles dispersed in water formed a localised ZrO_2 tribo-layer on the rubbing areas [37]. This layer could help relieve asperity welding (i.e., adhesive wear), thus leading to the reduced friction force and the improved worn surface quality. Compared to water, higher C residue is observed on the worn surface of the GO nanolubricant, as displayed in Table 1. The corresponding Raman spectrum in Figure 10e shows two notable peaks at 1345 cm^{-1} and 1615 cm^{-1}, representing D and G bands of GO, respectively [34]. This result indicates that the GO nanosheets dispersed in water indeed participated in the sliding. The GO-related film could serve as a lubrication layer to improve asperity contact, leading to the reduced friction and wear. Similarly, the worn surface of the ZrO_2/GO nanolubricant detected a higher C content than that using water, but has no detectable signs of Zr on the rubbing area, as shown in Figure 10d. The corresponding Raman spectrum shown in Figure 10f exhibits no peaks of ZrO_2 in the range 200 to 800 cm^{-1} [38], while the spectrum obtained from the worn surface presents the typical D and G bands of GO. This was most likely because during sliding the ZrO_2 nanoparticles, instead of generating a deposit layer on the rubbing area, rolled between the contact surfaces with the lubricant flow due to the reduced interfacial shear resistance resulting from the GO nanosheets, as reported in [16]. In addition, as shown in Figure 10f, the intensity ratio of D to G bands (I_D/I_G) is 0.92, smaller than that of 1.04 on the worn surface lubricated with the GO nanolubricant. The decrease in I_D/I_G indicated lower defects on GO surfaces, which was ascribed to the improved lubrication as a result of the combination of ZrO_2 and GO [34]. The synergistic lubricating effect of GO nanosheets and the rolling effect of ZrO_2 nanoparticles helped suppress asperity welding and relieve asperity ploughing, thus resulting in the significantly improved tribological performance. In contrast, water alone could not prevent surface asperities from direct contact, thus resulting in the severe wear and friction.

Table 1. EDS analysis results of the squared areas marked on the worn surfaces shown in Figure 7e,f, generated using water, and the 0.06 wt.% ZrO_2, 0.06 wt.% GO, and 0.03–0.03 wt.% ZrO_2/GO nanolubricants.

Analysed Area	Lubricant	C	O	Zr	Cr	Ni	Mn	Si	Fe
EDS 1	Water	1.6	0.5	/	14.7	5.6	0.9	0.6	76.1
EDS 2	ZrO_2	1.8	1.7	0.9	17.0	7.3	1.2	0.4	69.7
EDS 3	GO	4.4	3.5	/	14.7	6.5	0.9	0.6	69.4
EDS 4	ZrO_2/GO	3.6	2.8	/	14.8	5.9	0.9	0.6	71.4

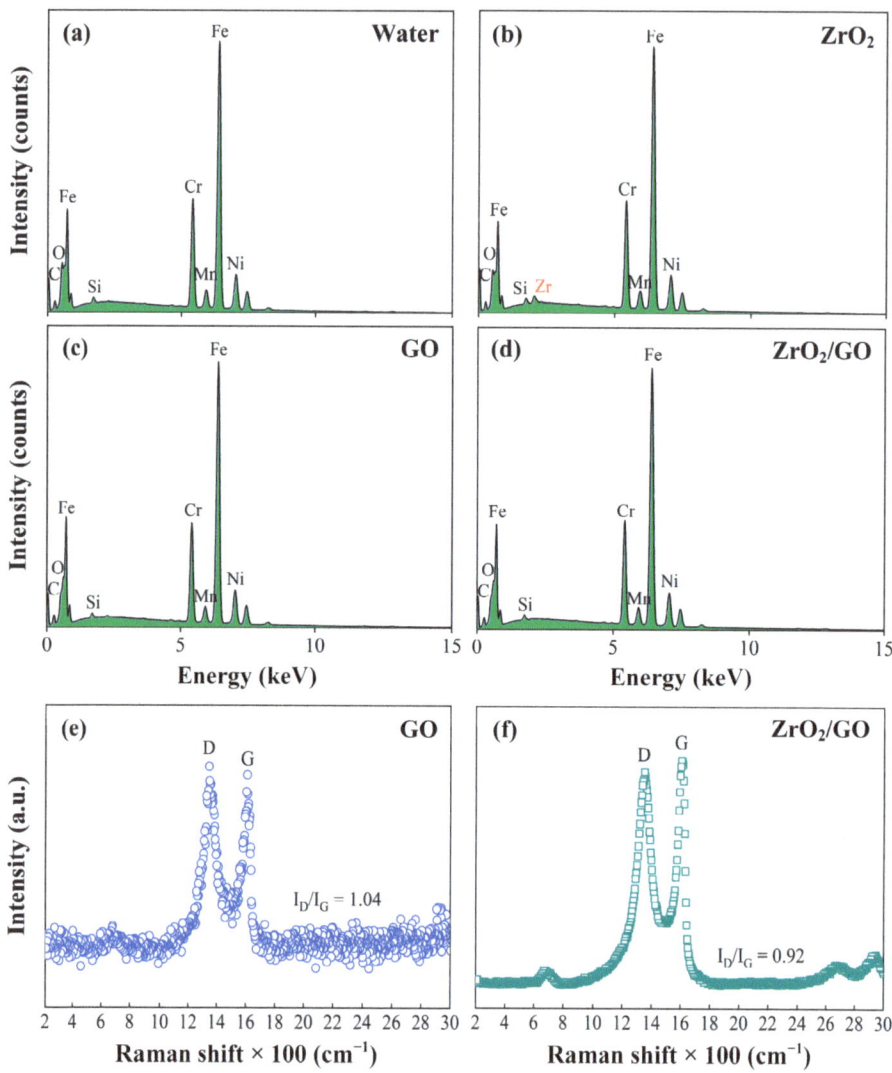

Figure 10. EDS spectra achieved with (**a**) water and (**b**) 0.06 wt.% ZrO$_2$, (**c**) 0.06 wt.% GO, and (**d**) 0.03–0.03 wt.% ZrO$_2$/GO nanolubricants, which were conducted on the squared areas marked in Figure 7e,f, respectively, and Raman spectra detected on the worn areas produced using (**e**) 0.06 wt.% GO and (**f**) 0.03–0.03 wt.% ZrO$_2$/GO nanolubricants.

4. Conclusions

Novel water-based hybrid nanolubricants were developed through ultrasonically dispersing ZrO$_2$ nanoparticles in GO aqueous suspensions, and their tribological performance was evaluated. The roles of nanoadditives in water-based nanolubrication were analysed and discussed. Key conclusions drawn from this study are shown below:

- Dimensional integration is an effective approach to improve nanoparticle dispersion in water.

- Addition of nanoadditives to water can improve its tribological performance, and there is an optimal concentration and mass ratio of nanoadditives for the best lubrication performance.
- In water-based lubrication, 0D-ZrO$_2$ nanoparticles are likely to generate a localised deposit layer on the rubbing area. This tribo-layer can help relieve asperity welding, i.e., adhesive wear, thus resulting in reduced friction and improved surface quality.
- In water-based lubrication, 2D-GO nanosheets tend to form a lubricating layer of low shear resistance on the sliding area. This self-lubricating layer can improve asperity contact and ploughing, thereby leading to reduced friction and wear.
- Hybrid nanoadditives of ZrO$_2$ nanoparticles and GO nanosheets can produce a synergistic friction-reduction and anti-wear effect in water-based lubrication, reliving abrasive wear (i.e., asperity ploughing) and suppressing adhesive wear (i.e., asperity welding), which thus enables the achievement of both improved tribological performance and enhanced surface quality.

Author Contributions: Conceptualization, S.H.; methodology, S.H.; validation, S.H.; formal analysis, S.H., Z.W., and L.X.; investigation, S.H; resources, S.H.; writing—original draft preparation, S.H.; writing—review and editing, S.H., Z.W., and L.X.; visualization, S.H.; supervision, C.H.; project administration, C.H.; funding acquisition, C.H. All authors have read and agreed to the published version of the manuscript.

Funding: This research was funded by the Baosteel-Australia Joint Research and Development Centre (BA17004), Australian Research Council (ARC) Industrial Transformation Research Hub for Computational Particle Technology (IH140100035), and Scientific Research Project for National High-level Innovative Talents of Hebei Province Full-time Introduction (2021HBQZYCXY004).

Institutional Review Board Statement: Not applicable.

Informed Consent Statement: Not applicable.

Data Availability Statement: The data presented in this study are available on request from the corresponding author.

Acknowledgments: S.H. would like to acknowledge the use of the facilities at, and the scientific and technical assistance of, the Nanomechanics and Advanced Manufacturing Laboratory at the University of Queensland.

Conflicts of Interest: The authors declare no conflict of interest.

References

1. Shokrani, A.; Dhokia, V.; Newman, S.T. Environmentally conscious machining of difficult-to-machine materials with regard to cutting fluids. *Int. J. Mach. Tool. Manu.* **2012**, *57*, 83–101. [CrossRef]
2. Zhang, Y.; Li, C.; Jia, D.; Zhang, D.; Zhang, X. Experimental evaluation of MoS$_2$ nanoparticles in jet MQL grinding with different types of vegetable oil as base oil. *J. Clean. Prod.* **2015**, *87*, 930–940. [CrossRef]
3. Reeves, C.J.; Siddaiah, A.; Menezes, P.L. Friction and wear behavior of environmentally friendly ionic liquids for sustainability of biolubricants. *J. Tribol.* **2019**, *141*, 051604. [CrossRef]
4. Siddaiah, A.; Kasar, A.K.; Manoj, A.; Menezes, P.L. Influence of environmental friendly multiphase lubricants on the friction and transfer layer formation during sliding against textured surfaces. *J. Clean. Prod.* **2019**, *209*, 1245–1251. [CrossRef]
5. Wu, H.; Jia, F.; Li, Z.; Lin, F.; Huo, M.; Huang, S.; Sayyar, S.; Jiao, S.; Huang, H.; Jiang, Z. Novel water-based nanolubricant with superior tribological performance in hot steel rolling. *Int. J. Extreme Manuf.* **2020**, *2*, 025002. [CrossRef]
6. Morshed, A.; Wu, H.; Jiang, Z. A comprehensive review of water-based nanolubricants. *Lubricants* **2021**, *9*, 89. [CrossRef]
7. Huang, S.; Wu, H.; Jiang, Z.; Huang, H. Water-based nanosuspensions: Formulation, tribological property, lubrication mechanism, and applications. *J. Manuf. Process.* **2021**, *71*, 625–644. [CrossRef]
8. Rahman, M.H.; Warneke, H.; Webbert, H.; Rodriguez, J.; Austin, E.; Tokunaga, K.; Rajak, D.K.; Menezes, P.L. Water-based lubricants: Development, properties, and performances. *Lubricants* **2021**, *9*, 73. [CrossRef]
9. Wu, H.; Kamali, H.; Huo, M.; Lin, F.; Huang, S.; Huang, H.; Jiao, S.; Xing, Z.; Jiang, Z. Eco-friendly water-based nanolubricants for industrial-scale hot steel rolling. *Lubricants* **2020**, *8*, 96. [CrossRef]
10. Sun, J.; Meng, Y.; Zhang, B. Tribological behaviors and lubrication mechanism of water-based MoO$_3$ nanofluid during cold rolling process. *J. Manuf. Process.* **2021**, *61*, 518–526. [CrossRef]
11. Meng, Y.; Sun, J.; He, J.; Yan, X.; Pei, Y. Recycling prospect and sustainable lubrication mechanism of water-based MoS$_2$ nano-lubricant for steel cold rolling process. *J. Clean. Prod.* **2020**, *277*, 123991.

12. Musavi, S.H.; Davoodi, B.; Niknam, S.A. Effects of reinforced nanoparticles with surfactant on surface quality and chip formation morphology in MQL-turning of superalloys. *J. Manuf. Process.* **2019**, *40*, 128–139. [CrossRef]
13. Wu, H.; Jia, F.; Zhao, J.; Huang, S.; Wang, L.; Jiao, S.; Huang, H.; Jiang, Z. Effect of water-based nanolubricant containing nano-TiO2 on friction and wear behaviour of chrome steel at ambient and elevated temperatures. *Wear* **2019**, *426*, 792–804. [CrossRef]
14. Wu, H.; Zhao, J.; Luo, L.; Huang, S.; Wang, L.; Zhang, S.; Jiao, S.; Huang, H.; Jiang, Z. Performance evaluation and lubrication mechanism of water-based nanolubricants containing nano-TiO2 in hot steel rolling. *Lubricants* **2018**, *6*, 57. [CrossRef]
15. He, A.; Huang, S.; Yun, J.-H.; Wu, H.; Jiang, Z.; Stokes, J.; Jiao, S.; Wang, L.; Huang, H. Tribological performance and lubrication mechanism of alumina nanoparticle water-based suspensions in ball-on-three-plate testing. *Tribol. Lett.* **2017**, *65*, 40. [CrossRef]
16. Huang, S.; Lin, W.; Li, X.; Fan, Z.; Wu, H.; Jiang, Z.; Huang, H. Roughness-dependent tribological characteristics of water-based GO suspensions with ZrO2 and TiO2 nanoparticles as additives. *Tribol. Int.* **2021**, *161*, 107073. [CrossRef]
17. He, A.; Huang, S.; Yun, J.-H.; Jiang, Z.; Stokes, J.R.; Jiao, S.; Wang, L.; Huang, H. Tribological characteristics of aqueous graphene oxide, graphitic carbon nitride, and their mixed suspensions. *Tribol. Lett.* **2018**, *66*, 42. [CrossRef]
18. Kinoshita, H.; Nishina, Y.; Alias, A.A.; Fujii, M. Tribological properties of monolayer graphene oxide sheets as water-based lubricant additives. *Carbon* **2014**, *66*, 720–723. [CrossRef]
19. Song, H.J.; Na, L. Frictional behavior of oxide graphene nanosheets as water-base lubricant additive. *Appl. Phys. A* **2011**, *105*, 827–832. [CrossRef]
20. Hu, Y.; Wang, Y.; Zeng, Z.; Zhao, H.; Ge, X.; Wang, K.; Wang, L.; Xue, Q. PEGlated graphene as nanoadditive for enhancing the tribological properties of water-based lubricants. *Carbon* **2018**, *137*, 41–48. [CrossRef]
21. Peña-Parás, L.; Maldonado-Cortés, D.; Rodríguez-Villalobos, M.; Romero-Cantú, A.G.; Montemayor, O.E. Enhancing tool life, and reducing power consumption and surface roughness in milling processes by nanolubricants and laser surface texturing. *J. Clean. Prod.* **2020**, *253*, 119836. [CrossRef]
22. Peña-Parás, L.; Maldonado-Cortés, D.; Rodríguez-Villalobos, M.; Romero-Cantú, A.G.; Montemayor, O.E.; Herrera, M.; Trousselle, G.; González, J.; Hugler, W. Optimization of milling parameters of 1018 steel and nanoparticle additive concentration in cutting fluids for enhancing multi-response characteristics. *Wear* **2019**, *426–427*, 877–886. [CrossRef]
23. Lin, C.-L.; Lin, W.; Huang, S.; Edwards, G.; Lu, M.; Huang, H. Tribological performance of zeolite/sodium dodecylbenzenesulfonate hybrid water-based lubricants. *Appl. Surf. Sci.* **2022**, *598*, 153764. [CrossRef]
24. Li, X.; Huang, S.; Wu, Y.; Huang, H. Performance evaluation of graphene oxide nanosheet water coolants in the grinding of semiconductor substrates. *Precis. Eng.* **2019**, *60*, 291–298. [CrossRef]
25. Li, C.; Li, X.; Huang, S.; Li, L.; Zhang, F. Ultra-precision grinding of Gd3Ga5O12 crystals with graphene oxide coolant: Material deformation mechanism and performance evaluation. *J. Manuf. Process.* **2021**, *61*, 417–427. [CrossRef]
26. Huang, H.; Li, X.; Mu, D.; Lawn, B.R. Science and art of ductile grinding of brittle solids. *Int. J. Mach. Tool. Manu.* **2021**, *161*, 103675. [CrossRef]
27. Wu, Y.; Mu, D.; Huang, H. Deformation and removal of semiconductor and laser single crystals at extremely small scales. *Int. J. Extreme Manuf.* **2020**, *2*, 012006. [CrossRef]
28. Wu, H.; Zhao, J.; Xia, W.; Cheng, X.; He, A.; Yun, J.H.; Wang, L.; Huang, H.; Jiao, S.; Huang, L. A study of the tribological behaviour of TiO2 nano-additive water-based lubricants. *Tribol. Int.* **2017**, *109*, 398–408. [CrossRef]
29. Yang, J.; Xia, Y.; Song, H.; Chen, B.; Zhang, Z. Synthesis of the liquid-like graphene with excellent tribological properties. *Tribol. Int.* **2017**, *105*, 118–124. [CrossRef]
30. He, A.; Huang, S.; Yun, J.-H.; Jiang, Z.; Stokes, J.; Jiao, S.; Wang, L.; Huang, H. The pH-dependent structural and tribological behaviour of aqueous graphene oxide suspensions. *Tribol. Int.* **2017**, *116*, 460–469. [CrossRef]
31. Wu, H.; Zhao, J.; Xia, W.; Cheng, X.; He, A.; Yun, J.H.; Wang, L.; Huang, H.; Jiao, S.; Huang, L. Analysis of TiO2 nano-additive water-based lubricants in hot rolling of microalloyed steel. *J. Manuf. Process.* **2017**, *27*, 26–36. [CrossRef]
32. Das, A.; Pradhan, O.; Patel, S.K.; Das, S.R.; Biswal, B.B. Performance appraisal of various nanofluids during hard machining of AISI 4340 steel. *J. Manuf. Process.* **2019**, *46*, 248–270. [CrossRef]
33. Peña-Parás, L.; Rodríguez-Villalobos, M.; Maldonado-Cortés, D.; Guajardo, M.; Rico-Medina, C.S.; Elizondo, G.; Quintanilla, D.I. Study of hybrid nanofluids of TiO2 and montmorillonite clay nanoparticles for milling of AISI 4340 steel. *Wear* **2021**, *477*, 203805. [CrossRef]
34. Huang, S.; He, A.; Yun, J.-H.; Xu, X.; Jiang, Z.; Jiao, S.; Huang, H. Synergistic tribological performance of a water based lubricant using graphene oxide and alumina hybrid nanoparticles as additives. *Tribol. Int.* **2019**, *135*, 170–180. [CrossRef]
35. He, J.; Sun, J.; Meng, Y.; Pei, Y. Superior lubrication performance of MoS2-Al2O3 composite nanofluid in strips hot rolling. *J. Manuf. Process.* **2020**, *57*, 312–323. [CrossRef]
36. Liu, Y.; Wang, X.; Pan, G.; Luo, J. A comparative study between graphene oxide and diamond nanoparticles as water-based lubricating additives. *Sci. China Technol. Sci.* **2013**, *56*, 152–157. [CrossRef]
37. Wu, H.; Zhao, J.; Cheng, X.; Xia, W.; He, A.; Yun, J.-H.; Huang, S.; Wang, L.; Huang, H.; Jiao, S. Friction and wear characteristics of TiO2 nano-additive water-based lubricant on ferritic stainless steel. *Tribol. Int.* **2018**, *117*, 24–38. [CrossRef]
38. Ji, P.; Mao, Z.; Wang, Z.; Xue, X.; Zhang, Y.; Lv, J.; Shi, X. Improved surface-enhanced raman scattering properties of ZrO2 Nanoparticles by Zn Doping. *Nanomaterials* **2019**, *9*, 983. [CrossRef]

 *lubricants*

Article

Effects of an Electrical Double Layer and Tribo-Induced Electric Field on the Penetration and Lubrication of Water-Based Lubricants

Zhiqiang Luan [1,2], Wenshuai Liu [1,2], Yu Xia [1,2], Ruochong Zhang [1,2], Bohua Feng [1,2], Xiaodong Hu [1,2], Shuiquan Huang [3,4,*] and Xuefeng Xu [1,2,*]

[1] College of Mechanical Engineering, Zhejiang University of Technology, Hangzhou 310023, China; 1112002019@zjut.edu.cn (Z.L.); 2112102109@zjut.edu.cn (W.L.); xiayu971007@163.com (Y.X.); zhangruochong@zjut.edu.cn (R.Z.); fengbohua@zjut.edu.cn (B.F.); hooxoodoo@zjut.edu.cn (X.H.)
[2] Key Laboratory of Special Purpose Equipment and Advanced Processing Technology, Ministry of Education and Zhejiang Province, Zhejiang University of Technology, Hangzhou 310023, China
[3] School of Mechanical Engineering, Yanshan University, Qinhuangdao 066004, China
[4] School of Mechanical and Mining Engineering, The University of Queensland, Brisbane, QLD 4072, Australia
* Correspondence: shuiquan.huang@uq.edu.au (S.H.); xuxuefeng@zjut.edu.cn (X.X.)

Abstract: Understanding the effects of electrical double layers (EDL) and tribo-induced electric fields on the electroosmotic behaviors of lubricants is important for developing high-performance water-based lubricants. In this study, EDL conductivities of aqueous lubricants containing a surfactant of 3-[(3-cholamidopropyl)-dimethylammonio]-1-propanesulfonate (CHAPS) or cetyltrimethylammonium bromide (CTAB) were analyzed. The interfacial zeta potentials of the synthesized lubricants and Al_2O_3 ceramic-alloy steel contacts were measured, and frictional potentials of ceramic and steel surfaces were determined using a modified ball-on-disc configuration. The distribution characteristics of the tribo-induced electric field of the ceramic-steel sliding contact were numerically analyzed. The electroosmotic behaviors of the lubricants were investigated using a four-ball configuration. It was found that an EDL and tribo-induced electric field was a crucial enabler in stimulating the electroosmosis of lubricants. Through altering EDL structures, CHAPS enhanced the electroosmosis and penetration of the water-based lubricant, thus resulting in improved lubrication.

Keywords: electroosmosis; electrical double layer; tribo-induced electric field; water-based lubricant; penetration; lubrication

Citation: Luan, Z.; Liu, W.; Xia, Y.; Zhang, R.; Feng, B.; Hu, X.; Huang, S.; Xu, X. Effects of an Electrical Double Layer and Tribo-Induced Electric Field on the Penetration and Lubrication of Water-Based Lubricants. *Lubricants* **2022**, *10*, 111. https://doi.org/10.3390/lubricants10060111

Received: 29 April 2022
Accepted: 30 May 2022
Published: 2 June 2022

1. Introduction

Water-based lubricants have attracted increasing academic and industrial interest in recent years due to their low-cost and environment-friendly nature [1,2]. They have been successfully applied to hot steel rolling and bearing lubrication [3–5]. It has been found that the penetration of water-based lubricants plays a critical role in their lubricating performance [6,7]. Capillary networks at the friction interface are the paths of penetration of lubricants, by which the lubricants can generate lubricating films on the sliding surface to reduce friction and wear [8,9]. Liquid head pressure under the action of the gravity field and capillary pressure caused by surface tension are the main driving forces of conventional capillary penetration [10]. The gradient of liquid surface tension caused by the temperature field makes the lubricant prefer to move away from the areas with the highest temperature [9,11]. Recent studies have proposed a lubricant electrokinetic effect under the electric field within the friction interface (i.e., electroosmosis) [12,13]. The penetrability of lubricants with different electroosmotic characteristics at the friction interface is significantly different [12]. However, in-depth investigations into essential prerequisites for the electroosmosis of lubricants, i.e., the effects of an electrical double layer (EDL) at the solid/liquid interface and a tribo-induced electric field on electroosmosis [14], are currently lacking, which has thus hindered the development of high-performance water-based lubricants.

An EDL is a phenomenon of uneven ion distribution of a liquid phase boundary and bulk caused by interface effects, such as charge transfer and charged particle adsorption [15]. The charged solid surface can attract the equivalent amount of counterions in the liquid bulk to aggregate at the interface, forming a stationary stern layer and a diffuse layer that allows ions to move, as shown in Figure 1. The degree of the uneven distribution of ions is related to the ion concentration in the liquid phase. The lower the ion concentration, the more obvious the uneven distribution, which can be reflected in the difference between the EDL conductivity and the bulk conductivity of the liquid. The adsorption of ionic surfactants on the charged surface can change the structure of the EDL. The electroosmotic velocity and direction can be significantly affected by adding only 10^{-1} mmol/L surfactant to the base fluid [16–18]. Generally, the zeta potential at the shear plane between the two layers reflects the distribution of ions in the EDL [19]. The tangential streaming potential method is a simple and widely used method for measuring the zeta potential [20]. It is worth noting that the contributions to the formation of streaming potential come entirely from the movement of free ions in the diffusion layer [21]. For the interface between an insulating material and the liquid with an extremely low ion concentration, it is important to clarify the EDL conductivity to calculate the zeta potential using the streaming potential. If the liquid bulk conductivity is used for the calculation, the zeta potential will be underestimated. The method of the galvanostatic four-electrodes system combined with electrochemical impedance spectroscopy is usually used to measure the pore conductivity of membranes to obtain information about the membrane structure [22]. The approximate conductivity of the EDL can be calculated when a membrane with a pore diameter closed to the EDL thickness is selected. For the interface between a conductor (e.g., metals) and liquid, a large number of ions will flow through the conductor bulk, and the zeta potential is also easily underestimated. This problem can be effectively solved by measuring the total conductance of the streaming potential measurement cell as a substitute for liquid conductivity [23].

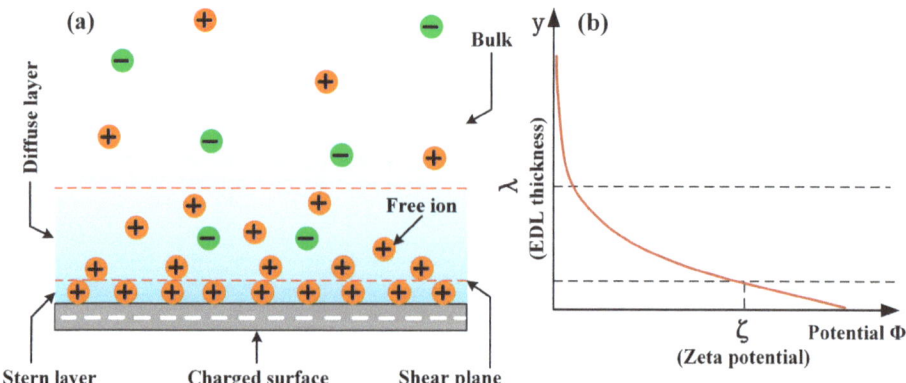

Figure 1. (a) Schematic illustration of EDL produced at a solid-liquid interface and (b) its corresponding potential distribution.

Researchers have shown that triboelectrification can lead to the formation of a high-strength tribo-induced electric field in narrow slits of friction interfaces, which is sufficient to discharge the air in the slits even under boundary lubrication conditions, resulting in an electron avalanche and tribo-plasma phenomena [24,25]. The electrical conductivity of material has a great influence on this electric field. A conductor contains a large number of mobile electrons, and the static surface electricity is easy to dissipate, resulting in a low electric field intensity [26]. Once static electricity is charged on an insulator surface, it is difficult to dissipate, causing a high electric field intensity [27]. The tribo-induced electric field may have a significant effect on the properties of water-based lubricants, such as penetrability and stability [28–30]. The electric field distribution at the friction interface can be calculated by Laplace's equation on the basis of determining the triboelectrification situation of the material [31]. If the electric field intensity within the lubricant penetrated into a capillary is 100–1000 kV/m, the lubricant will penetrate the friction interface by electroosmosis [15,32].

This paper systematically investigates the effects of the EDL and tribo-induced electric field at the interface of alumina ceramic sliding against AISI 52100 steel on the penetration and lubrication of water-based lubricants. The galvanostatic four-electrodes system was established, and the EDL conductivities of dilute lubricants with the highest surfactant CHAPS (or CTAB) concentration of 0.2 mmol/L were measured using the electrochemical impedance spectroscopy. A tangential streaming potential measurement platform was built, and the zeta potentials at friction material/lubricant interfaces were calculated according to the formation principle of the streaming potential. The distribution of the tribo-induced electric field in a capillary at the friction interface was studied by numerical simulation. Finally, the tribological behaviors of lubricants at the ceramic-steel interface were investigated using the four-ball tribometer. The penetrability of lubricants was evaluated by elemental analysis of worn surfaces. The electroosmotic mechanism of lubricants with different electroosmotic characteristics at the ceramic-steel interface was revealed.

2. Experimental Details

2.1. Preparation of Water-Based Lubricants

Potassium chloride (KCl, AR Grade, Sinopharm Chemical Reagent Co., Ltd., Shanghai, China) was added into deionized water to prepare KCl solutions with concentrations of 0.00001–1 mol/L for investigating the influence of ion concentration on the difference of the solution bulk conductivity and EDL conductivity. To avoid the interference of conventional additives, such as rust inhibitors and extreme pressure agents, the lubricants used in this study were only composed of pure water and a surfactant, 3-[(3-cholamidopropyl)-dimethylammonio]-1-propanesulfonate (CHAPS, Biotech Grade, 98%, Shanghai Macklin Biochemical Co., Ltd., Shanghai, China) or cetyltrimethylammonium bromide (CTAB, Biotech Grade, 99%, Shanghai Macklin Biochemical Co., Ltd., Shanghai, China). The concentrations of CHAPS (or CTAB) were 0.0125, 0.05, 0.1 and 0.2 mmol/L. A drop weight method was used to measure the surface tension of the synthesized lubricants. Their wetting angle on the alumina ceramic and AISI 52100 steel surfaces was measured using a goniometry method [33,34]. The effects of CHAPS and CTAB on the capillary penetration of the lubricants were then analyzed.

2.2. Measurement of EDL Conductivity and Zeta Potential

The EDL is a phenomenon in which the net charge concentration near the solid/liquid interface is much higher than the liquid phase bulk, as shown in Figure 1. If the solid phase is an annular structure (such as pores in a membrane) and the intracavity diameter is about twice larger than the EDL thickness, it can be considered that the intracavity ions are distributed in the EDL structure; that is, the intracavity liquid conductivity is approximate to the EDL conductivity. The EDL thicknesses of different liquids were firstly calculated with the Poisson–Boltzmann equation. In this paper, the liquid EDL thicknesses are within 50 nm. To avoid the overlap of the EDLs in a membrane pore, the alumina ceramic flat membrane with a pore diameter of about 100 nm was selected for the measurement. The measuring liquid was then pumped through the ceramic membrane for 30 min. Next, the membrane was soaked in this liquid for 24 h to ensure that all pores were filled with the liquid. The membrane resistance was measured using electrochemical impedance spectroscopy on the galvanostatic four-electrodes system to calculate the liquid conductivity within a pore, i.e., the approximate conductivity of the EDL (the deduction details are presented in Appendix A), as shown in Figure 2a. The resistance measurement was carried out in the cell with a membrane immersed in the measuring liquid and the cell with a lone liquid, respectively, to obtain the total resistance of the liquid and membrane ($R_m + R_{sol}$) and the resistance of the liquid (R_{sol}). The membrane resistance R_m can be obtained by subtracting the two values. The dimension "D" in Figure 2a was the distance between the two voltage electrodes when the membrane was immersed in the cell, and "d" was the membrane thickness. In addition, the liquid bulk conductivities were measured using a model DDS-307A conductivity meter.

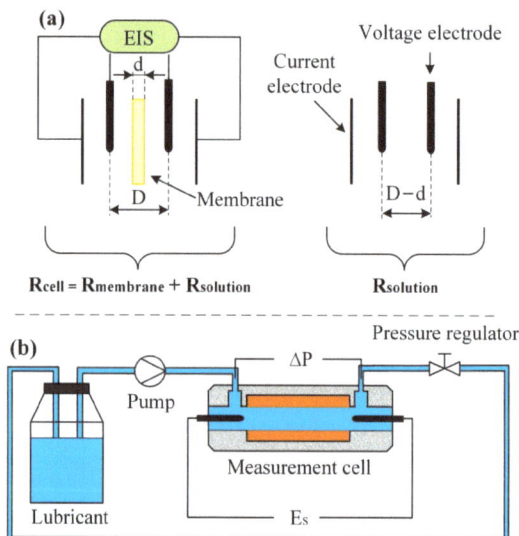

Figure 2. Schematic illustrations of (**a**) galvanostatic four-electrodes system and (**b**) tangential streaming potential measurement platform.

Figure 2b presents the schematic illustration of the tangential streaming potential measurement platform. During the test, the measuring lubricant was circularly forced through the measurement cell under the drive of a pump. The slit channel within the cell consisted of two alumina ceramic (or AISI 52100 steel) slices, and its dimensions were $160 \times 10 \times 0.3$ mm. The hydrodynamic pressure difference ΔP between the two sides of the channel was regulated by a pressure regulator, which increased from 0.02 to 0.1 Mpa. An Agilent 34420A voltmeter was used to measure the potential difference between the upstream and downstream under different ΔP, i.e., streaming potential E_s. The zeta potential at the material/lubricant interface was calculated with the values of $E_s/\Delta P$ of the lubricant on the ceramic and steel surfaces obtained from linear regression (details are shown in Appendix B).

2.3. Analysis of Tribo-Induced Electric Field

A W1 three-dimensional optical profilometer (CHOTEST Instrument Co., Ltd., Shenzhen, China) was used to analyze the worn surface morphologies of the ceramic and steel balls lubricated by water. The geometric parameters of the surface profile were used as the basis for establishing the capillary model at the friction interface. A ball-on-disc material friction device was developed to explore the tribo-electrification between ceramic and steel, as shown in Figure 3. The diameters of the ball and disc were 6 and 75 mm, respectively. The friction load was 1 N, and the sliding velocity was 157 mm/s. Once the disc slipped out of the friction zone, its potential was detected by an EST102 electrometer placed 10 mm above the disc surface. The triboelectrification characteristics of the ceramic (or the steel) were obtained by switching the paired combinations of ball/disc materials.

According to the capillary model and the ceramic/steel triboelectrification results, the distribution of tribo-induced electric field within the friction interface was simulated and analyzed with the finite element software ANSYS.

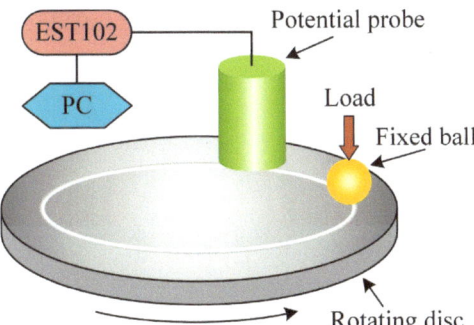

Figure 3. Schematic diagram of triboelectrification electrostatic potential measurement configuration.

2.4. Tribological Tests

The four-ball module of the MMW-1 tribometer was used to investigate the tribological behaviors and penetrability of the lubricants at the ceramic/steel friction interfaces. The pairing mode of one alumina ceramic ball and three AISI 52100 steel balls was adopted. Before the test, the ball pot, lock ring, ceramic and steel balls were cleaned with acetone. Considering the relatively weak load-carrying capacity of water-based lubricants, the test was conducted under a load of 49 N and a rotation speed of 1000 r/min to avoid the seizure of friction pairs [35]. The test was repeated three times for each lubricant, and the duration was 30 min. The wear scar diameters (WSD) were measured using a VW-6000 microscope system (Keyence Co., Ltd., Osaka, Japan). A Sigma HV-01-43 scanning electron microscope (SEM) (Zeiss Co., Ltd., Oberkochen, Germany) equipped with an angle-selective backscattered electron detector (AsB) and an energy-dispersive spectrometer (EDS) was adopted to analyze the element components on the worn surfaces.

3. Results and Discussion

3.1. Electroosmotic Prerequisites

Table 1 presents the bulk conductivities of KCl solutions with different concentrations and the corresponding membrane resistances. The solution EDL conductivities were calculated using Equation (A4), and their proportional relationships (k) with bulk conductivities were obtained and are shown in Figure 4a. The difference between the EDL conductivity and bulk conductivity gradually increases with the decrease in the KCl concentration. The EDL conductivity of water is about 21 times higher than its bulk conductivity. This is attributed to the ion aggregation at the solid/liquid interface induced by the electrostatic force and van der Waals force between ions and solid surface, which forms a completely different ion distribution from solution bulk, i.e., EDL. Moreover, the lower the ion concentration (lower bulk conductivity), the more obvious this phenomenon is, resulting in the EDL conductivity being much higher than the bulk conductivity. As noted in Appendix B, the contribution of generating streaming potential mainly comes from the ion movement within the diffuse layer of the EDL. Therefore, it is of great significance to clarify the EDL conductivities of extremely dilute solutions to characterize zeta potentials at solid/liquid interfaces.

Table 1. Membrane resistances and bulk conductivities of KCl solutions.

C (mmol/L)	1	0.1	0.01	0.001	0.0001	0.00001	Pure water
λ_0 (μS/cm)	109,600	11,080	1428	161.6	22.8	8.56	5.26
R_m ($\Omega\cdot$cm^2)	0.1	0.93	6.8	24	51	89	99

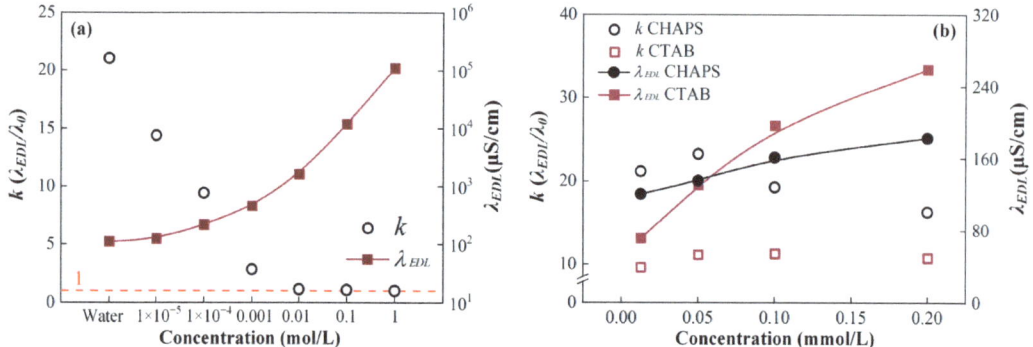

Figure 4. EDL conductivities of (**a**) KCl solutions and (**b**) CHAPS and CTAB lubricants, plotted against concentrations and their relationships with bulk conductivities.

The bulk conductivities of the diluted lubricants prepared in this paper and the corresponding membrane resistances are listed in Table 2. Figure 4b shows the relationships of CHAPS and CTAB concentrations with the EDL conductivity and k. The EDL conductivities of both types of lubricants are higher than their bulk conductivities, and the k of CHAPS lubricants is larger than that of CTAB lubricants. This might be related to the different effect mechanisms of those two surfactants on the EDL, resulting in differences in the ion concentration within the EDL. The EDL conductivities of these two diluted lubricants were therefore used for the subsequent calculation of the zeta potentials at the ceramic/lubricant interfaces.

Table 2. Membrane resistances and bulk conductivities of lubricants.

Parameters	CHAPS Lubricants				CTAB Lubricants			
Concentration (mmol/L)	0.0125	0.05	0.1	0.2	0.0125	0.05	0.1	0.2
λ_0 (μS/cm)	5.69	5.82	8.36	11.20	7.39	11.71	17.39	24.03
R_m ($\Omega\cdot$cm^2)	91	81	68	60	154	84	55.8	42.3

Figure 5a,b depict the streaming potentials of water and lubricants on the ceramic and steel surfaces under different hydrodynamic pressures. It is seen that the streaming potentials of pure water and CHAPS lubricants on both surfaces are negative, and the amplitudes increase linearly with the increase in the pressure, indicating that the ceramic and steel surfaces are negatively charged in those liquids. However, CTAB lubricants show positive potential, suggesting that CTAB can change the charge characteristics of the two materials in liquids, and the surfaces are positively charged. In addition, due to the excellent electrical conductivity of the steel, some of the ions accumulated downstream flow back through the material bulk, exhibiting lower streaming potential amplitudes than ceramics. These two material surfaces can be charged by the effects of charge transfer and ion adsorption at the solid/liquid interface. The surface charges can attract the counter ions of the lubricant bulk to form an EDL, one of the electroosmotic prerequisites.

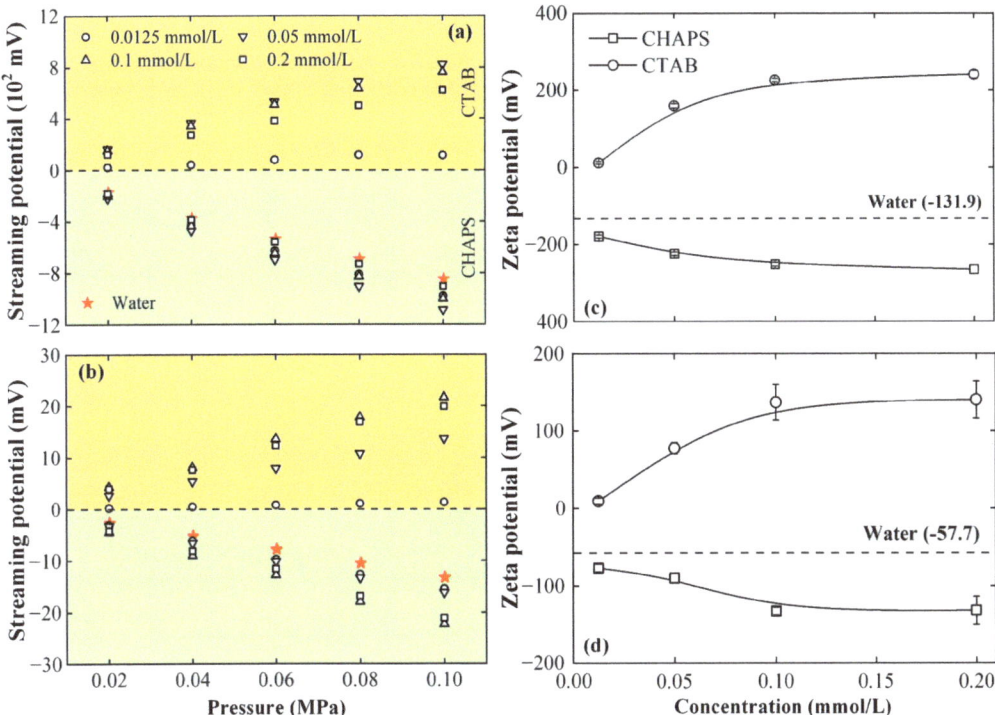

Figure 5. The streaming potential of lubricants on (**a**) ceramic and (**b**) steel surfaces plotted against pressure values (data in the yellow (or green) area was obtained using lubricants prepared with CTAB (or CHAPS)); Zeta potential of at the (**c**) ceramic-lubricant and (**d**) steel-lubricant interfaces, plotted as a function of concentration.

The relationships between the streaming potentials and the hydrodynamic pressures, $E_s/\Delta P$, were obtained by linear regression, and the zeta potentials at the ceramic/lubricant and steel/lubricant interfaces were calculated using Equations (A12) and (A15), respectively, as shown in Figure 5c,d. With the increase in CHAPS (or CTAB) concentration, the zeta potentials of the two materials show the same change trend: CHAPS does not change the polarity of the zeta potential, while CTAB can change that with the concentration of only 0.0125 mmol/L. Increasing the concentrations of both surfactants can increase the zeta potential amplitude. It is well known from the Helmholtz–Smoluchowski equation that the polarity and amplitude of the zeta potential affect the direction and velocity of electroosmosis, respectively, as shown below,

$$v_{eo} = -\frac{\varepsilon_r \cdot \varepsilon_0 \cdot \zeta}{\eta} \cdot E, \qquad (1)$$

where v_{eo} is the velocity of electroosmosis, ε_r is the liquid permittivity, ε_0 is the permittivity of free space, ζ is the Zeta potential, and E is the applied electric field intensity [15]. This indicates that the CHAPS (or CTAB) concentration possesses an important effect on the electroosmotic properties of lubricants. Furthermore, it can be found from Figure 5c,d that the zeta potentials of ceramics are higher than those of steels, indicating that the charging ability of ceramics in lubricants is stronger than that of steels.

Figure 6a,b present the profile curves of the ceramic and steel worn surfaces. It is shown that there are peaks and valleys on the ceramic surface, and the maximal vertical spacings between peaks and valleys are about 0.5 μm. Compared with the ceramic surface, the steel surface profile is smoother, with the largest vertical spacings of about 0.08 µm. The surface roughness of the worn ceramic and steel balls is summarized in Table 3. Referring to the above profile features and roughness information of the ceramic and steel worn surfaces, the geometric model of the lubricant penetrating a capillary at the friction interface was established for simulation analysis of the tribo-induced electric field shown in Figure 6d. Since the capillary is thought to be formed by the peaks of the harder ceramic surface ploughing the softer steel surface, the Rz (maximum height of profile) of the worn ceramic surface is used to define the capillary thickness. The 2Ra (arithmetical mean deviation of the profile) and Rsm (average width of profile unit) of the ceramic and steel surface are used to define the heights and widths of the bulges of the two materials' walls within the capillary. The specific parameters are shown in Figure 6d.

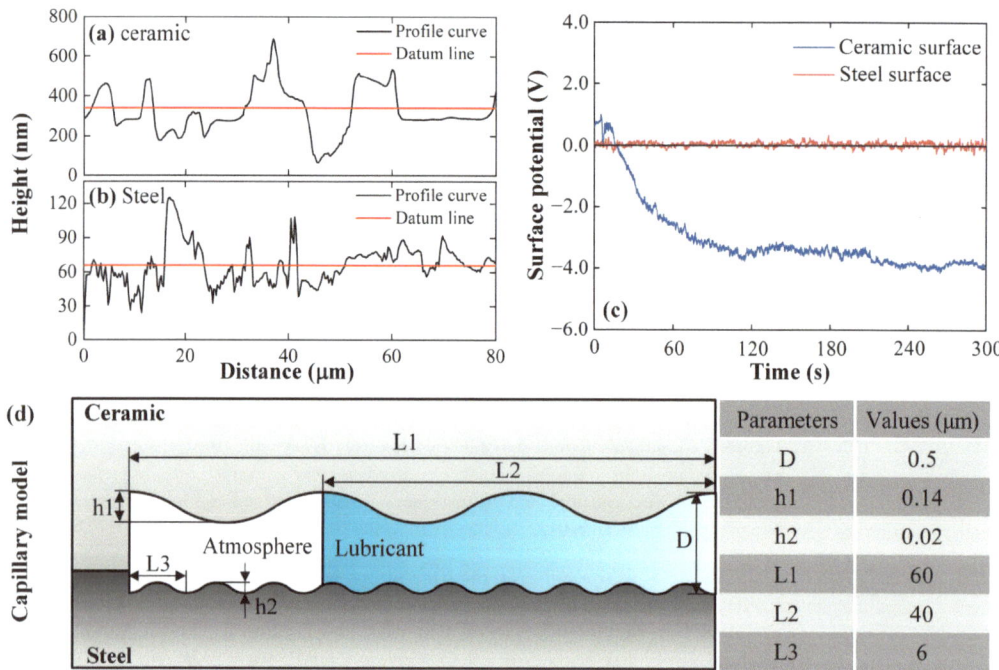

Figure 6. Profiles of (**a**) ceramic and (**b**) steel worn surfaces, (**c**) triboelectrification electrostatic potentials of steel and ceramic surfaces, and (**d**) capillary model of friction interface.

Table 3. Roughness Evaluation Parameters of the Ceramic and Steel Worn Surfaces.

Worn Surface	Ra (µm)	Rz (µm)	Rsm (µm)
Alumina ceramic	0.07	0.51	20.05
AISI 52100 steel	0.01	0.08	5.68

Figure 6c presents the triboelectrification electrostatic potentials of the worn surfaces of the ceramic sliding against the steel. The potential of the ceramic surface reduces continuously with the increase in friction time and reaches a steady status after 60 s. The averaged potential after stabilization is −3.53 V. This is because the alumina ceramic is ranked after most metals in the triboelectric series, which is negatively charged when rubbed with a metal [36]. Due to the excellent electrical conductivity of the steel, the surface charges are easy to dissipate into the bulk [26], showing a weak positive potential. The average potential within 5 min was 0.04 V.

Laplace's equation was used to describe the electric field distribution within the capillary to explore the effect of the electric field generated by the material triboelectrification on the lubricant penetration behavior at the ceramic/steel interface. According to the results of triboelectrification between the ceramic and steel, high potentials were loaded on the capillary wall bulges that were not submerged by the lubricant, which was −3.53 and 0.04 V, respectively. The lubricant was loaded with zero potential because it was placed in the grounding ball pot. The simulation results are displayed in Figure 7. Figure 7a depicts the potential distribution within the capillary. The potential in the lubricant gradually increases from the capillary inside to the outside, indicating that the electric field direction within the lubricant points to the inner end of the capillary. Figure 7b shows the electric field intensity values at the capillary center. It is seen that due to the periodic distribution of the micro-bulges on the capillary walls, the electric field intensity within the capillary inner end fluctuates around 8×10^4 V/cm, which is sufficient to induce gas discharge. This is close to the results observed in Nakayama's experiments [37], indicating the rationality of the finite element model used in this paper. Due to the higher dielectric constant of the lubricant compared to the air, a sudden drop in electric field intensity occurs at the interface of gas and liquid when it decreases from the inside to the outside. In addition, the electric field intensities within the lubricant are about 300–600 V/cm, meeting the strength condition for inducing the lubricant electroosmosis [15,38].

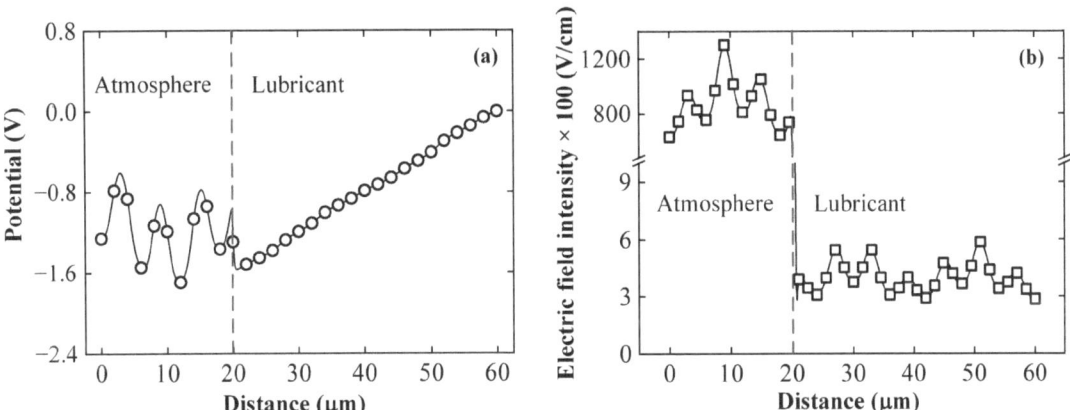

Figure 7. Simulation results of the tribo-induced electric field: values of (**a**) potential and (**b**) electric field intensities at the capillary center.

3.2. Tribological Behaviors

The effects of surfactant concentration on the tribological performance of the lubricants are shown in Figure 8. The coefficient of friction (COF) and WSD gradually decreased with the increase in CHAPS concentration but increased when CTAB was used. The COF and WSD of 0.2 mmol/L CHAPS lubricant were 12.5% and 6.8% lower than water, whereas 0.2 mmol/L CTAB lubricant produced 15.5% and 6.2% increases, respectively. The difference in tribological performance might be related to the penetrability of the lubricants and the lubricity of the surfactants.

Figure 9 shows the SEM-AsB images of the worn surface and the corresponding EDS results. Sparse dark areas are observed on the steel worn surface produced using pure water. The corresponding EDS mapping detected high O content in those areas, suggesting that a lubricating film of oxides was likely formed due to the tribochemical reaction between the steel and water [39]. Figure 9b,c show that as the CHAPS concentration increased, more surface areas were oxidized, and O content increased to 6.32 wt.%. This indicates that the lubricant penetrability was improved by increasing the CHAPS concentration. As shown in Figure 9d,e, the worn surfaces lubricated by the 0.2 mmol/L CTAB lubricant present no significant oxide films, and the O content is only 0.55 wt.%. This suggests that the penetrability of the lubricant decreases with increasing CTAB concentration. In addition, a small quantity of S and Br exist on the worn surface, with the contents of 0–0.11 wt.% and 0–0.19 wt.%, respectively. According to the molecular structures of CHAPS and CTAB, element S is derived from the sulfonic group in CHAPS, and element Br is from the bromine ion in CTAB. The result suggests that only a small quantity of CHAPS and CTAB participated in the lubrication during the friction. The lubricity of surfactants possesses a little effect on the anti-friction and anti-wear properties of the lubricants. The difference in the tribological performance between the two lubricants is thus mainly caused by the penetrability at the friction interface.

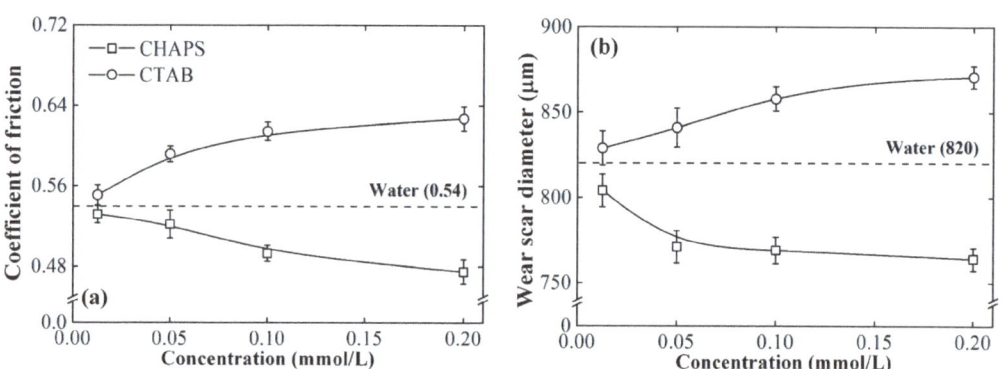

Figure 8. (**a**) Coefficient of friction values and (**b**) wear scar diameters of ceramic/steel friction pairs using different lubricants.

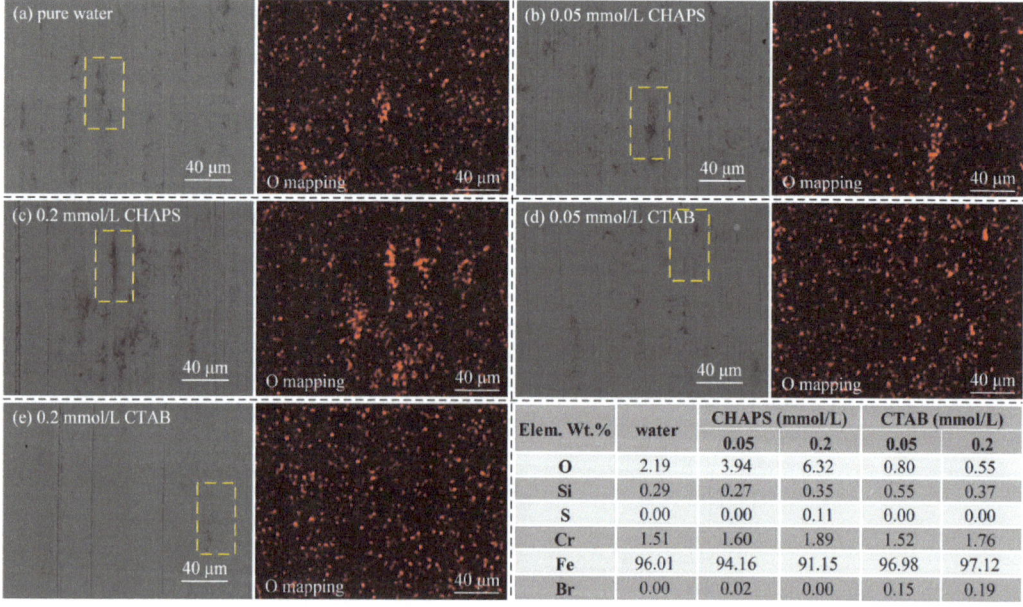

Elem. Wt.%	water	CHAPS (mmol/L)		CTAB (mmol/L)	
		0.05	0.2	0.05	0.2
O	2.19	3.94	6.32	0.80	0.55
Si	0.29	0.27	0.35	0.55	0.37
S	0.00	0.00	0.11	0.00	0.00
Cr	1.51	1.60	1.89	1.52	1.76
Fe	96.01	94.16	91.15	96.98	97.12
Br	0.00	0.02	0.00	0.15	0.19

Figure 9. SEM-AsB images and EDS analyses of the worn surfaces lubricated by (**a**) pure water and lubricants with different additive concentrations: (**b**) 0.05 mmol/L CHAPS, (**c**) 0.2 mmol/L CHAPS, (**d**) 0.05 mmol/L CTAB, and (**e**) 0.2 mmol/L CTAB.

3.3. Mechanism Discussion

Figure 10a,b exhibit the contact angles of lubricants on ceramic and steel surfaces. The increase in surfactant concentrations leads to a reduction in contact angle for both CHAPS and CTAB lubricants. The contact angles of the 0.2 mmol/L CHAPS on the ceramic and steel surfaces were 5.6% and 4.1% lower compared with water, respectively. However, the contact angles produced using the 0.2 mmol/L CTAB were 9.3% and 7.8% lower than those of water. Figure 10c shows the effects of the surfactants on the lubricant surface tension. The surface tension of the lubricant decreases continuously with increasing CHAPS or CTAB concentration. When the surfactant concentrations increase to 0.2 mmol/L, the surface tension decreases by 9.5% and 13.1%, respectively. This is because the positive adsorptions of CHAPS and CTAB on the liquid surface can reduce the liquid surface tension [40]. It is known from the capillary force equation that the conventional capillary penetration is mainly affected by the liquid surface tension and contact angle, as shown below:

$$F_{cap} = 2\pi r\gamma \cos\theta, \tag{2}$$

where F_{cap} is the capillary force, r is the capillary radius, γ is the surface tension, and θ is the contact angle [41]. Assuming that the contributions of the ceramic and steel surfaces to the capillary force are equal and the capillary diameter is 3 μm, the capillary forces of lubricants at the ceramic/steel friction interfaces were calculated, as shown in Figure 10d. It is shown that with the increase in the CHAPS concentration, the capillary force first increases and then decreases, and the variation range is within 4.0%. As the CTAB concentration increases, the capillary force shows an increasing trend, with an increase of 4.5%. The above results are inconsistent with the lubricant penetrations at friction interfaces, indicating that CHAPS and CTAB present little effect on the lubricant conventional capillary penetration and the differences in the penetration behaviors of lubricants are mainly related to their electroosmotic characteristics.

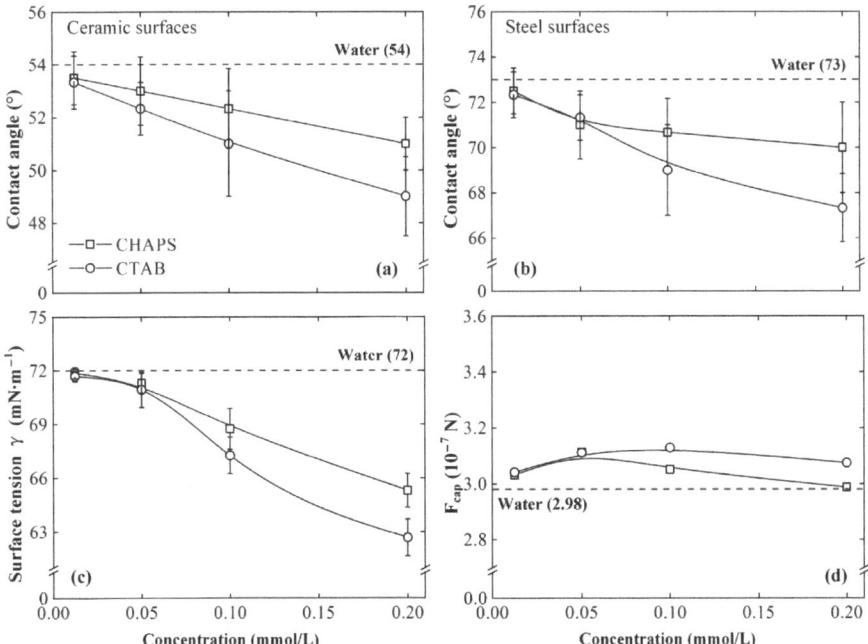

Figure 10. Lubricant contact angles on (**a**) ceramic and (**b**) steel surfaces. Effects of CHAPS and CATB on (**c**) lubricant surface tensions and (**d**) capillary forces at ceramic/steel friction interfaces.

The investigation into the EDL and tribo-induced electric field characteristics shows that the ceramic/steel friction interface has the prerequisites for inducing lubricant electroosmosis. The ceramic and steel surfaces are negatively charged in pure water through the interface effects of the charge transfer and ion adsorption at the solid/liquid interface, attracting the equivalent amount of the counterions in liquid bulk to form the EDL. Free ions in the diffuse layer of the EDL drive liquid molecules to form an electroosmotic flow (EOF) that flows into the capillary under the tribo-induced electric field, as shown in Figure 11a.

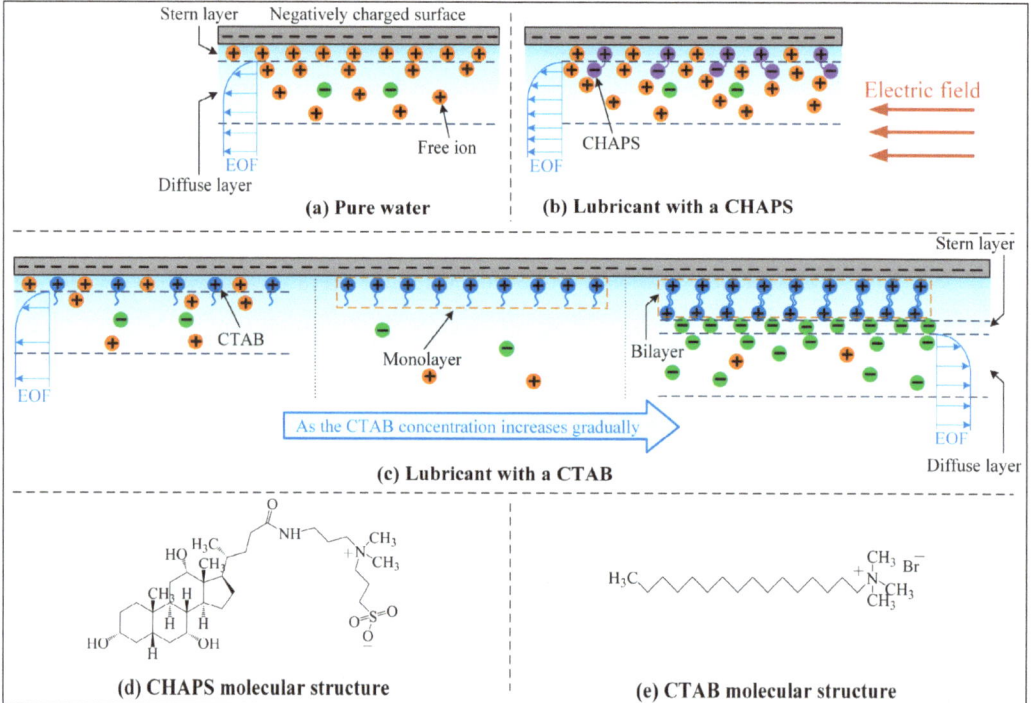

Figure 11. Illustrations of (**a**) pure water EOF and impacts of (**b**) CHAPS and (**c**) CTAB on the EOF. Molecular structures of (**d**) CHAPS and (**e**) CTAB.

Surfactants impact the EOF rate and orientation by dynamically coating the solid surface [42]. As a zwitterionic surfactant, CHAPS can be absorbed on the negatively charged surface through electrostatic interaction between its positive amino group and the surface. Moreover, its negative sulfonic group attracts the cations in liquid bulk, causing an increase in the number of the free ions in the diffuse layer, thereby increasing the electroosmotic velocity [18,43], as shown in Figure 11b. Therefore, adding the appropriate amount of CHAPS can improve the lubricant penetrability at the friction interface, showing a better tribological performance. As a cationic surfactant, CTAB's positive group is absorbed on the charged surface. With increasing concentration, the surface was first gradually neutralized by forming a CTAB monolayer and then positively charged by forming a CTAB bilayer, as shown in Figure 11c. As a result, the electroosmotic velocity first decreases to zero and then increases in reverse [16,42]. Thus, adding the appropriate amount of CTAB can suppress the penetrability of the lubricant, presenting worse tribological properties.

4. Conclusions

The major conclusions are as follows:

- The difference between the EDL conductivity and the bulk conductivity of a liquid (λ_{EDL}/λ_0, k) is sensitive to the change in the ion concentration. The lower the ion concentration, the more obvious the difference. The k of 0.01 mol/L KCl solution is 1.13, while that of pure water is 21.05;

- Due to the charge transfer and ion adsorption at the solid/liquid interface, the alumina ceramic and AISI 52100 steel surfaces are negatively charged in the prepared lubricants. The charged surface attracts the counterions within the lubricant bulk to form an EDL, one of the electroosmotic prerequisites. Cause of the special molecular structure of surfactants CHAPS and CTAB, their adsorption at the solid/liquid interface changes the EDL structure, showing the different zeta potentials;

- The triboelectrification electrostatic potential during ceramic/steel friction was measured. The potential of the ceramic surface is −3.53 V, and that of the steel surface is 0.04 V. The distribution characteristics of the tribo-induced electric field within the capillary at the friction interface were analyzed using the numerical simulation method. The results show that the electric field direction in the lubricant is directed to the capillary inner end, and the intensity is maintained at about 300–600 V/cm, which satisfies the strength condition for driving the lubricant electroosmosis;

- The ceramic/steel friction interface possesses the conditions for inducing the capillary electroosmosis of the lubricant. CHAPS can promote the capillary penetration of the lubricant by improving its electroosmotic properties, presenting satisfactory anti-friction and anti-wear performances. CTAB can reverse the lubricant electroosmosis, thus suppressing its penetrability, showing poor tribological performance.

Author Contributions: Conceptualization, Z.L., X.H. and X.X.; methodology, Z.L., R.Z. and X.X.; validation, Z.L., W.L. and Y.X.; formal analysis, B.F. and S.H.; investigation, Z.L., W.L. and X.X.; resources, R.Z., X.H. and X.X.; writing—original draft preparation, Z.L.; writing—review and editing, Z.L., Y.X., S.H. and X.X.; visualization, Z.L., S.H. and B.F.; supervision, R.Z., X.H. and X.X.; project administration, X.X.; funding acquisition, X.H. and X.X. All authors have read and agreed to the published version of the manuscript.

Funding: This research was funded by the National Key Research and Development Program of China, grant number 2020YFB2010600, the National Natural Science Foundation of China, grant number 51775507, and the Natural Science Foundation of Zhejiang Province, grant number LY19E050006.

Informed Consent Statement: Not applicable.

Data Availability Statement: Not applicable.

Conflicts of Interest: The authors declare no conflict of interest.

Appendix A. Calculation of EDL Conductivity from Membrane Resistance

Assuming that the pores in the membrane are ideal parallel capillaries, the membrane resistance R_m can be considered as the resistance of the parallel circuit of the liquid within pores and the pore walls. Since the liquid resistance is much lower than the pore wall resistance, the relationship between the membrane resistance R_m and the liquid resistance within a pore R_{pore} is as follows:

$$R_m = \frac{R_{pore}}{N}, \tag{A1}$$

where N is the number of the pores within the membrane. The relationship between the conductivity and resistance of the liquid within a pore is shown below:

$$\frac{\lambda_{pore}^l}{\lambda_{pore}^h} = \frac{G_{pore}^l}{G_{pore}^h} = \frac{R_{pore}^h}{R_{pore}^l}, \tag{A2}$$

where λ^l_{pore} is the measuring liquid conductivity within a pore, λ^h_{pore} is the conductivity of a high concentration electrolyte solution within a pore (i.e., the liquid conductivity within a pore can be assumed to be equal to its bulk conductivity [44]. 1 mol/L KCl solution is used in this paper), G^l_{pore} and G^h_{pore} are the pore conductance of the measuring liquid and the high concentration electrolyte solution, and R^l_{pore} and R^h_{pore} are the pore resistances of these two liquids. Substituting for Equation (A2) from Equation (A1):

$$\lambda^l_{pore} = \frac{R^h_{pore}}{R^l_{pore}} \cdot \lambda^h_{pore} \approx \frac{R^h_m}{R^l_m} \cdot \lambda^h_0, \tag{A3}$$

where R^l_m and R^h_m are the membrane resistances when the pores are filled with the measuring liquid and the high concentration electrolyte solution, and λ^h_0 is the solution bulk conductivity. Since the selected membrane pore size is equivalent to the EDL thickness of the measuring liquid, it can be considered that:

$$\lambda^l_{EDL} \approx \lambda^l_{pore} \approx \frac{R^h_m}{R^l_m} \cdot \lambda^h_0, \tag{A4}$$

where λ^l_{EDL} is the *EDL* conductivity of the measuring liquid.

Appendix B. Calculation of Zeta Potential from Streaming Potential

The streaming potential is an electrokinetic effect at the solid/liquid interface, resulting from the relative movement of the *EDL* under external pressure, as shown in Figure A1. The free ions in the diffuse layer move downstream to form a streaming current, I_s:

$$I_s = \int_0^{\frac{h}{2}} \frac{4w}{h} y v_z(y) \rho(y) dy, \tag{A5}$$

where h and w are the height and width of the slit channel, and $v_z(y)$ is the linear velocity of the liquid at a distance y from the axis of the channel, which is given by Hagen-Poiseuille's equation:

$$v_z(y) = \Delta p \frac{\left(\left(\frac{h}{2}\right)^2 - y^2\right)}{2\eta l}, \tag{A6}$$

Since the EDL is a thin region near the solid surface, only the movement of the free ions in the diffuse layer near $y = h/2$ is important in determining the streaming current [21]. Substituting for Equation (A6) from $y = (h/2 - x)$, hence,

$$v_z(y) = v_z(h/2 - x) = \Delta p \frac{(hx - x^2)}{2\eta l} \approx \frac{\Delta p h x}{2\eta l}, \tag{A7}$$

Since the channel height is much larger than the EDL thickness, the upper and lower EDLs do not overlap. The ion density within the channel can be described by Poisson's equation:

$$\frac{d^2\psi}{dx^2} = -\frac{\rho}{\varepsilon}, \tag{A8}$$

Substituting for Equation (A5) from Equations (A7) and (A8):

$$I_s = -\frac{\varepsilon\zeta}{\eta l} wh\Delta p, \tag{A9}$$

The accumulation of ions downstream results in the formation of an electric field, causing a backflow of ions, i.e., a conduction current, I_b. When the channel is composed of insulating materials with low conductivities, the downstream ions only flow back through the liquid phase, I_b is as follows:

$$I_b = I_l = \frac{whE_s \cdot \lambda_0}{l},\tag{A10}$$

where I_l is the backflow of ions through the liquid phase, l is the channel length, λ_0 is the liquid conductivity. Note that when the liquid phase is a dilute solution with an extremely low ion concentration, due to the large difference between the bulk conductivity and the EDL conductivity, the λ_0 in Equation (A10) should be replaced by the liquid EDL conductivity λ_{EDL}. When the streaming potential, Es, reaches a steady-state, $I_s + I_b = 0$, hence,

$$\frac{whE_s \cdot \lambda_{EDL}}{l} - \frac{\varepsilon\zeta}{\eta l}wh\Delta p = 0,\tag{A11}$$

$$\zeta = \frac{E_s}{\Delta p} \cdot \frac{\eta}{\varepsilon} \cdot \lambda_{EDL},\tag{A12}$$

where ΔP the is the hydrodynamic pressure difference between the two ends of the channel, η is the liquid viscosity, and ε is the liquid permittivity. When the channel is composed of materials with excellent electrical conductivities, such as metals, a large number of ions flow back through the material bulk; hence,

$$I_b = I_l + I_v = E_s \cdot G_t,\tag{A13}$$

where I_v is the backflow of ions through the material bulk, G_t is the total conductance within the channel, and so,

$$E_s \cdot G_t - \frac{\varepsilon\zeta}{\eta l}wh\Delta p = 0,\tag{A14}$$

$$\zeta = \frac{E_s}{\Delta p} \cdot \frac{\eta}{\varepsilon} \cdot \frac{l}{wh} \cdot G_t,\tag{A15}$$

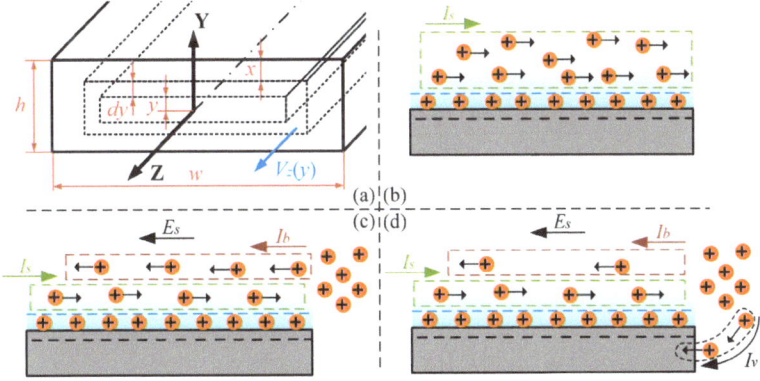

Figure A1. (a) Illustration of the slit channel for the calculation of streaming current, I_s; **(b)** pressure drives ions in diffuse layers to generate a streaming current; **(c)** the accumulation of ions downstream results in streaming potential E_s and an ion backflow to generate a back current, I_b; **(d)** a backflow through the metal bulk.

References

1. Morshed, A.; Wu, H.; Jiang, Z. A Comprehensive Review of Water-Based Nanolubricants. *Lubricants* **2021**, *9*, 89. [CrossRef]
2. Rahman, M.H.; Warneke, H.; Webbert, H.; Rodriguez, J.; Austin, E.; Tokunaga, K.; Rajak, D.K.; Menezes, P.L. Water-Based Lubricants: Development, Properties, and Performances. *Lubricants* **2021**, *9*, 73. [CrossRef]
3. Wu, H.; Zhao, J.; Luo, L.; Huang, S.; Wang, L.; Zhang, S.; Jiao, S.; Huang, H.; Jiang, Z. Performance Evaluation and Lubrication Mechanism of Water-Based Nanolubricants Containing Nano-TiO$_2$ in Hot Steel Rolling. *Lubricants* **2018**, *6*, 57. [CrossRef]
4. Wu, H.; Kamali, H.; Huo, M.; Lin, F.; Huang, S.; Huang, H.; Jiao, S.; Xing, Z.; Jiang, Z. Eco-Friendly Water-Based Nanolubricants for Industrial-Scale Hot Steel Rolling. *Lubricants* **2020**, *8*, 96. [CrossRef]
5. Sagraloff, N.; Winkler, K.J.; Tobie, T.; Stahl, K.; Folland, C.; Asam, T. Investigations on the Scuffing and Wear Characteristic Performance of an Oil Free Water-Based Lubricant for Gear Applications. *Lubricants* **2021**, *9*, 24. [CrossRef]
6. Lin, W.; Klein, J. Control of surface forces through hydrated boundary layers. *Curr. Opin. Colloid Interface Sci.* **2019**, *44*, 94–106. [CrossRef]
7. Huang, S.; Wu, H.; Jiang, Z.; Huang, H. Water-based nanosuspensions: Formulation, tribological property, lubrication mechanism, and applications. *J. Manuf. Processes* **2021**, *71*, 625–644. [CrossRef]
8. Godlevskiy, V.A. Technological lubricating means: Evolution of materials and ideas. *Front. Mech. Eng.* **2016**, *11*, 101–107. [CrossRef]
9. Brinksmeier, E.; Meyer, D.; Huesmann-Cordes, A.G.; Herrmann, C. Metalworking fluids-Mechanisms and performance. *CIRP Ann.* **2015**, *64*, 605–628. [CrossRef]
10. Smith, T.; Naerheim, Y.; Lan, M.S. Theoretical analysis of cutting fluid interaction in machining. *Tribol. Int.* **1988**, *21*, 239–247. [CrossRef]
11. Zeytounian, R.K. *Interfacial Phenomena and the Marangoni Effect*, 1st ed.; Springer: Vienna, Austria, 2002; pp. 123–190.
12. Xu, X.; Luan, Z.; Zhang, T.; Liu, J.; Feng, B.; Lv, T.; Hu, X. Effects of electroosmotic additives on capillary penetration of lubricants at steel/steel and steel/ceramic friction interfaces. *Tribol. Int.* **2020**, *151*, 106441. [CrossRef]
13. Feng, B.; Luan, Z.; Zhang, T.; Liu, J.; Hu, X.; Guan, J.; Xu, X. Capillary electroosmosis properties of water lubricants with different electroosmotic additives under a steel-on-steel sliding interface. *Friction* **2021**, *10*, 1019–1034. [CrossRef]
14. Chen, L.X.; Ma, J.P.; Tan, F.; Guan, Y.F. Generating high-pressure sub-microliter flow rate in packed microchannel by electroosmotic force: Potential application in microfluidic systems. *Sens. Actuators B Chem.* **2003**, *88*, 260–265. [CrossRef]
15. Li, D. EDL Potential. In *Encyclopedia of Microfluidics and Nanofluidics*; Li, D., Ed.; Springer: West Lafayette, IN, USA, 2008; pp. 444–453.
16. Lucy, C.A.; Underhill, R.S. Characterization of the cationic surfactant induced reversal of electroosmotic flow in capillary electrophoresis. *Anal. Chem.* **1996**, *68*, 300–305. [CrossRef]
17. MacDonald, A.M.; Sheppard, M.A.W.; Lucy, C.A. Enhancement of electroosmotic flow using zwitterionic additives. *Electrophoresis* **2005**, *26*, 4421–4428. [CrossRef]
18. Buchberger, W.; Winna, K. Determination of free fatty acids by capillary zone electrophoresis. *Mikrochim. Acta* **1996**, *122*, 45–52. [CrossRef]
19. Möckel, D.; Staude, E.; Dal-Cin, M.; Darcovich, K.; Guiver, M. Tangential flow streaming potential measurements: Hydrodynamic cell characterization and zeta potentials of carboxylated polysulfone membranes. *J. Membr. Sci.* **1998**, *145*, 211–222. [CrossRef]
20. Fievet, P.; Sbaï, M.; Szymczyk, A.; Vidonne, A. Determining the ζ-potential of plane membranes from tangential streaming potential measurements: Effect of the membrane body conductance. *J. Membr. Sci.* **2003**, *226*, 227–236. [CrossRef]
21. Hunter, R.J. Chapter 3-The Calculation of Zeta Potential. In *Zeta Potential in Colloid Science*; Hunter, R.J., Ed.; Academic Press: Cambridge, MA, USA, 1981; pp. 59–124.
22. Szymczyk, A.; Fievet, P.; Aoubiza, B.; Simon, C.; Pagetti, J. An application of the space charge model to the electrolyte conductivity inside a charged microporous membrane. *J. Membr. Sci.* **1999**, *161*, 275–285. [CrossRef]
23. Exartier, C.; Maximovitch, S.; Baroux, B. Streaming potential measurements on stainless steels surfaces: Evidence of a gel-like layer at the steel/electrolyte interface. *Corros. Sci.* **2004**, *46*, 1777–1800. [CrossRef]
24. Nakayama, K. Triboemission of charged particles from various solids under boundary lubrication conditions. *Wear* **1994**, *178*, 61–67. [CrossRef]
25. Nakayama, K. The plasma generated and photons emitted in an oil-lubricated sliding contact. *J. Phys. D Appl. Phys.* **2007**, *40*, 1103–1107. [CrossRef]
26. Chang, Y.P.; Chu, H.M.; Chou, H.M. Effects of mechanical properties on the tribo-electrification mechanisms of iron rubbing with carbon steels. *Wear* **2007**, *262*, 112–120. [CrossRef]
27. Charlson, E.M.; Charlson, E.J.; Burkett, S.; Yasuda, H.K. Study of the contact electrification of polymers using contact and separation current. *IEEE Trans. Electr. Insul.* **1992**, *27*, 1144–1151. [CrossRef]
28. He, J.; Sun, J.; Meng, Y.; Pei, Y. Superior lubrication performance of MoS$_2$-Al$_2$O$_3$ composite nanofluid in strips hot rolling. *J. Manuf. Processes* **2020**, *57*, 312–323. [CrossRef]
29. He, J.; Sun, J.; Meng, Y.; Tang, H.; Wu, P. Improved lubrication performance of MoS$_2$-Al$_2$O$_3$ nanofluid through interfacial tribochemistry. *Colloids Surf. A Physicochem. Eng. Asp.* **2021**, *618*, 126428. [CrossRef]
30. Xiong, S.; Zhang, B.; Luo, S.; Wu, H.; Zhang, Z. Preparation, characterization, and tribological properties of silica-nanoparticle-reinforced B-N-co-doped reduced graphene oxide as a multifunctional additive for enhanced lubrication. *Friction* **2021**, *9*, 239–249. [CrossRef]

31. Pérez, A.T.; Fernández-Mateo, R. Electric force between a dielectric sphere and a dielectric plane. *J. Electrostat.* **2021**, *112*, 103601. [CrossRef]
32. Tatsumi, K.; Nishitani, K.; Fukuda, K.; Katsumoto, Y.; Nakabe, K. Measurement of electroosmotic flow velocity and electric field in microchannels by micro-particle image velocimetry. *Meas. Sci. Technol.* **2010**, *21*, 11. [CrossRef]
33. Gao, S.; Liu, H. *Capillary Mechanics*, 1st ed.; Science Press: Beijing, China, 2010; pp. 121–123.
34. Huang, S.; Li, Z.; Yao, W.; Hu, J.; Xu, X. Tribological performance of charged vegetable lubricants. *Tribology* **2014**, *34*, 371–378.
35. Wang, J.; Li, C.; Wang, J.; Zhao, G.; Wang, X. Synthesis and tribological properties of a water-soluble lubricant additive. *Lubr. Eng.* **2012**, *37*, 1–6.
36. Zou, H.; Guo, L.; Xue, H.; Zhang, Y.; Wang, Z.L. Quantifying and understanding the triboelectric series of inorganic non-metallic materials. *Nat. Commun.* **2020**, *11*, 2093. [CrossRef] [PubMed]
37. Nakayama, K.; Hashimoto, H. Effect of surrounding gas-pressure on triboemission of charged-particles and photons from wearing ceramic surfaces. *Tribol. Trans.* **1995**, *38*, 35–42. [CrossRef]
38. Zhao, W.; Liu, X.; Yang, F.; Wang, K.G.; Bai, J.T.; Qiao, R.; Wang, G.R. Study of Oscillating Electroosmotic Flows with High Temporal and Spatial Resolution. *Anal. Chem.* **2018**, *90*, 1652–1659. [CrossRef]
39. Liu, J.Y.; Liu, H.P.; Han, R.D.; Wang, Y. The study on lubrication action with water vapor as coolant and lubricant in cutting ANSI 304 stainless steel. *Int. J. Mach. Tool Manuf.* **2010**, *50*, 260–269.
40. Wang, S.; Li, X.; Liu, D. *Surfactant Chemistry*, 1st ed.; Chemical Industry Press: Beijing, China, 2005; pp. 8–13.
41. Xu, X.F.; Lv, T.; Luan, Z.Q.; Zhao, Y.Y.; Wang, M.H.; Hu, X.D. Capillary penetration mechanism and oil mist concentration of Al_2O_3 nanoparticle fluids in electrostatic minimum quantity lubrication (EMQL) milling. *Int. J. Adv. Manuf. Technol.* **2019**, *104*, 1937–1951. [CrossRef]
42. Chen, Y. *Capillary Electrophoresis Technology and Its Application*, 2nd ed.; Chemical Industry Press: Beijing, China, 2005; pp. 9–91.
43. Hines, H.B.; Brueggemann, E.E. Factors affecting the capillary electrophoresis of ricin, a toxic glycoprotein. *J. Chromatogr. A* **1994**, *670*, 199–208. [CrossRef]
44. Sbai, M.; Fievet, P.; Szymczyk, A.; Aoubiza, B.; Vidonne, A.; Foissy, A. Streaming potential, electroviscous effect, pore conductivity and membrane potential for the determination of the surface potential of a ceramic ultrafiltration membrane. *J. Membr. Sci.* **2003**, *215*, 1–9. [CrossRef]

 *lubricants*

Article

Friction and Wear Properties of a Nanoscale Ionic Liquid-like GO@SiO₂ Hybrid as a Water-Based Lubricant Additive

Liang Hao [1,†], Wendi Hao [1], Peipei Li [2,†], Guangming Liu [3], Huaying Li [3,*], Abdulrahman Aljabri [4] and Zhongliang Xie [5,*]

[1] School of Mechano-Electronic Engineering, Xidian University, Xi'an 710071, China; haoliang@xidian.edu.cn (L.H.); xdhaowendi@163.com (W.H.)
[2] School of Advanced Materials and Nanotechnology, Xidian University, Xi'an 710126, China; lip@xidian.edu.cn
[3] School of Materials Science and Engineering, Taiyuan University of Science and Technology, Taiyuan 030024, China; brightliu2008@126.com
[4] Department of Mechanical Engineering, Islamic University of Madinah, Medina 42351, Saudi Arabia; aaljabri@iu.edu.sa
[5] Department of Engineering Mechanics, Northwestern Polytechnical University, Xi'an 710072, China
*   Correspondence: huayne@163.com (H.L.); zlxie@nwpu.edu.cn (Z.X.)
†   These authors contributed equally to this work.

Abstract: In this study, a nanoscale ionic liquid (NIL) GO@SiO₂ hybrid was synthesized by attaching silica nanoparticles onto graphene oxide (GO). It was then functionalized to exhibit liquid-like behavior in the absence of solvents. The physical and chemical properties of the synthesized samples were characterized by means of a transmission electron microscope, X-ray diffraction, Fourier transform infra-red, Raman spectroscopy, and thermogravimetric analysis. The tribological properties of the NIL GO@SiO₂ hybrid as a water-based (WB) lubricant additive were investigated on a ball-on-disk tribometer. The results illustrate that the NIL GO@SiO₂ hybrid demonstrates good dispersity as a WB lubricant, and can decrease both the coefficient of friction (COF) and wear loss.

Keywords: tribological tests; GO@SiO₂ hybrid; water-based lubricant; additive

Citation: Hao, L.; Hao, W.; Li, P.; Liu, G.; Li, H.; Aljabri, A.; Xie, Z. Friction and Wear Properties of a Nanoscale Ionic Liquid-like GO@SiO₂ Hybrid as a Water-Based Lubricant Additive. *Lubricants* **2022**, *10*, 125. https://doi.org/10.3390/lubricants10060125

Received: 29 April 2022
Accepted: 7 June 2022
Published: 13 June 2022

Publisher's Note: MDPI stays neutral with regard to jurisdictional claims in published maps and institutional affiliations.

1. Introduction

Lubricants have become essential in the modern manufacturing industry, reducing energy consumption and improving the surface finish of products and the reliability of the production process. Petroleum derivatives and functional additives, such as extreme pressure agents, antioxidants, detergents, dispersants, etc., constitute traditional lubricants which are environmentally unfriendly and detrimental to human health. With growing concerns of energy crises and environmental issues, "Green Manufacturing", or "Environmentally Conscious Manufacturing", has gained considerable attention [1,2]. Therefore, novel lubricants, which are environmentally friendly and effective, are imperative to be developed to substitute traditional oil-based lubricants.

Since water is of low cost with a high cooling capacity, water-based lubricants are potential candidates for novel lubricants. However, the weaknesses of water, such as low viscosity, its corrosive properties, and especially the low strength of water films, constitute major barriers for tribological applications [3,4]. In order to adjust and improve water-based lubricants, high-quality functional additives are of great significance. Among these functional additives, nano-materials are extensively investigated due to their distinctive physiochemical and mechanical properties [5–7]. In addition, nanoparticles can improve tribological properties by improving the viscosity of water and nanolubrication mechanisms [8,9]. Nanoparticles added, such as metals (e.g., Cu [10,11] and Cu–Al alloy [12]), metal oxides (e.g., CuO [13], Fe₃O₄ [14], Al₂O₃ [15,16], TiO₂ [17–19], and ZnO [20,21]), non-metal oxides (e.g., SiO₂ [22–24]), sulfides (e.g., MoS₂ [25,26] and WS₂ [27]) and rare earth compounds (e.g., CeO_2 [28] and $BCeO_3$ [29]), offer anti-wear properties and lower friction due to the formation of tribofilms, in addition to micro-bearing, polishing and mending [30]. However, the poor dispersity of these nanoparticles in base stocks fails to enhance their tribological properties. A typical technique to stabilize nanoparticles in base stocks is to use physical methods, chemical methods, and self-dispersed methods. Traditional physical suspension processes include mechanical stirring, ultrasonication ball mill, and high-pressure homogenization [31,32]. Chemical methods mean modifying the inorganic–organic interface by attaching different functional groups on the surface of the nanoparticles [33]. Current studies have also revealed that the dispersion of the stability of nanoparticles can also be enhanced via the structure regulation of nanoparticles [34].

He et al. [16] mechanically dispersed different sizes and concentrations of Al_2O_3 nanoparticles in glycerol water-based lubricants using an ultrasonic probe. The synthesized suspensions were found to be stable for only 3 days. Wu et al. [35] prepared TiO_2 nano-additive water-based lubricants by modifying TiO_2 nanoparticles with polyethyleneimine (PEI), and the suspensions were only stable for 7 days. Gup et al. [36] used oleic acid and ionic liquid to engineer ZnO and WS_2 nanoparticles as oil additives, and the nanolubricants stratified after 10 days. In addition, silane coupling agents, commonly employed to modify nanoparticles, failed to form enough steric repulsion to stabilize the nanolubricants for a long period due to their light molecular weights [14,37,38]. Man et al. synthesized novel CuO@Graphene and added PAO-6 oil, reducing COF by more than 50% with 0.5 wt.% [39]. Surface-functionalized nanoparticles with liquid-like behavior in the absence of solvent have been coined nanoscale ionic liquids, which are organic–inorganic hybrids comprising a nanoparticle core functionalized with a covalently tethered ionic corona and oppositely charged canopy. The physical properties (rheological and solubility) of the nanoscale ionic liquids can be engineered over a broad range by adjusting the chemical character-istics of the corona and canopy [40–42]. Li et al. [43] synthesized a nanoscale liquid-like graphene@Fe_3O_4 hybrid according to the nanoscale ionic liquid method, and identified excellent amphiphilicity. On the other hand, graphene-based nanolubricant additives have gained increasing attention because of their superior lubricating performance, as well as their green and dashless properties. Graphene-based nanocomposites reveal good lubrication properties due to their synergistic effects [44,45].

In this work, a new kind of graphene-based composite was synthesized and modified for a water-based nanolubricant additive with enhanced stability. Following Section 1, the rest of the paper is organized as follows. The detailed synthesis and functionalization pro-cess, tribological tests, and characterizations are given in Section 2. In Section 3, synthetic, modification and tribological results, as well as lubrication mechanisms, are presented. Finally, Section 4 provides conclusions.

2. Experimental Section

2.1. Materials

GO solutions (purity: >99%; content: 1.55%; thickness: 0.55–2 nm; size: 1–5 μm) were purchased from Best Material Co., Ltd. (Chengdu, China). Tetraethyl orthosilicate (TEOS, 99% purity), ammonia (25% aqueous solution) and ethanol (99% purity) were purchased from Tianjin Organics (Tianjin, China). 3-(Trihydroxysilyl)-1-propanesulfonic acid (30–35% in water) (SIT8378.3, $(CH_3)_3Si(CH_2)_3HSO_3$) was obtained from Gelest Inc. (Shanghai, China), while Jeffamine M-2070 Polyetheramine $(CH_3-(OCH_2CH_2)_6-(OCH_2CH-CH_3)_{35}-NH_2)$ was from HengYu Trading Co., Ltd. (Guangzhou, China). A ferritic stainless steel (FSS 444) was used as a disk material (Taiyuan, China). All the disks were cut to 28 mm in diameter, 2 mm in thickness, and surfaces were ground to a roughness of 0.6 μm. GCr15 steel balls with a diameter of 6 mm and an identical Ra of 0.02 mm were employed for the ball-on-disk tests. The Vickers hardness of the ball and disk were 790 HV and 168 HV0.1. The main chemical compositions of FSS 444 (wt.%) were as follows: C 0.0094, Si 0.084, Mn 0.064, Cr 18.4, Mo 1.81, Nb 0.22, and Fe balance.

2.2. Synthesis of the GO@SiO₂ Compound

Firstly, 400 mL ethanol, 25 mL GO solution, 21 mL ammonia were sequentially poured into a three-necked, round-bottomed flask and stirred at 700 rpm in a 50 °C water bath for 30 min. Secondly, 16.1 mL TEOS was added, and the chemical reaction (700 rmp stirring and 50 °C water bath) occurred for 2 h. Then, the black precipitate was collected by centrifugation and washed with ethanol three times. Finally, the wet precipitate was free-dried for 24 h to obtain the GO@SiO₂ compound.

2.3. Preparation of the Nanoscale Liquid-like GO@SiO₂ Hybrid

To begin with, 500 mg GO@SiO₂ was dispersed in 10 mL deionized water under sonication to obtain the GO@SiO₂ suspension, followed by dropwise adding 5 mL SIT8378.3 solution and stirring for 30 min. Then, the NaOH solution (1 mol L⁻¹) was added until the pH became 7, and the solution was stirred at room temperature for 24 h. In order to remove the residual SIT 8378.3, the solution was dialyzed using a dialysis tube for 48 h and the deionized water was exchanged every 8 h. Next, a cation exchange resin was employed to remove the Na⁺ ions to protonate the sulfonate group. Finally, the polymer chains were attached onto the functionalized GO@SiO₂ compound by dropwise injecting 10 wt.% Jeffamine M-2070 solution to neutralize all sulfonate groups connected to the surface of the GO@SiO₂ compound. Finally, the solution was dried to a constant weight under vacuum at 50 °C, and the nanoscale liquid-like GO@SiO₂ hybrid was obtained. A synthetic framework is illustrated in Figure 1.

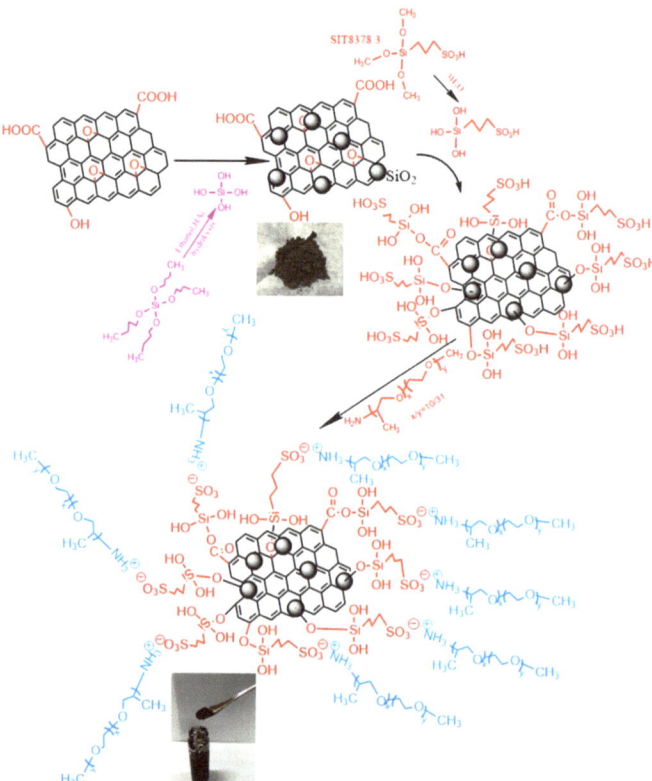

Figure 1. Synthetic framework of the nanoscale liquid-like GO@SiO₂ hybrid.

2.4. Tribological Tests

The first step was to prepare the water-based (WB) lubricant (base fluid) and WB nanolubricants with different concentrations of the NIL GO@SiO$_2$ hybrid (1.0 wt.%, 2.0 wt.%, 4.0 wt.% and 8.0 wt.%). The detailed preparation process of all suspensions was as follows: (a) the WB lubricant consisted of 10.0 wt.% glycerol and 90.0 wt.% deionized water; (b) the base fluid was kept at 60 °C with an electromagnetic stirring heater, and proper amounts of NIL GO@SiO$_2$ were added into the base fluids under stirring for 30 min; (c) the suspensions were ultrasonicated for 30 min to obtain homogeneous WB nanolubricants.

Ball-on-disk tests were conducted on an Rtec MFT (multi-functional tribometer) 5000. Moving 10 mm away from the center of the disk, a normal force of 15 N was applied to the ball (equal to 1.6 GPa of the maximum Hertz contact stress) against the rotating disk at 200 rpm. The normal force employed to the ball holder was measured by an F_z load cell installed above a spring. The friction force was generated by the combination of the rotating motion and the normal load. The coefficient of friction (COF) was calculated by the ratio between an F_x load cell attached onto the ball holder and F_z. The tribological trials were performed under different lubrication conditions for 20 min., and each test was repeated at least three times. During the tests, the COF variations were recorded automatically every two seconds. In addition, the wear of the balls after tests was evaluated by measuring the worn surface areas.

2.5. Characterizations

The samples were prepared by placing a few drops of the GO and GO@SiO$_2$ dispersions onto a copper grid and then evaporating the solvent. Afterwards, transmission electron microscope (TEM) images were examined using JEM-2100F. The X-ray diffraction (XRD) analysis was conducted on a D8 Advance using Cu Kα radiation. The measured 2θ values ranged from 10° to 80° and the scan step was 0.02. The chemical composition and physical properties of the GO, GO@SiO$_2$ compound and NIL GO@SiO$_2$ hybrid were investigated using a Fourier transform infra-red (FTIR, Nicolet iS50) spectra in the range of 4000 to 500 cm^{-1}, and Raman spectroscopy (inVia). A Zeta potential analyzer (Zen3690) was used to characterize the dispersion stability. Thermogravimetric analysis (TGA) measurements were taken under Argon flow at a heating rate of 10 °C min^{-1} using a STA 449F5 instrument.

The wear scars on the balls and the worn tracks of the disks were characterized using a Leica optical microscope (OM) and JSM-7800F field emission scanning electron microscope (FE-SEM) and EDS. In addition, atomic force microscopy (AFM) was carried out to measure the surface morphology of the wear tracks.

3. Results and Discussion

3.1. Structural Analysis

The microstructures of the GO and GO@SiO$_2$ compounds were characterized by TEM, as shown in Figure 2. Figure 2a reveals that the GO nanosheets were efficiently exfoliated to form separate and transparent sheets. In addition, the dark color in the picture indicates that the GO nanosheets were folded. Compared with the GO (Figure 2a), the GO@SiO$_2$ compound was decorated with SiO$_2$ nanoparticles with a diameter of approximately 100 nm, in which the SiO$_2$ nanoparticles were sparsely attached onto the lamellae of GO, displaying no apparent nano-SiO$_2$ agglomeration.

The crystalline structures of GO, the GO@SiO$_2$ compound and the NIL GO@SiO$_2$ hybrid were analyzed by XRD. As shown in Figure 3, the diffraction of the GO peaks occurred at $2\theta = 11.4°$, corresponding to the (0 0 2) plane. In addition, the characteristic diffraction peak at $2\theta = 22°$ can be assigned to the (1 1 1) plane reflection of SiO$_2$ according to PAN-ICSD NO.01-089-3435, which suggests that SiO$_2$ nanoparticles were successfully attached to the GO. Moreover, the XRD pattern of the NIL GO@SiO$_2$ hybrid revealed that GO@SiO$_2$ compounds retain their structural and size integrity after surface functionalization.

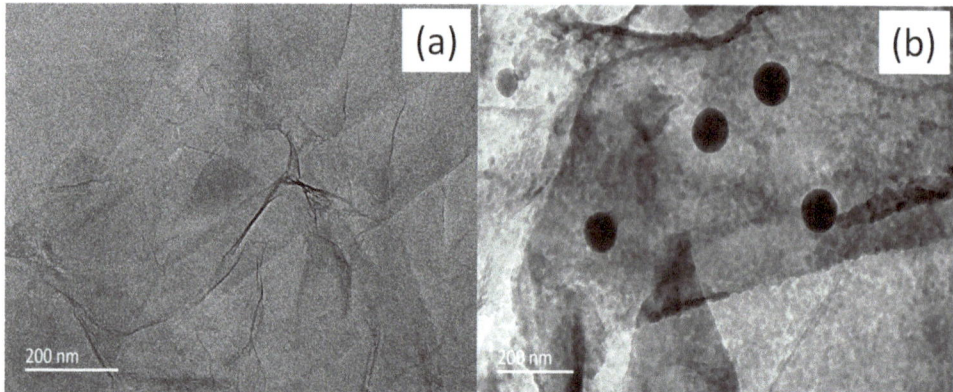

Figure 2. TEM images of the GO (**a**) and the GO@SiO$_2$ compound (**b**).

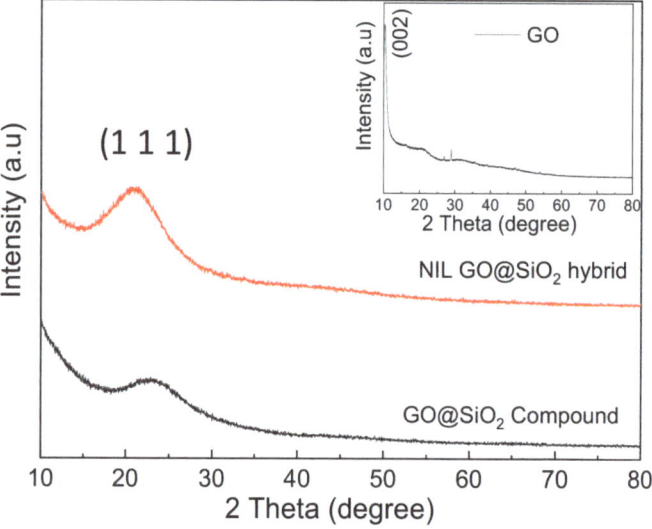

Figure 3. XRD patterns of GO and the GO@SiO$_2$ compound.

The FTIR spectra of GO, the GO@SiO$_2$ compound, and the NIL GO@SiO$_2$ hybrid are presented in Figure 4. The FTIR spectrum of GO reveals the presence of hydroxyl (~3400 cm^{-1}), epoxy (~1177 cm^{-1}), carboxyl (~1733 cm^{-1}), the O-H deformation (~1049 cm^{-1}), and the oxygenous groups (~1624 cm^{-1}) [46]. The intensities of these IR peaks dropped considerably after the attachment of the silica nanoparticles. In addition, the characteristic absorption peaks of Si-O (~1105 cm^{-1}) and Si-O-Si (~1100 cm^{-1}) were both observed, which confirms that silica was successfully deposited on the surface of the GO. As for the NIL GO@SiO$_2$ hybrid, the bands at ~2920 cm^{-1} and 674 cm^{-1} were due to the C-H vibrations from the Jeffamine M-2070 and SO$_3{}^{2-}$ from SIT 8378.3, respectively. Owing to the relatively small number of terminal ammonium groups present in the high molecular weight of the the Jeffamine M-2070, it is difficult to detect them in the FTIR spectra [47].

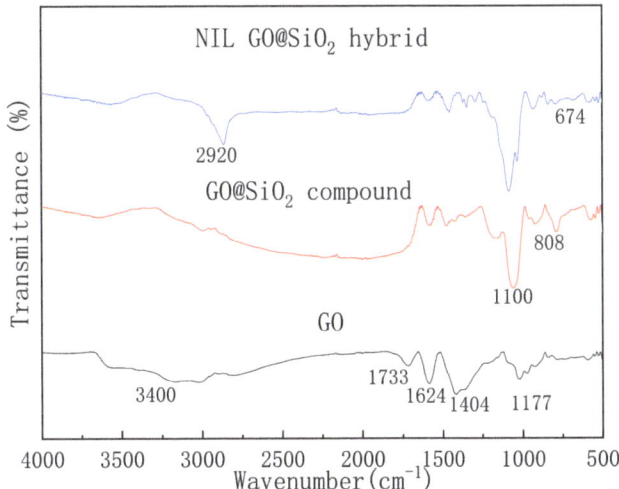

Figure 4. FTIR spectra of GO, GO@SiO$_2$ compound and NIL GO@SiO$_2$ hybrid.

To further investigate the layered and defect structure of carbonaceous materials, Raman spectroscopy was employed to explore the GO, the GO@SiO$_2$ compound and the NIL GO@SiO$_2$ hybrid. As shown in Figure 5, two bands around 1350 cm^{-1}(D) and 1595 cm^{-1} (G) represent the disorder of symmetry and crystallization, respectively [48], and are attributed to the activation of the first-order scattering process of sp^3 carbon and sp^2-bonded carbon atoms in graphene sheets, respectively. The ratio of the D-band and G-band I(D)/I(G) is correlated to the ratio of disordered sp^3 and ordered sp^2 carbon domains. The higher the ratio of the D-band to G-band, the more defects in the carbon materials. Compared with GO (I_D/I_G = 0.96) and the GO@SiO$_2$ compound (I_D/I_G = 0.94), the NIL GO@SiO2 hybrid increased to 1.02, implying that the sp^2 carbon domain decreased and new defects and polar groups were produced. In addition, the molecular charge transferred between M-2070 and the graphene also led to the increase in the I(D)/I(G) ratio [43].

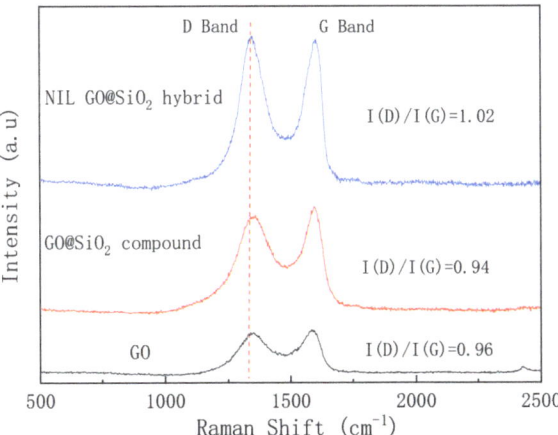

Figure 5. Raman spectra of GO, GO@SiO$_2$ compound and NIL GO@SiO$_2$ hybrid.

The content of the organic canopy attached to the NIL GO@SiO$_2$ hybrid affects the properties of the material. Therefore, TGA was conducted to evaluate the thermal stability and organic quantity of the material (Figure 6). Jeffamine M-2070 exhibited a relatively higher decomposition at the temperature range of 300~420 °C, while the NIL GO@SiO$_2$ hybrid mainly underwent weight loss at the range of 280~360 °C. In addition, the GO@SiO$_2$ compound showed gradual weight loss up to approximately 14.5% until 600 °C, while the decomposition residual of the M-2070 was roughly 6.9%. Most especially, the TGA trace under Argon flow demonstrated that the NIL GO@SiO$_2$ hybrid is virtually solvent-free and a new hybrid, and is not a simple mixture of the M-2070 and GO@SiO$_2$ compound. As a result, it can be estimated that the organic component (surface functionalization groups) in the hybrid accounted for roughly 77.9%.

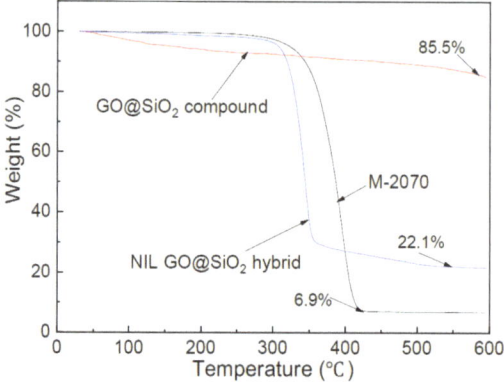

Figure 6. TGA curves of M-2070, GO@SiO$_2$ compound and NIL GO@SiO$_2$ hybrid.

3.2. Dispersion Stability

Since nanoparticles have a tendency to agglomerate, stability is a key issue for nanolubricants. The state of the nanoparticles results from a combined effect of the van der Waals attraction force and the electrical double-layer repulsive force [49]. The zeta potential (ZP) is an important and measurable indicator of the stability of colloids, and its magnitude reveals the degree of electrostatic repulsion between adjacent charged particles. High ZP implies highly charged particles, which prevents the aggregation of the particles due to electric repulsion, while low ZP implies that attraction overcomes repulsion, leading to coagulation. Figure 7 shows the ZP of nanolubricants with the GO@SiO$_2$ compound and the NIL GO@SiO$_2$ hybrid, respectively. The absolute ZP value (-45.8 mV) of the GO@SiO$_2$ hybrid dispersion was greater than that (-28.2 mV) of the GO@SiO$_2$ compound dispersion. Therefore, the surface functionalization of GO@SiO$_2$ by the NIL method improved the dispersion stability in the water-based lubricant.

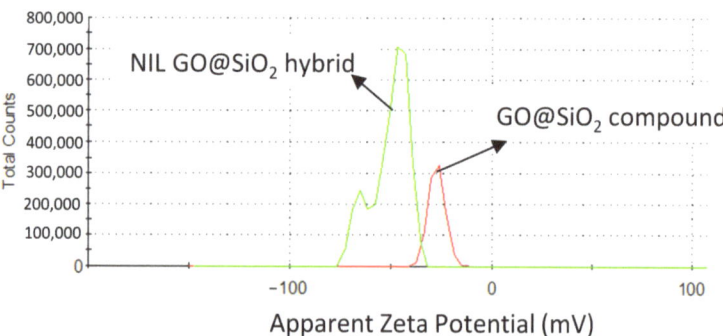

Figure 7. Zeta potential of nanolubricants with the GO@SiO$_2$ compound and the NIL GO@SiO$_2$ hybrid.

3.3. Tribological Properties

The friction and wear properties of the NIL GO@SiO$_2$ hybrid as the additive in the water-based lubricant were investigated by ball-on-disk tests. From Figure 8a, the in situ COF curves versus time revealed that the COF slightly varied from the start to the end in each scenario. It can be seen that the COF was relatively high under the lubrication of the water-based stock; however, 1.0 wt.% NIL GO@SiO$_2$ hybrid lubricant made the COF slightly fluctuate and hardly improved the lubrication performance. As the concentration of the water-based lubricant continuously increased, although the COF curve varied, it proceeded to be lower than that of the water-based stock. In particular, the COF values of 4.0 and 8.0 wt.% concentrations presented the comparatively lowest level of all scenarios. The AFC (average friction coefficient) and AWS (area of wear scar) of the balls are shown in Figure 8b. It can be seen that the addition of the NIL GO@SiO$_2$ hybrid was able to ameliorate the tribological performance by reducing the AFC and AWS. In contrast, the best lubrication performance was obtained at 4.0 wt.% hybrid concentration, in which the AFC and AWS were 0.33 and 0.084 µm^2, respectively. Compared with the WB lubricant, these two values were reduced by 20.7% and 36.6%, respectively.

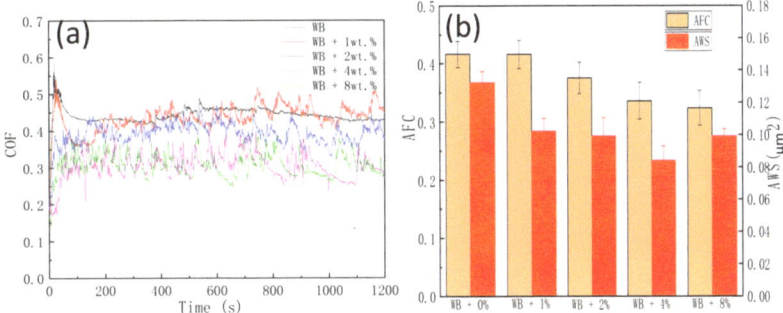

Figure 8. (**a**) COF versus time and (**b**) AFC and AWS lubricated by different concentrations.

As shown in Figure 9, wear tracks lubricated by the different concentrations of WB nanolubricants were characterized by AFM. It is evident that the addition of NIL GO@SiO$_2$ hybrid flattened the surface roughness. Deep grooves were generated under the lubrication of the WB lubricant with the surface roughness (R_a) at about 700 nm (Figure 9a). Additionally, the addition of the NIL GO@SiO$_2$ hybrid decreased the track roughness at all concentrations tested (Figure 9b–e), which was about 100 nm. As a result, abrasive polishing was ascribed to be one of the lubrication forms of the NIL GO@SiO$_2$ hybrid [50].

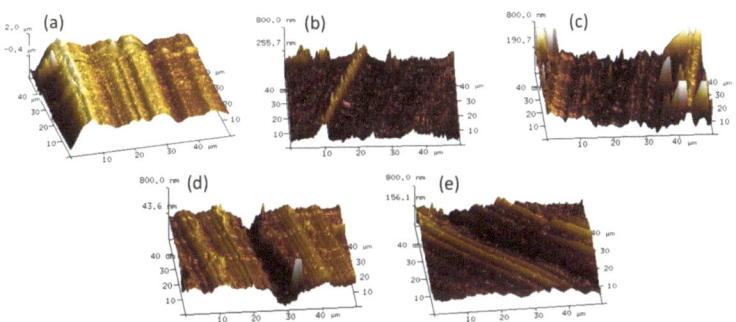

Figure 9. AFM images of wear tracks lubricated by (**a**) the WB lubricant, (**b**) 1.0 wt.%, (**c**) 2.0 wt.%, (**d**) 4.0 wt.%, and (**e**) 8.0 wt.% NIL GO@SiO$_2$.

3.4. Inquiry of Lubrication Mechanisms

The lubrication enhancement of nanoparticles can be primarily divided into four mechanisms, including the micro-bearing effect, the self-repairing effect, tribo-film, and the polishing effect [51]. As the ball was pressed against the rotating disk under the nanolubricants, the GO@SiO$_2$ nanoparticles were dragged into the engaging surfaces with the base fluid, leaving deposits on the mating surfaces.

In order to understand the mechanisms of friction reduction and anti-wear for the NIL GO@SiO$_2$ hybrid, FE-SEM morphologies of the worn track lubricated under the different concentrations are displayed in Figure 10. Spherical nanoparticles can be discerned at all concentrations; these are the silica nanoparticles attached to the GO. The friction reduction was less obvious at low concentrations (1.0 wt.% and 2.0 wt.%), because there were limited nanoparticles deposited on the mating surfaces (Figure 10a,b). When the concentration increased to 4.0 wt.% and 8.0 wt.%, both friction reduction and anti-wear were reduced more than 20%. Furthermore, the shape and size of the nanoparticles were well maintained after the wear tests, which implies that the NIL GO@SiO$_2$ hybrid may have a typical micro-bearing effect in common cases [52]. Moreover, graphene also promotes relative sliding between the mating surfaces.

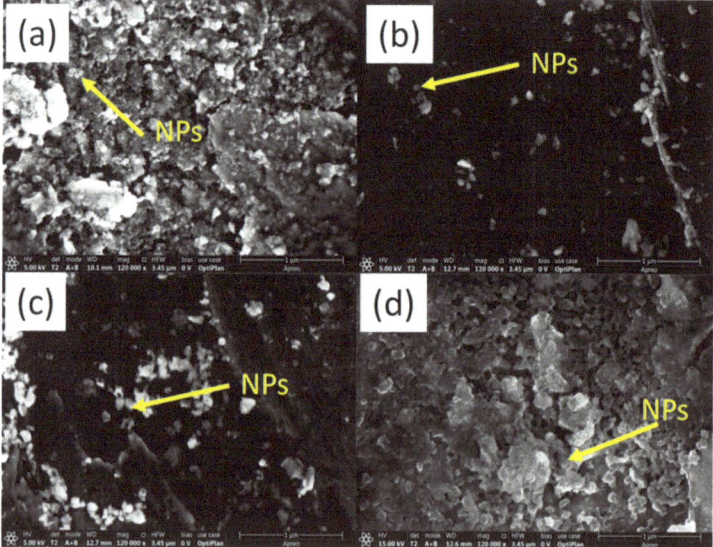

Figure 10. FE-SEM images of the worn surfaces on disks lubricated by (**a**) 1.0 wt.%, (**b**) 2.0 wt.%, (**c**) 4.0 wt.%, and (**d**) 8.0 wt.% NIL GO@SiO$_2$.

FE-SEM image of the wear track lubricated by the 4.0% NIL GO@SiO$_2$ hybrid and a point element analysis are given in Figure 11. The spherical particles (bright color) can be discerned, and the EDS spectra analysis (Figure 11b) reveals Si (wt.% 2.76) and C (wt.% 2.49) elements on the worn track, which illustrates that the NIL GO@SiO$_2$ hybrid was embedded onto the steel substrate under high contact pressure. This indicates that the spherical SiO$_2$ nanoparticles in the lubricant can roll between the rubbing surfaces during the friction process. The graphene deposited onto the rubbing surfaces also plays a role in anti-wear and friction reduction. At the later stage, the hybrid could take effect and avoid the direct steel-to-steel contact to reduce wear. In future, more exquisite techniques (focused ion beam—FIB) should be applied to characterize the lubrication films formed due to nanoparticles. Furthermore, Bao and our former findings [24,53] demonstrate that there is no chemical reaction of SiO$_2$ nanoparticles in the tribofilms, and the lubrication effect is only a physical effect.

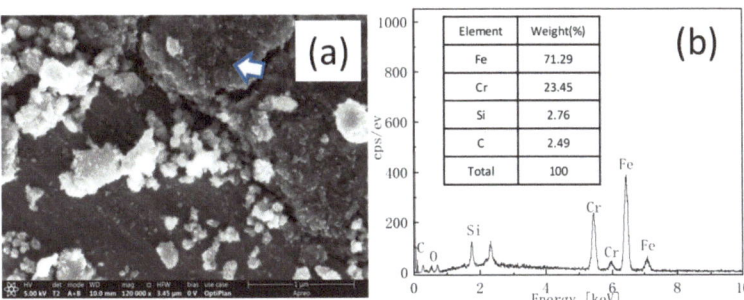

Figure 11. (**a**) FE-SEM image of the worn surfaces lubricated by 4.0 wt.% NIL GO@SiO$_2$ and (**b**) the point element analysis labeled in (**a**).

Therefore, the mechanisms responsible for the NIL GO@SiO$_2$ hybrid's ability to reduce friction and wear can be attributed to a synergy of mechanisms. Firstly, the nanoscale ionic liquid-functionalized GO@SiO$_2$ compound restrained the agglomeration of GO@SiO$_2$ in the water-based fluid; secondly, when the NIL GO@SiO$_2$ hybrid WB nanolubricants were used, some nanoparticles were embedded onto the disk surface, acting as micro-bearings. Additionally, the NIL GO@SiO$_2$ hybrid could be pressed to form tribofilms, preventing direct steel-to-steel contact. Thirdly, during the stable period, the nanoparticles refilled the friction pair surfaces (so-called "mending"), and the replenishment and loss of nanoparticle obtained a balance. The nanolubricants also took away debris and friction heat to avoid welding between asperities. Thus, the NIL GO@SiO$_2$ hybrid exhibits a good tribological performance.

4. Conclusions

In this work, silica nanoparticles with a diameter of 100 nm were successfully attached onto graphene oxide. Then, based on the nanoscale ionic liquid method, the GO@SiO$_2$ compound was functionalized to obtain a liquid-like GO@SiO$_2$ hybrid. The lubrication performance of the NIL GO@SiO$_2$ hybrid as a water-based lubricant was investigated using a ball-on-disk tribometer. Based on the aforementioned analysis, the following conclusions were obtained:

1. The as-synthesized NIL GO@SiO2 hybrid consisted of approximately 77.9% organic components and 22.1% inorganic components, exhibiting good dispersity and stability as a WB lubricant;
2. The addition of the NIL GO@SiO2 hybrid reduced the COF and AWS at all tested concentrations. Compared with the WB lubricant, the 4.0 wt% hybrid nanolubricant lowered COF and AWS by 20.7% and 36.6%, respectively;
3. The tribological enhancement of the NIL GO@SiO2 hybrid can be explained by the synergistic mechanisms of micro-rolling, polishing and mending in the GO@SiO2 compound.

Author Contributions: Conceptualization, L.H. and P.L.; methodology, G.L.; software, W.H.; validation, H.L.; formal analysis, A.A.; investigation, Z.X.; resources, L.H.; data curation, W.H.; writing—original draft preparation, L.H.; writing—review and editing, P.L.; visualization, W.H.; supervision, G.L.; project administration, L.H.; funding acquisition, L.H., H.L. and Z.X. All authors have read and agreed to the published version of the manuscript.

Funding: This research was funded by National Natural Science Foundation of China (Grant No. 51904217), Natural Science Foundation of Shaanxi Province (Grant No. 2020JQ-294), Science and Technology Innovation Project of Educational Commission of Shanxi Province (Grant No. 2020L0333) and Natural Science Basic Research Program of Shaanxi (Program No. 2022JM-003).

Institutional Review Board Statement: Not applicable.

Informed Consent Statement: Not applicable.

Lubricants **2022**, *10*, 125

Data Availability Statement: Not applicable.

Conflicts of Interest: The authors declare no conflict of interest.

Nomenclature

NIL	nanoscale ionic liquid
English Symbols	
FSS	Ferritic Stainless Steel
GO	Graphene Oxide
TEO	Tetraethyl orthosilicate
WB	water-based
MFT	multi-functional tribometer
COF	coefficient of friction
TEM	Transmission Electron Microscope
XRD	X-ray diffraction
FTIR	Fourier Transform Infra-Red
TGA	Thermogravimetric analyzer
OM	Optical Microscope
AFM	Atomic Force Microscopy

References

1. Wu, H.; Zhao, J.; Luo, L.; Huang, S.; Wang, L.; Zhang, S.; Jiao, S.; Huang, H.; Jiang, Z. Performance Evaluation and Lubrication Mechanism of Water-Based Nanolubricants Containing Nano-TiO$_2$ in Hot Steel Rolling. *Lubricants* **2018**, *6*, 57. [CrossRef]
2. Rahman, M.H.; Warneke, H.; Webbert, H.; Rodriguez, J.; Austin, E.; Tokunaga, K.; Rajak, D.K.; Menezes, P.L. Water-Based Lubricants: Development, Properties, and Performances. *Lubricants* **2021**, *9*, 73. [CrossRef]
3. Tomala, A.; Karpinska, A.; Werner, W.S.M.; Olver, A.; Störi, H. Tribological properties of additives for water-based lubricants. *Wear* **2010**, *269*, 804–810. [CrossRef]
4. Xie, Z.; Zhu, W. An investigation on the lubrication characteristics of floating ring bearing with consideration of multi-coupling factors. *Mech. Syst. Signal Process.* **2022**, *162*, 108086. [CrossRef]
5. Khalid Shafi, W.; Charoo, M.S. NanoLubrication Systems: An Overview. *Mater. Today Proc.* **2018**, *5*, 20621–20630. [CrossRef]
6. Jia, X.; Huang, J.; Li, Y.; Yang, J.; Song, H. Monodisperse Cu nanoparticles @ MoS$_2$ nanosheets as a lubricant additive for improved tribological properties. *Appl. Surf. Sci.* **2019**, *494*, 430–439. [CrossRef]
7. Darminesh, S.P.; Sidik, N.A.C.; Najafi, G.; Mamat, R.; Ken, T.L.; Asako, Y. Recent development on biodegradable nanolubricant: A review. *Int. Commun. Heat Mass Transf.* **2017**, *86*, 159–165. [CrossRef]
8. Wang, L.; Tieu, A.K.; Zhu, H.; Deng, G.; Cui, S.; Zhu, Q. A study of water-based lubricant with a mixture of polyphosphate and nano-TiO$_2$ as additives for hot rolling process. *Wear* **2021**, *477*, 203895. [CrossRef]
9. Thampi, A.D.; Prasanth, M.A.; Anandu, A.P.; Sneha, E.; Sasidharan, B.; Rani, S. The effect of nanoparticle additives on the tribological properties of various lubricating oils—Review. *Mater. Today Proc.* **2021**, *47*, 4919–4924. [CrossRef]
10. Guo, Z.; Zhang, Y.; Wang, J.; Gao, C.; Zhang, S.; Zhang, P.; Zhang, Z. Interactions of Cu nanoparticles with conventional lubricant additives on tribological performance and some physicochemical properties of an ester base oil. *Tribol. Int.* **2020**, *141*, 105941. [CrossRef]
11. Chen, X.; Han, Z.; Li, X.; Lu, K. Lowering coefficient of friction in Cu alloys with stable gradient nanostructures. *Sci. Adv.* **2016**, *2*, e1601942. [CrossRef] [PubMed]
12. Chen, X.; Han, Z.; Lu, K. Enhancing wear resistance of Cu–Al alloy by controlling subsurface dynamic recrystallization. *Scr. Mater.* **2015**, *101*, 76–79. [CrossRef]
13. Alves, S.M.; Mello, V.S.; Faria, E.A.; Camargo, A.P.P. Nanolubricants developed from tiny CuO nanoparticles. *Tribol. Int.* **2016**, *100*, 263–271. [CrossRef]
14. Atila Dinçer, C.; Yıldız, N.; Aydoğan, N.; Çalımlı, A. A comparative study of Fe$_3$O$_4$ nanoparticles modified with different silane compounds. *Appl. Surf. Sci.* **2014**, *318*, 297–304. [CrossRef]
15. Luo, T.; Wei, X.; Huang, X.; Huang, L.; Yang, F. Tribological properties of Al$_2$O$_3$ nanoparticles as lubricating oil additives. *Ceram. Int.* **2014**, *40*, 7143–7149. [CrossRef]
16. He, A.; Huang, S.; Yun, J.H.; Wu, H.; Jiang, Z.; Stokes, J.; Jiao, S.; Wang, L.; Huang, H. Tribological Performance and Lubrication Mechanism of Alumina Nanoparticle Water-Based Suspensions in Ball-on-Three-Plate Testing. *Tribol. Lett.* **2017**, *65*, 40. [CrossRef]
17. Ingole, S.; Charanpahari, A.; Kakade, A.; Umare, S.S.; Bhatt, D.V.; Menghani, J. Tribological behavior of nano TiO$_2$ as an additive in base oil. *Wear* **2013**, *301*, 776–785. [CrossRef]
18. Wu, H.; Zhao, J.; Xia, W.; Cheng, X.; He, A.; Yun, J.H.; Wang, L.; Huang, H.; Jiao, S.; Huang, L.; et al. A study of the tribological behaviour of TiO$_2$ nano-additive water-based lubricants. *Tribol. Int.* **2017**, *109*, 398–408. [CrossRef]

19. Kong, L.; Sun, J.; Bao, Y.; Meng, Y. Effect of TiO$_2$ nanoparticles on wettability and tribological performance of aqueous suspension. *Wear* **2017**, *376*, 786–791. [CrossRef]
20. Gara, L.; Zou, Q. Friction and Wear Characteristics of Oil-Based ZnO Nanofluids. *Tribol. Trans.* **2013**, *56*, 236–244. [CrossRef]
21. Javed, R.; Usman, M.; Tabassum, S.; Zia, M. Effect of capping agents: Structural, optical and biological properties of ZnO nanoparticles. *Appl. Surf. Sci.* **2016**, *386*, 319–326. [CrossRef]
22. Peng, D.X.; Chen, C.H.; Kang, Y.; Chang, Y.P.; Chang, S.Y. Size effects of SiO$_2$ nanoparticles as oil additives on tribology of lubricant. *Ind. Lubr. Tribol.* **2010**, *62*, 111–120. [CrossRef]
23. Kumar, R.S.; Sharma, T. Stability and rheological properties of nanofluids stabilized by SiO$_2$ nanoparticles and SiO$_2$-TiO$_2$ nanocomposites for oilfield applications. *Colloids Surf. A Physicochem. Eng. Asp.* **2018**, *539*, 171–183. [CrossRef]
24. Bao, Y.; Sun, J.; Kong, L. Effects of nano-SiO$_2$ as water-based lubricant additive on surface qualities of strips after hot rolling. *Tribol. Int.* **2017**, *114*, 257–263. [CrossRef]
25. Wu, H.; Johnson, B.; Wang, L.; Dong, G.; Yang, S.; Zhang, J. High-efficiency preparation of oil-dispersible MoS$_2$ nanosheets with superior anti-wear property in ultralow concentration. *J. Nanoparticle Res.* **2017**, *19*, 339. [CrossRef]
26. Forsberg, V.; Zhang, R.; Bäckström, J.; Dahlström, C.; Andres, B.; Norgren, M.; Andersson, M.; Hummelgård, M.; Olin, H. Exfoliated MoS$_2$ in Water without Additives. *PLoS ONE* **2016**, *11*, e0154522. [CrossRef]
27. Aldana, P.U.; Vacher, B.; Le Mogne, T.; Belin, M.; Thiebaut, B.; Dassenoy, F. Action Mechanism of WS$_2$ Nanoparticles with ZDDP Additive in Boundary Lubrication Regime. *Tribol. Lett.* **2014**, *56*, 249–258. [CrossRef]
28. Loya, A.; Stair, J.L.; Ren, G. Simulation and experimental study of rheological properties of CeO$_2$–water nanofluid. *Int. Nano Lett.* **2015**, *5*, 1–7. [CrossRef]
29. Boshui, C.; Kecheng, G.; Jianhua, F.; Jiang, W.; Jiu, W.; Nan, Z. Tribological characteristics of monodispersed cerium borate nanospheres in biodegradable rapeseed oil lubricant. *Appl. Surf. Sci.* **2015**, *353*, 326–332. [CrossRef]
30. Shahnazar, S.; Bagheri, S.; Abd Hamid, S.B. Enhancing lubricant properties by nanoparticle additives. *Int. J. Hydrogen Energy* **2016**, *41*, 3153–3170. [CrossRef]
31. Chakraborty, S.; Panigrahi, P.K. Stability of nanofluid: A review. *Appl. Therm. Eng.* **2020**, *174*, 115259. [CrossRef]
32. Yu, F.; Chen, Y.; Liang, X.; Xu, J.; Lee, C.; Liang, Q.; Tao, P.; Deng, T. Dispersion stability of thermal nanofluids. *Prog. Nat. Sci. Mater. Int.* **2017**, *27*, 531–542. [CrossRef]
33. Wu, W.; Liu, J.; Li, Z.; Zhao, X.; Liu, G.; Liu, S.; Ma, S.; Li, W.; Liu, W. Surface-Functionalized NanoMOFs in Oil for Friction and Wear Reduction and Antioxidation. *Chem. Eng. J.* **2021**, *410*, 128306. [CrossRef]
34. Zhao, J.; Huang, Y.; He, Y.; Shi, Y. Nanolubricant additives: A review. *Friction* **2021**, *9*, 891–917. [CrossRef]
35. Wu, H.; Zhao, J.; Cheng, X.; Xia, W.; He, A.; Yun, J.H.; Huang, S.; Wang, L.; Huang, H.; Jiao, S.; et al. Friction and wear characteristics of TiO$_2$ nano-additive water-based lubricant on ferritic stainless steel. *Tribol. Int.* **2018**, *117*, 24–38. [CrossRef]
36. Guo, J.; Barber, G.C.; Schall, D.J.; Zou, Q.; Jacob, S.B. Tribological properties of ZnO and WS$_2$ nanofluids using different surfactants. *Wear* **2017**, *382*, 8–14. [CrossRef]
37. Sonn, J.S.; Lee, J.Y.; Jo, S.H.; Yoon, I.-H.; Jung, C.-H.; Lim, J.C. Effect of surface modification of silica nanoparticles by silane coupling agent on decontamination foam stability. *Ann. Nucl. Energy* **2018**, *114*, 11–18. [CrossRef]
38. Kang, T.; Jang, I.; Oh, S.-G. Surface modification of silica nanoparticles using phenyl trimethoxy silane and their dispersion stability in N-methyl-2-pyrrolidone. *Colloids Surf. A Physicochem. Eng. Asp.* **2016**, *501*, 24–31. [CrossRef]
39. Man, W.; Huang, Y.; Gou, H.; Li, Y.; Zhao, J.; Shi, Y. Synthesis of novel CuO@Graphene nanocomposites for lubrication application via a convenient and economical method. *Wear* **2022**, *498–499*, 204323. [CrossRef]
40. Rodriguez, R.; Herrera, R.; Archer, L.A.; Giannelis, E.P. Nanoscale Ionic Materials. *Adv. Mater.* **2008**, *20*, 4353–4358. [CrossRef]
41. Rodriguez, R.; Herrera, R.; Bourlinos, A.B.; Li, R.; Amassian, A.; Archer, L.A.; Giannelis, E.P. The synthesis and properties of nanoscale ionic materials. *Appl. Organomet. Chem.* **2010**, *24*, 581–589. [CrossRef]
42. Jespersen, M.L.; Mirau, P.A.; von Meerwall, E.; Vaia, R.A.; Rodriguez, R.; Giannelis, E.P. Canopy Dynamics in Nanoscale Ionic Materials. *ACS Nano* **2010**, *4*, 3735–3742. [CrossRef] [PubMed]
43. Li, P.; Zheng, Y.; Wu, Y.; Qu, P.; Yang, R.; Wang, N.; Li, M. A nanoscale liquid-like graphene@Fe$_3$O$_4$ hybrid with excellent amphiphilicity and electronic conductivity. *New J. Chem.* **2014**, *38*, 5043–5051. [CrossRef]
44. Huang, J.; Li, Y.; Jia, X.; Song, H. Preparation and tribological properties of core-shell Fe$_3$O$_4$@C microspheres. *Tribol. Int.* **2019**, *129*, 427–435. [CrossRef]
45. Min, C.; He, Z.; Song, H.; Liang, H.; Liu, D.; Dong, C.; Jia, W. Fluorinated graphene oxide nanosheet: A highly efficient water-based lubricated additive. *Tribol. Int.* **2019**, *140*, 105867. [CrossRef]
46. He, D.; Peng, Z.; Gong, W.; Luo, Y.; Zhao, P.; Kong, L. Mechanism of a green graphene oxide reduction with reusable potassium carbonate. *RSC Adv.* **2015**, *5*, 11966–11972. [CrossRef]
47. Fernandes, N.; Dallas, P.; Rodriguez, R.; Bourlinos, A.B.; Georgakilas, V.; Giannelis, E.P. Fullerol ionic fluids. *Nanoscale* **2010**, *2*, 1653–1656. [CrossRef] [PubMed]
48. Zhang, J.; Li, P.; Zhang, Z.; Wang, X.; Tang, J.; Liu, H.; Shao, Q.; Ding, T.; Umar, A.; Guo, Z. Solvent-free graphene liquids: Promising candidates for lubricants without the base oil. *J. Colloid Interface Sci.* **2019**, *542*, 159–167. [CrossRef] [PubMed]
49. Kong, L.; Sun, J.; Bao, Y. Preparation, characterization and tribological mechanism of nanofluids. *RSC Adv.* **2017**, *7*, 12599–12609. [CrossRef]

50. Ali, M.K.A.; Xianjun, H.; Turkson, R.F.; Peng, Z.; Chen, X. Enhancing the thermophysical properties and tribological behaviour of engine oils using nano-lubricant additives. *RSC Adv.* **2016**, *6*, 77913–77924.
51. Azman, N.F.; Samion, S. Dispersion Stability and Lubrication Mechanism of Nanolubricants: A Review. *Int. J. Precis. Eng. Manuf. Green Technol.* **2019**, *6*, 393–414. [CrossRef]
52. Ranga Babu, J.A.; Kumar, K.K.; Srinivasa Rao, S. State-of-art review on hybrid nanofluids. *Renew. Sustain. Energy Rev.* **2017**, *77*, 551–565. [CrossRef]
53. Hao, L.; Wang, Z.; Zhang, G.; Zhao, Y.; Duan, Q.; Wang, Z.; Chen, Y.; Li, T. Tribological evaluation and lubrication mechanisms of nanoparticles enhanced lubricants in cold rolling. *Mech. Ind.* **2020**, *21*, 108–113. [CrossRef]

 *lubricants*

Article

Lubrication Mechanisms of a Nanocutting Fluid with Carbon Nanotubes and Sulfurized Isobutylene (CNTs@T321) Composites as Additives

Jiju Guan [1,*], Chao Gao [1], Zhengya Xu [1], Lanyu Yang [1] and Shuiquan Huang [2]

[1] College of Mechanical Engineering, Changshu Institute of Technology, Suzhou 215500, China
[2] School of Mechanical Engineering, Yanshan University, Qinhuangdao 066004, China
*   Correspondence: guanjiju@cslg.edu.cn

Abstract: Developing high-efficiency lubricant additives and high-performance green cutting fluids has universal significance for maximizing processing efficiency, lowering manufacturing cost, and more importantly reducing environmental concerns caused by the use of conventional mineral oil-based cutting fluids. In this study, a nanocomposite is synthesized by filling sulfurized isobutylene (T321) into acid-treated carbon nanotubes (CNTs) with a liquid-phase wet chemical method. The milling performance of a nanocutting fluid containing CNTs@T321 composites is assessed using a micro-lubrication technology in terms of cutting temperature, cutting force, tool wear, and surface roughness. The composite nanofluid performs better than an individual CNT nanofluid regarding milling performance, with 12%, 20%, and 15% reductions in the cutting force, machining temperature, and surface roughness, respectively. The addition of CNTs@T321 nanocomposites improves the thermal conductivity and wetting performance of the nanofluid, as well as produces a complex lubricating film by releasing T321 during machining. The synergistic effect improves the cutting state at the tool–chip interface, thereby resulting in improved machining performance.

Keywords: carbon nanotube; composites; nanofluids; milling; lubrication mechanism

Citation: Guan, J.; Gao, C.; Xu, Z.; Yang, L.; Huang, S. Lubrication Mechanisms of a Nanocutting Fluid with Carbon Nanotubes and Sulfurized Isobutylene (CNTs@T321) Composites as Additives. *Lubricants* **2022**, *10*, 189. https://doi.org/10.3390/lubricants10080189

Received: 22 July 2022
Accepted: 14 August 2022
Published: 19 August 2022

Publisher's Note: MDPI stays neutral with regard to jurisdictional claims in published maps and institutional affiliations.

1. Introduction

Milling is an important processing method in mechanical manufacturing. Modern milling technology has increasingly higher requirements for efficiency, quality, and environmental protection. In the high-speed milling of difficult-to-machine materials, the reasonable selection of cooling and lubrication methods can improve the tool/workpiece friction state and suppress tool wear, which has become a key factor to be considered in cutting process design [1]. Cutting fluid is mainly composed of a base fluid and oily agent, an extreme pressure agent, emulsifier, rust inhibitor, etc., and has the functions of cooling, lubricating, and cleaning. The cost of cutting fluid accounts for 7–17% of the part processing costs, while the cost of cutting tools accounts for 2–4%, according to the investigation in [2]. Various harmful additives appended in cutting fluid have caused many problems, such as environmental pollution, resource consumption, and harm to human health [3]. As a result, continuously developing high-efficiency lubricating additives and new green cutting fluid has universal meaning for maximizing processing efficiency, lowering manufacturing costs, and reducing environmental pollution.

Choi et al. proposed the concept of nanofluid in 1995, which is a suspension formed by dispersing various nanoparticles into a base fluid in a certain proportion [4]. Nanoparticles can put a positive spin on the heat transfer and lubrication performance of the base fluid. As nanotechnology has introduced a new era of rapid development, researchers have recently begun to use nanofluids to achieve efficient cooling and nanotechnology. Nanofluid possesses better thermal conductivity, lower cutting temperature and better tribological conductivity, which can upgrade the lubrication performance of the base fluid compared to traditional cutting fluid [5,6]. Therefore, nanofluid-based milling processing technology, as an environmentally friendly machining method that achieves cooling and lubrication effects, has good prospects in terms of application. Lee P H and Nam T S used MoS$_2$ nanofluid in face milling to prove that the use of MoS$_2$ nanofluid in minimal quantity lubrication (MQL) processing could minimize the cutting force and achieve a better surface quality of a workpiece than the use of conventional cooling methods and cold air cutting [7]. Rahamti et al. used MoS$_2$ nanofluid to mill an A16061-T6 alloy in a vertical machining center and studied its surface morphology, and the concentration of nanoparticles in basic oil were 0.0%, 0.2%, 0.5% and 1.0 wt%, respectively. The result showed that the surface quality of workpieces was enhanced in regard to the rolling, filling, and polishing of nanoparticles in the processing area. When the concentration of MoS$_2$ nanoparticles was 0.5 wt%, the surface quality was optimal, while the surface quality declined above 1 wt% [8].

Marcon A and Melkote S prepared water-based graphite nanofluid for the micro-milling of steel H13 under MQL conditions. The results indicated that nanofluid equipped with cooling and lubrication functions could minimize the milling force and had a good polishing effect on the workpiece [9]. Sarah AAD et al. prepared an oil-based nanocutting fluid of onion-like fullerene C60 and carried out research on milling hard aluminum alloy aviation parts. This illustrated that the addition of C60 could reduce the milling force by 21.99% and improve the surface roughness by 46.32%. They believe that C60 acted as a nanoball at the cutting friction interface, thus reducing the cutting force and the tool wear [10]. Sayuti identified the surface morphology of milling A16061-T6 under new SiO$_2$ nanofluid lubrication, and the concentrations of SiO$_2$ were 0.0, 0.2, 0.5, and 1.0 wt%. The study showed that the increase in the concentration of SiO$_2$ nanoparticles promoted the growth of the tool–chip interface protective film, and the film that formed on the machined surface improved the surface quality of the workpiece and reduced the cutting temperature and cutting force [11]. Long T M et al. used MoS$_2$ nanosheet nanofluid and Al$_2$O$_3$/MoS$_2$ hybrid nanofluid in minimum quantity cooling lubrication (MQCL) hard milling, and the obtained results indicated that better cooling/better effects and good surface quality were achieved [12,13].

Carbon nanotubes (CNTs) are seamless nanotubes formed by monolayer or multi-layer graphite sheets that centrally rotate at a particular helical angle. Every layer of the nanotubes—which are in the shape of a cylinder—consists of a flat hexagon carbon atom attached to three carbon atoms with sp2 hybridization [14]. CNTs are an excellent thermally conductive material with good lubrication performance. Their thermal conductivity can reach 6000 W/mK [15], which can improve the heat transfer property of the base fluid and play a similar role in the friction areas of "micro-bearings" [16,17]. Therefore, CNTs are also more suitable as cutting fluid additives. Puneet Sharma et al. used carbon nanotubes to prepare oil-based nanocutting fluids and carried out a study on turning AISI D2 (American Iron and Steel Institute D2) steel under MQL conditions. They identified that, under the action of CNTs, the cutting temperature decreased significantly, the quality of the surface was enhanced, and the tool wear was minimized [18]. The turning of AISI 1040 steel under the lubrication of CNTs nanofluid was analyzed by Rao S N, and the cutting temperature and cutting force were measured, indicating that while the CNT content was below 2%, the temperature and tool wear could be reduced [19]. Sougata Roy et al. studied the turning performance of Al$_2$O$_3$ and CNT nanocutting fluids, proving that CNT nanofluids could significantly improve roughness under high-speed turning conditions [20].

As an additive to nanofluids, CNTs have some limitations due to their molecular structure. For instance, they tend to agglomerate and form sediment in the base fluid where the dispersion stability performs inferiorly, incurring the instability of the prepared nanofluid [21]. Though used as a lubricating additive for the deficiency of lubricous molecular groups between molecules, the lubricating efficacy under severe friction conditions is limited. However, CNTs have a special cavity structure with a cavity between 5 and 50 nm. Other substances are permitted to be infused into the cavities of CNTs to structure a composite under suitable conditions [22,23]. To prepare various composites, various lubricants are filled into CNT cavities, as proposed by the research group. With the introduction of these lubricating compositions into the CNT molecules, the internal filling method improves its lubricating properties compared to the external chemical modification method. In the filling phase, the "end cap" of the CNTs needs to be opened before filling, and hydroxyl and carboxyl groups are bonded to the surface of CNTs at the same time as opening the "end cap" [24,25], fortifying CNTs' dispersion stability in basic solution. In this paper, the team first fill a typical extreme pressure additive isobutylene sulfide (T321) into CNTs to prepare a CNTs@T321 composite and use it to prepare nanocutting fluid samples and carry out milling processing tests. In the area of cutting, the essence of force and temperature, the workpiece surface roughness and tool wear are comparatively studied, and the machinery of the composite is analyzed. The research findings provide the theoretical basis for the application of CNT nanocutting fluids for milling flat and thin plate parts.

2. Materials and Methods

2.1. Preparation and Characterization of CNTs@T321 Composites

CNTs were purchased from Shanghai Aladdin Reagent Company with 99.9% purity and an inside diameter of 20–50 nm. On account of the large slenderness ratio of these CNTs, which is unfavorable for filling, the CNTs were acidified and shortened first. At the time of acid treating, we first put 20 g of CNTs into 800 mL acid treatment solution (consisting of nitric acid, hydrochloric acid and sulphuric acid; the volume ratio was 3:1:1, with water subdilution 1:1). Then, placing the mixture in a three-necked flask and heating the reflux at 85 °C for 12 h, its magnetic stirring was utilized at a rate of 500 r/min. Then, it was filtered by vacuum and fried at 80 °C; it was then ball grinded for 6 h and acid-treated CNTs were acquired.

Sulfurized isobutylene (T321) was provided by Kangtai Lubricant Additive Co., Ltd. We first dissolved 30 g of T321 with 500 mL of acetone, and then added 90 g acid-treated CNTs. The mixture was placed in a bottle evacuating into a −0.06 Mpa vacuum, and the ultrasonic treatment was applied for 8 h. Then, the filter cake was collected by suction filtration and repeatedly cleaned with acetone to eliminate the T321 that had not been filled, and then the resultant was filtered and the cake was dried at 80 °C. The composites could be obtained by 8 h of ball milling. FEI TECNAI G20 TEM (FEI Company, Hillsboro, OR, USA) was used to observe the microstructure of the nanoparticles, with a 200 VK acceleration voltage in the test. A TG 6300 comprehensive thermal analyzer (Seiko Instruments Inc-SII, Nagano, Japan) was used to test the thermal weight loss characteristic of the composite with a temperature rise of 100–800 °C at a heating rate of 20 °C/min.

2.2. Preparation and Performance Testing of Nanofluids

For nanofluid preparation, surfactants served as dispersers. In this paper, through the preliminary test, a mixture of Tween-80 (TW-80) and dodecylbenzene sulfonate (SDBS) (with mass ratio of 6:4) was used as the surface dispersant. The formulation of water-based nanofluids can be seen in Table 1 below. The nanofluid preparation steps included mixing, mechanically stirring at 50 °C for 30 min, and ultrasonically dispersing for 1 h. After a complete preparation, there were tests for key performances during which three groups were applied to each sample:

Table 1. Formulation of the nanocutting fluids.

Ingredient	Mass Content	Function
Deionization water	97.3 wt%–98.4 wt%	Base Fluid
MWCNTs/composites	0.1 wt%–1.2 wt%	Lubricant additives
Tw-80/SDBS surfactant	0.5 wt%	Activator
Triethanolamine	0.5 wt%	Antirust

The first was the dispersing stability test. We kept every group stationary for 30 days, took 1 mL of the supernatant liquid daily and diluted it by 2 times, and then put it into a cuvette and used a 752 N ultraviolet–visible light photometer to test the absorbency performance of nanofluids at a 500 nm wavelength. The absorbency stability value was used as the basis for estimating the dispersing stability of the nanofluids.

The second was thermal conductivity test where the thermal conductivity of each sample was measured by using a TC3010 L thermal conductivity meter (Xi'an Xiaxi Electronic Technology Co., Ltd., Xi'an, China). Each injection was around 30 mL. After the completion of the sample, the instrument automatically determined the thermal conductivity.

Finally, the wettability of the nanofluid was evaluated based on C45 steel surface's contact angle, with measurements from the Kruss DSA25 (KRÜSS Scientific, Hamburg, Germany) instrument. The injected quantity of every drop in the test was 2 μL. For a stable 15 s contact with the surface of the C45 steel, the droplets were in optimum display status throughout the movement test bench and measurement reading.

2.3. Milling Test

The milling test was carried out at the Fadal VMC 3016 L machining center (Fadal Machining Centers, LLC, Murray, GA, USA), as shown in Figure 1a. C45 steel was used as the machining workpiece with the dimensions of 120 mm × 120 mm × 80 mm for the rectangular blank; the tool was a TiAlN-coated milling cutter for the test and the milling insert specification was APMT1135PDER-M2/H2 (Chengdu Bangpu Cutting Tools Co., Ltd., Chengdu, China). We used a Kistler 9129A dynamometer (Kistler, Barcelona, Spain) to measure the cutting force, which is a criterion for the resultant force of a three-way cutting force. An RX4006D thermometer (Hangzhou Meacon Automation technology Co., Ltd., Hangzhou, China) was employed to measure the cutting temperature, and a K-type thermocouple (Elitech Co., Ltd., Xuzhou, China) with a diameter of 0.5 mm was buried in the workpiece with small holes, recording the temperature when the machining state was stable.

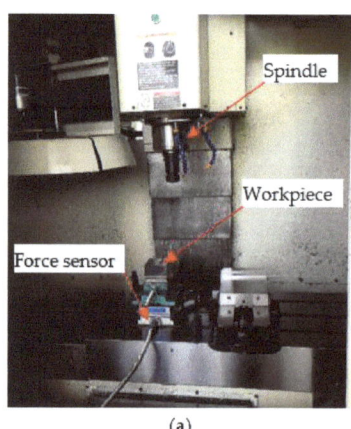 

(a)   (b)

Figure 1. The milling test: (**a**) equipment used in the test; (**b**) tool being cut.

First, we investigated the cutting performance of common emulsion, acid-treated CNT nanofluid, and composite nanofluid under MQL conditions, as shown in the A1–A3 groups in Table 2. Second, the milling performance of the nanofluids was investigated when the content of the nanoparticles was 0.02%, 0.04%, 0.06%, 0.08%, 0.1%, 0.3%, 0.5%, 0.7%, 0.9%, and 1.1 wt% under MQL conditions, as shown in the B1–B10 and C1–C10 groups. Moreover, a series of tests of dry cutting in identical cutting conditions were executed, in which the spindle speed was 5000 r/min, the back cutting depth a_p was 0.5 mm, the side cutting depth a_e was 1 mm, the feed per tooth f_z was 0.05 mm, and the total cutting thickness was 10 mm. We performed three parallel tests for each group of nanofluids. A surface roughness measuring instrument (Beijing Time Yuanfeng Technology Co., Ltd., Beijing, China) was used to measure the surface quality of the workpiece, and the Nikon microscope (Nikon Corporation, Tokyo, Japan) was employed to observe the wear of the tool (*VB*).

Table 2. The cutting test scheme.

Groups	Fluids	Cooling Methods	MQL Conditions
A1	Emulsion (1 wt%)		
A2	CNTs nanofluids (0.1 wt%)	MQL	Pressure: 0.5 MPa
A3	Composites nanofluids (0.1 wt%)		Rate: 40 mL/h
B1–B10	CNTs nanofluids (0.02 wt%–1.1 wt%)	MQL	Spray distance:
C1–C10	Composites nanofluids (0.02 wt%–1.1 wt%)		10 mm

A metastable lubricating layer may form on the tool face during the milling process, and it is necessary to analyze the composition of the surface lubricating layer. The tools were cut for the analysis under the conditions of dry cutting and emulsifying, CNT nanofluids, and composite nanofluid lubrication, and the thickness of the piece was roughly 1 mm, as shown in Figure 1b. X-ray photoelectron spectroscopy (XPS), an analysis of the wear area blinding energy of the essential elements, used the internal standard of 80 eV for the electron flux and 284.8 eV for the binding energy of the contaminated carbon C_{1s}.

3. Results and Discussion

3.1. Characterization of Composites

Figure 2 displays the transmission electron microscopy (TEM) images of the CNTs and composites. Figure 2a shows the hollow tubular structure of the CNTs. Figure 2b presents that several segments of the CNTs@T321 composite sample are filled with T321 in the observed area, proving a successful filling of T321 in the CNTs on the microstructure. The dominating driving force of capillary action for the filling of T321 molecules into the CNT cavity proves that an adequately large force exists in the filler and the nanotube, which can wet each other [26]. Formula (1) is derived from Laplace Equation (2) which gives the relationship of the surface tension and the contact angle at the liquid–solid interface with the pressure difference at the gas–liquid interface [27]:

$$cos\theta = (\gamma_{sV} - \gamma_{sL})/\gamma \tag{1}$$

$$\Delta P = 2\gamma cos\theta / r \tag{2}$$

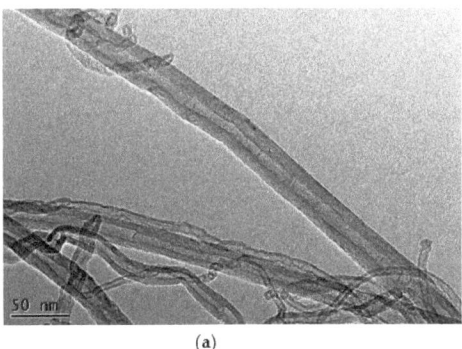

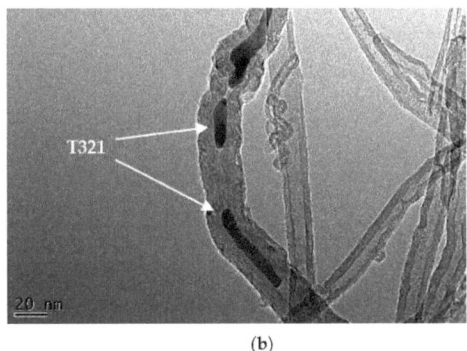

(a)   (b)

Figure 2. TEM images of nanoparticles: (**a**) CNTs; (**b**) composites.

In the formulas, θ is the contact angle of the liquid–solid interface, γ_{sV} is the surface tension of the solid–air interface, γ_{sL} is the surface tension of the solid–liquid interface, γ is the liquid surface tension, and r is the radius of curvature. The determinant of the wetting of CNTs by lubricants is the contact angle θ as capillary action takes place. When $\theta > 90°$, the pressure difference ΔP at the gas–liquid interface is negative, with no wetting occurring. When $\theta < 90°$, only then can infiltration and filling occur. Substances such as water, ethanol, acids, and so forth, with surface tension below 100–200 mN/m are capable of filling in CNTs in specific conditions [28]. The surface tension of T321 is around 97 mN/m, which is further reduced when dissolved in acetone, making it easier to enter CNTs by capillary action. The subsequent drying process evaporates anhydrous ethanol while the T321 in the tube is retained.

In addition, thermal analysis was performed on CNTs, acid-treated CNTs and CNTs@T321 composites, as given in Figure 3. Figure 3a demonstrates that the thermal weight loss properties of CNTs and acidified CNTs were similar with a similar thermal weight loss procedure. However, the thermal weight loss curves of CNT composites had a distinct loss process at a lower temperature as the result of T321 escaping from the object during the heating process. This also proves that T321 exists in CNTs. In accordance with the formula of the latent heat of phase change, the T321 fill rate can be calculated by the following mathematical calculation [29]:

$$\eta = \frac{H_f}{H_p} \times 100\% \qquad (3)$$

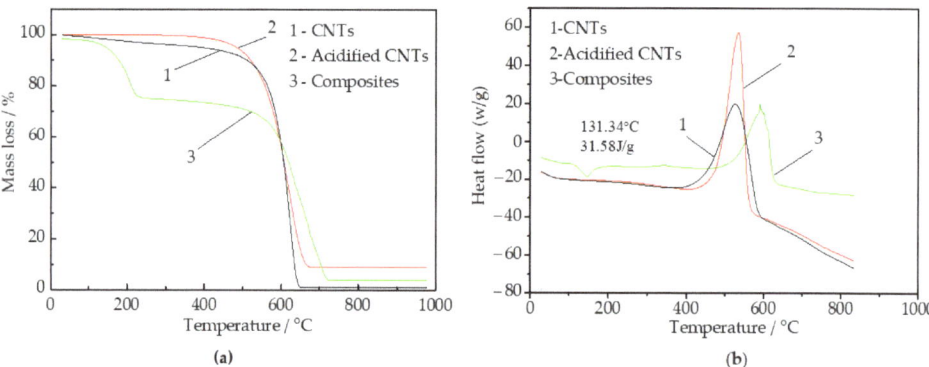

Figure 3. Thermal analysis of CNTs, acidified CNTs and the nanocomposites: (**a**) TG; (**b**) DSC.

In this formula, H_f is the phase change heat of T321 that fills in CNTs, J/g, and H_p is the phase change heat of T321 itself, J/g. As the composite in the differential scanning calorimetry (DSC) curve illustrates in Figure 3b, the calculated result of the phase change heat of T321 that fills in CNTs is 31.58 J/g, and the phase change heat of T321 itself is 123.24 J/g under the same heating conditions, so it can be calculated that the filling rate of T321 in CNTs is 25.6%. In addition, the acetone in the filtrate extracted during the composite was evaporated to dryness, and the mass of the remaining T321 was weighed. It can be calculated that the filling rate of the carbon tube was also about 25%, which was basically consistent with the DSC analysis result.

3.2. Physical Performance Test Results of Nanofluids

Figure 4a shows nanofluids with various composite contents, and Figure 4b shows the scanning electron microscope(SEM) image of the nanofluid. It can be seen that the composite had no agglomeration and may have stable dispersion performance. Figure 5a shows the effect of the acid-treated CNTs and composite content on the absorbance of the nanofluids. It serves to show that a gradual increase in absorbance follows the increase in content with the condition of small nanoparticle content. When the content of the carbon tubes reached a roughly "saturated" 0.3% content, there was an unapparent increase in absorbance, attributed to the deposition of carbon nanotubes that were not dispersed. The dispersion stability of the composite under the same content was better than that of the acid-treated CNTs, which should have been due to the exposure of part of the T321 molecular groups outside the tube. These groups can be combined with surfactants in order to more deeply reinforce the effect of dispersion [30]. Figure 5b illustrates the influence of nanoparticle content on thermal conductivity. The thermal conductivity increases with the increase in content while there is low content, and the base fluid thermal conductivity can be increased by 105% at most. When the content continues increasing, there is a relatively slow increasing trend of thermal conductivity owing to the agglomeration and sedimentation caused by plus nanoparticles with excessive content, which contributes to the decrease in nanoparticles with an obvious effect on heat transfer.

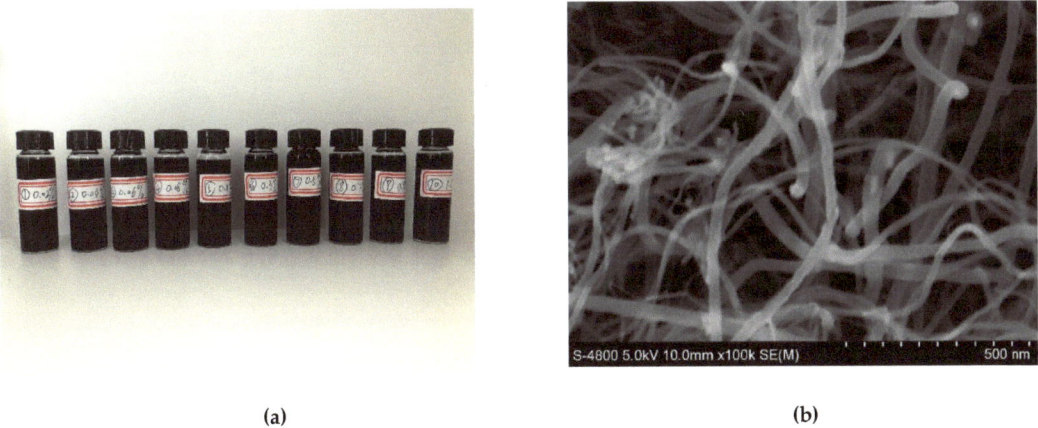

(a) (b)

Figure 4. Preparation and SEM image of the nanofluids: (**a**) nanofluid samples; (**b**) SEM image of nanofluid.

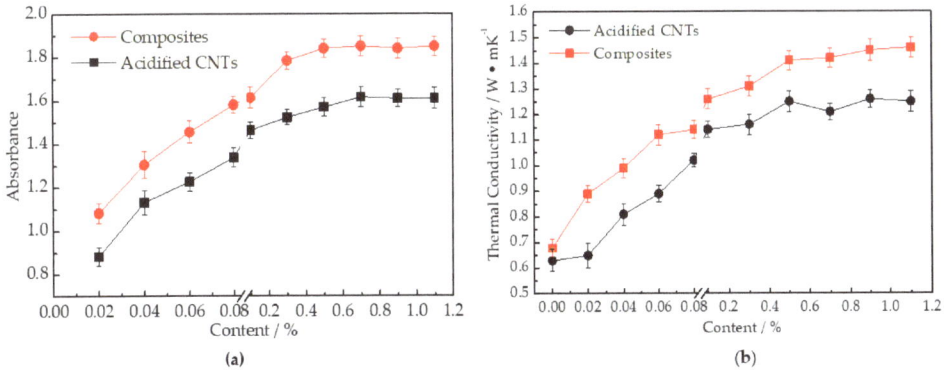

(a) (b)

Figure 5. Effects of nanoparticle content on: (**a**) absorbance; (**b**) thermal conductivity.

Figure 6a demonstrates the effect of the content on the nanofluids' wettability when the content of surfactant is 0.5%. Where there is a decrease first, there is an increase later to the contact angle of nanofluids following the increasing nanoparticle content as perceived. Since definite surface activity may be observed with CNTs after treatment with acid, when they are thoroughly separated in the base fluid, which molecules are repulsed with the nanoparticles and the increasing molecular spacing leads to a reduction in surface tension [31]. In the event that there is an excessively high composite content, worse nanofluid dispersion and viscosity results in an increasing intermolecular force of components, which leads to an increase in contact angle. In addition, in a variety of composite contents, the performance of the composites, compared with those of the acid-treated CNTs in the aspects of wettability and thermal conductivity in relation to surface activity, is stronger and shows more stability in the base solution. Figure 6b shows the nanofluids' viscosity increasing continuously with the increase in content.

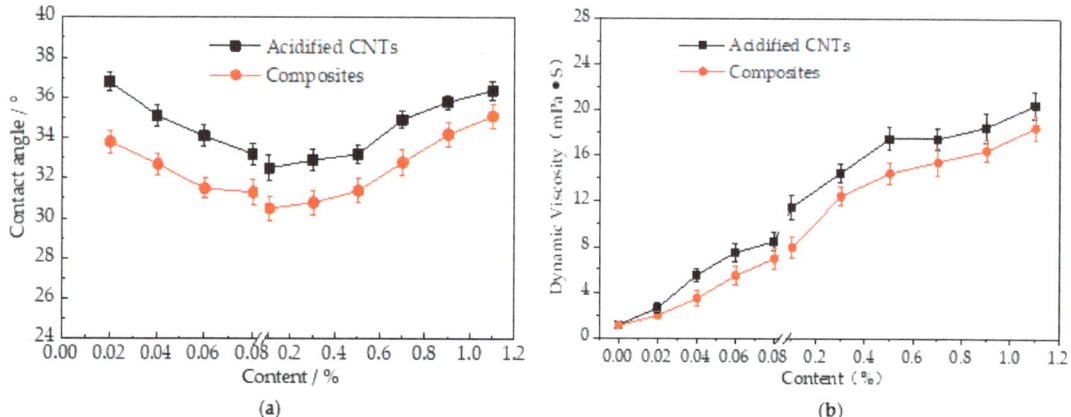

Figure 6. Effects of nanoparticles content on: (**a**) wettability; (**b**) viscosity.

3.3. Milling Performance of Different Fluids under MQL Conditions
3.3.1. Force and Temperature in the Cutting Area

The experiment in this paper uses MQL as the supply method of cutting fluid. This is mainly because the provision of cutting fluid presents in mist form in the conditions of MQL with high speed, pressure, and directionality during supply. Penetrating into the cutting area more conductively, the nanofluid has a better anti-friction effect [32,33]. Other studies show an unchanged or increased cutting force in the case of MQL, largely relating to the object of processing and cutting [34,35].

The cutting force is higher when emulsion and CNT nanofluids act in Table 3. An effect of lubrication is produced in the emulsion in which the anti-friction agent is capable of penetrating the cutting area. CNTs can also play a similar role as "micro-bearings" in the cutting area to achieve an effect of definite lubrication [36–38]. However, the molecular structures of CNTs lack effective lubricating groups, and its chemical properties are stable, making a hard reaction in the cutting area. Therefore, its cutting force is relatively high under the same cutting conditions. When the composite nanofluid acts, its cutting force is around 14% lower than that of the emulsion. This is mainly because the composite contains T321 inside, and the composite penetrates it at the time of the cutting process. Additionally, in that case, shear and destroy occur for the composite in the cutting area. The T321 molecules in the composites are released to the cutting area, where the composite is promoted for its anti-friction effect and a cutting force reduction during cutting.

Table 3 demonstrates a comparison between ordinary emulsions and a composite nanofluid whose cutting temperature is lower by almost 33%, chiefly owing to the good thermal conductivity of CNTs. It can be used as an additive that gives a remarkable improvement to the nanofluid's heat transfer capacity and plays an important role in the better heat conduction of the cutting area, which is the reason for the lower nanofluid temperature than the ordinary emulsion. The prepared composite nanofluid compared with the prepared acid-treated CNT nanofluid possesses a better cooling effect and a better wettability and thermal conductivity of the nanofluid prepared by the composite.

Table 3. Variation in cutting force (*F*), cutting temperature (*T*), surface roughness (*Ra*) and tool wear (*VB*) under different cutting fluids.

Cutting Fluids	Main Testing Items			
	F/N	*T*/°C	*Ra*/μm	*VB*/μm
1 wt% Emulsion	218	105	0.939	158
0.1 wt% CNTs	201	85	0.701	134
0.1 wt% Composites	185	71	0.674	105

3.3.2. Tool Wear and Surface Roughness

Table 3 also illustrates the change in the surface quality of the workpiece and the *VB* value of the tool under the action of various cutting fluids. It points out that the surface roughness under the action of the composite is reduced by 30% and the tool wear is reduced by 34% compared with the ordinary emulsion. It is the extraordinary anti-friction performance of the CNTs that partly separates the tool and workpiece for a cutting force reduction, and the definite polishing on the surface of the workpiece that reduces the surface roughness and retards the tool wear. Under the action of composite nanofluids, the workpiece has the best surface quality and retards the tool wear preferably, which is also due to the better permeability of the composite nanofluid which increases the nanometer entering into the cutting area. In the cutting process, the T321 molecule is released while destruction occurs to the structure of the composite molecule in the cutting area, which brings better lubrication to the composite to obtain a better cutting effect [39].

3.4. Effect of Concentration of Nanoparticle on Milling Properties

3.4.1. Force and Temperature in Cutting Area

The effectiveness of acid-treated CNTs, composite nanoparticle content and cutting area on the cutting force and cutting temperature is demonstrated in Figure 7. It can be seen that with the increase in nanoparticle content, the force and temperature of cutting first decrease and then slowly increase. This is the minimum force and temperature of cutting that can be acquired in the case of the nearly 0.1–0.3% content of the carbon tubes. The lower the nanoparticle content is, the fewer composite particles are in the cutting friction area. This means the relatively modest function of lubrication and heat conduction of the cutting fluid has a limited impact on friction and temperature, with rising improvement in the cutting area. Within a large proportion of nanoparticles, an unsteady dispersion is shown in the base fluid, and the agglomeration and precipitation of the composite are produced, reducing the actual number of nanoparticles that enter the cutting area, leading to the friction increase in the cutting area. Therefore, a more appropriate content of carbon tubes will have a better generation role. Under different contents, a superior effect of lubricating and cooling is generated in the nanofluid that is prepared by the composite. This is because the composite in the base fluid shows better dispersion stability and is more uniform, and consistent prepared nanofluids make for an easier penetration into the cutting area.

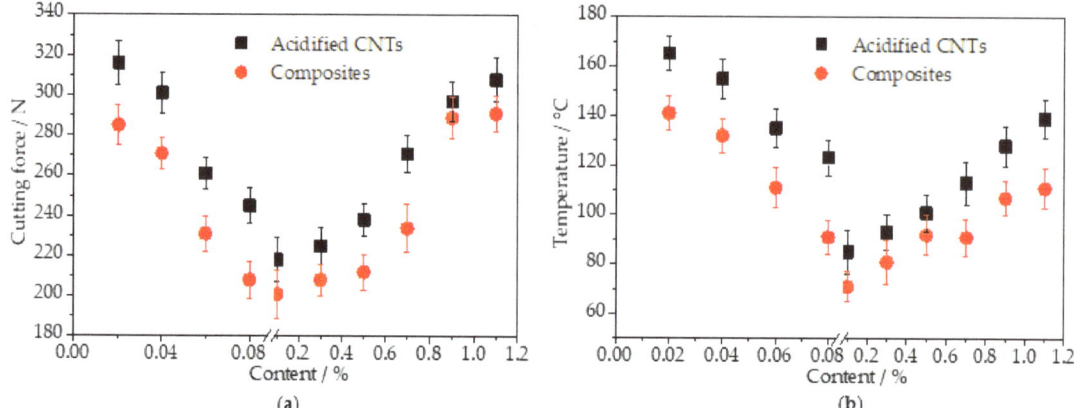

Figure 7. Effect of nanoparticles content: (**a**) cutting force; (**b**) cutting temperature.

3.4.2. Tool Wear and Surface Roughness of Workpiece

Figure 8 demonstrates the effect of nanoparticle content on the tool wear and the surface quality of the workpiece. It points out that with the increase in the content of nanoparticles, the surface roughness and the VB value of tool wear also show a tendency to first increase and then decrease. When the content of carbon tubes is about 0.3%, nanofluid can build up workpiece surface quality and slow down the tool wear. With a small proportion of nanoparticle content, there are fewer nanoparticles with limited effects of lubricating and polishing in the cutting area. Nanoparticle composites, with the agglomeration and precipitation that is caused by an excessively large proportion of them, result in fewer nanoparticles entering the cutting area. The nanoparticle decrease causes insufficient lubrication in the cutting area and also brings about the poor surface quality and increased tool wear. Compared with the nanofluid prepared by the acid-treated CNTs, the nanofluid prepared by the composite had a more obvious improvement impact on the tool wear and workpiece surface quality. Primarily, compared with ordinary CNT nanofluids, it is easier to penetrate the cutting processing area with composites. After entering the cutting processing area, nanoparticle composites can have a more sufficient lubrication effect, thereby leading to a reduction in tool wear and an improvement in workpiece surface quality. Next, in the process of cutting, destruction occurs to the composite's molecular structure, and a release of T321 molecules occurs at the cutting area, forming a more sufficient lubrication film. This results in a lower coefficient of friction than ordinary CNTs, and thus the composite has a more obvious lubrication effect on the cutting area, leading to an obvious reduction in tool wear and an improvement in the workpiece surface quality. Lastly, the composite nanofluid has a modified thermal conductivity compared to the CNT nanofluid, and the heat is more easily transferred out during processing, which reduces the temperature of the tool, protects the tool to a certain extent, and reduces wear.

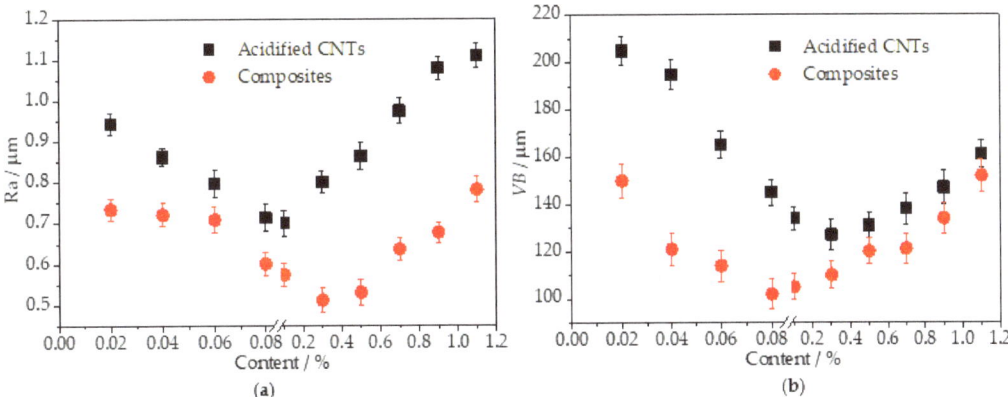

Figure 8. Effect of nanoparticle content: (**a**) surface roughness; (**b**) tool wear.

4. Discussion on the Lubrication Mechanism

An analysis has been given by some scholars about the determination of the composition of the stable lubricating film that forms on the worn tool surface [40]. This paper aims to find the major elements of the wear surface of the tool surface under the circumstances of dry cutting or emulsion; CNT nanofluids and composite nanofluids were analyzed by an XPS spectrum to try to give a determination of the process of forming the lubricating layer in the cutting area. C peak separation results of elements, O elements, and S elements are demonstrated in Figure 9. Measurements for every surface of the relative contents of elements C, O, and Fe are listed in Table 4. From Figure 9a,b, we can see that at the time of dry cutting, the C_{1s} and O_{1s} peak shapes on the wear surface of the tool are comparatively simple, mainly belonging to the contaminated carbon, Fe-C bond, and Fe-O bond element types. This surface does not form an effective lubricating layer during dry cutting.

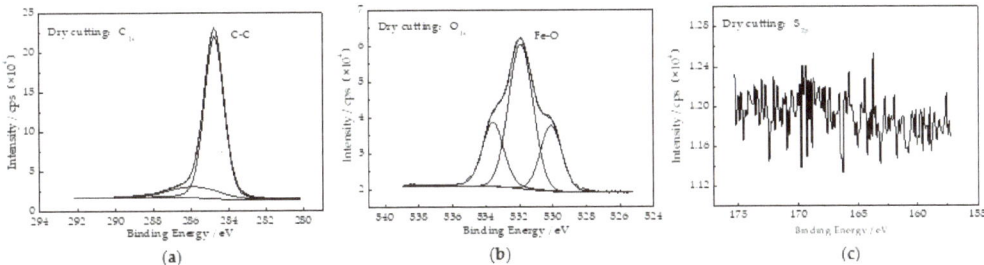

Figure 9. XPS analysis on the worn surfaces of flanks under the condition of dry cutting: (**a**) C_{1s}; (**b**) Fe_{2p}; (**c**) S_{2p}.

Table 4. Relative concentration of the main elements on wear surface of the tool.

Elements	Concentration of Main Elements			
	Dry Cutting	Emulsion	CNTs Nanofluids	Composite Nanofluids
C	24.41	41.34	49.47	55.11
O	43.57	40.05	32.69	35.97
Fe	32.02	18.61	17.84	8.92

As shown in Figure 10a,b, under the condition of emulsion lubrication, new peaks of C_{1s} and O_{1s} energy spectrum appear at 289.1 eV and 533.2 eV discriminably, basically belonging to C-O bond and C=O bond [41,42], which shows the effect of the lateral adsorption of the active lubricating components in the emulsion. Visible peaks of the S element appear at the tool wear surface with the micro-lubricating emulsion in comparison to the condition of dry cutting. This indicates that the emulsion is sulfur-added, which is related to the appearance of S_{2p} peaks at around 162.2 eV and 168.4 eV, showing that they remain with the Fe-S bond and S-O bond in several instances. The adsorption of active ingredients in the emulsion during the cutting is proven in Table 4, which illustrates a 70% content increase in C in comparison with the wear surface of the tool under the condition of dry cutting.

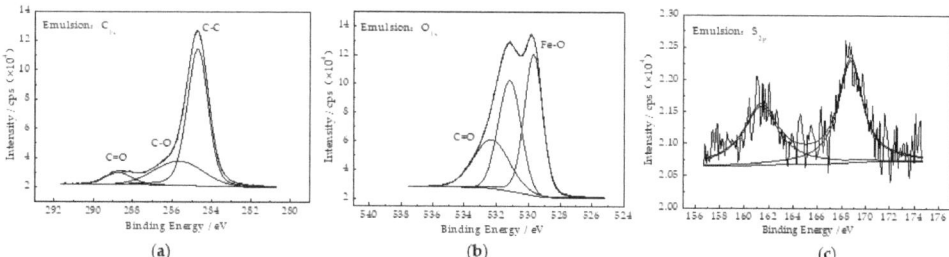

Figure 10. XPS analysis on the worn surfaces of flanks under the lubrication of emulsion: (a) C_{1s}; (b) Fe_{2p}; (c) S_{2p}.

As shown in Figure 11a,b, with the CNT nanofluid function, the C_{1s} and O_{1s} energy spectra of the flank wear area are split. It may be inferred that 284.6 eV, 285.8 eV, 288.2 eV and 291.2 eV are the energy spectra of four small peak shapes attributed to C_{1s}. Being part of the carbon species, the first three fitting peaks belong to C-C, C-O and C=O, comparatively, and the last one of 289.1 eV is more intricate, such as polymer-[C-C]n-, and so forth [42]. A small peak appears near 534.3 eV on the O_{1s} energy spectrum, proving that CNTs form a more complicated lubricating ingredient when acting. Table 4 gives a comparison with the worn surface of dry cutting; the content of the C element on the CNT nanocutting fluid surface has doubled, proving that CNTs also have components of lubrication that form in the cutting area. The "micro-bearings" role is predominantly played by CNTs in the cutting friction process, while some of them break with the frictional shear action. A visible increase can be seen in C element's relative content, which is led by the result of CNT depositing and adsorption affecting the enhancement of the anti-friction effect of the composite.

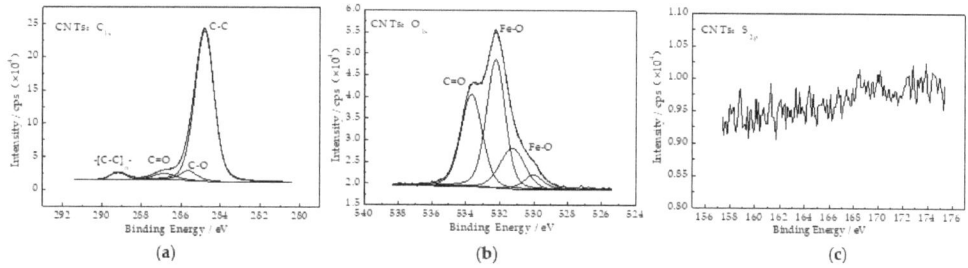

Figure 11. XPS analysis on the worn surfaces of flanks under the lubrication of CNT nanofluids: (a) C_{1s}; (b) Fe_{2p}; (c) S_{2p}.

Figure 12b exhibits that when the nanofluid prepared by the composite acts as lubrication, the peak shape of the C_{1s} energy spectrum is similar to that in Figure 11a, but the relative area of the peak shape of the fitting peak C=O is larger, with which -[C-C]n- is smaller. The O_{1s} spectrum is relatively similar to that in Figure 11b, but with a wider peak shape, and the sub-peaks are attributed to more prominent C-O and C=O, which proves that a similarity exists between the composite lubricating layer and CNTs in the process of forming. Then, a certain influence on the forming lubrication layer of CNTs may be affected by the discharge of T321 in the process of cutting. Being exposed to the nanofluid of the composite, the S_{2p} spectrum on the worn surface also has two peaks appearing at 162.4 eV and 168.2 eV distinctly, demonstrating that a vulcanized film is formed. Table 4 figures that the C element on the worn surface of the composite nanofluid lubrication is further increasing, the relative Fe element content is minimum, and the O content is between that of T321 and CNTs separately. This proves that a more abundant lubricating film is formed on the wear surface when the CNTs@T321 composite acts. It may be seen that the lubricating layer, which is composed of a vulcanized film, is formed between T321 and CNTs when the action happens separately. Figure 13 shows a model illustrating the lubrication mechanism of the composite at the cutting zone [43]. During cutting, T321 is released from the composites, and the active groups in T321 form a lubricating film through physical or chemical adsorption, relieving the friction at both the tool–chip and tool–workpiece interfaces. This reduces the cutting force and cutting temperature, and hence improves the surface quality. Meanwhile, CNTs are broken due to the action of cutting, and the generated fragment groups such as $R-CH_2-$ and $R-O-$ are adsorbed on the frictional interfaces, generating a lubricative film to help improve the interfacial lubrication. In addition, the composites can also act like micro bearings to lower the sliding friction, resulting in improved cutting performance.

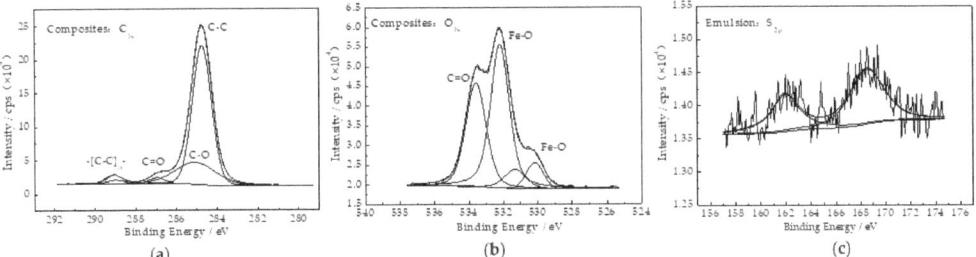

Figure 12. XPS analysis on the worn surfaces of flanks under the lubrication of composite nanofluids: (a) C_{1s}; (b) Fe_{2p}; (c) S_{2p}.

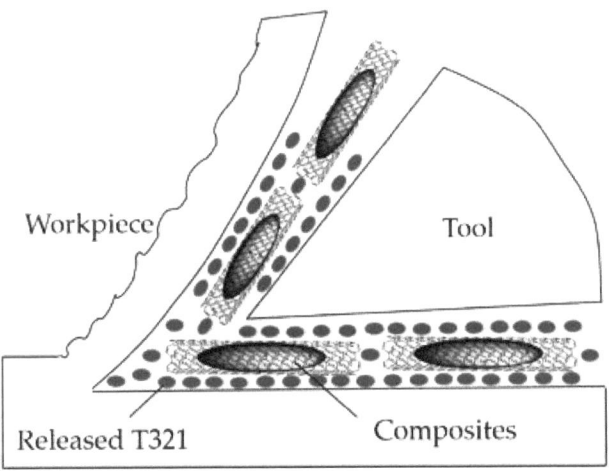

Figure 13. A model of the lubrication mechanism of the composites.

5. Conclusions

(1) CNTs@T321 nanocomposites were synthesized with an approximate fill rate of T321 of 25%. The addition of the synthesized nanocomposites significantly improved the dispersion stability, heat transferability and wettability of the nanofluid;

(2) The application of the developed CNTs@T321 nanofluid generated lower cutting force and cutting temperature, as well as better machined surface quality and tool life, with 12%, 20%,15%, and 15% improvements compared with the commercially available emulsion;

(3) During cutting, CNTs@T321 composite nanofluids produced a complex protective film containing a T321-based lubricating layer and a CNT-related tribo-layer. The synergistic lubrication effect of the protective film improved the friction state at both the tool–chip and tool–workpiece interfaces, thereby resulting in improved machining performance;

(4) This research offers an effective and practical way to synthesize metalworking fluids for the optimal performance and sustainable machining of difficult-to-machine materials.

Author Contributions: Conceptualization, J.G. and S.H.; methodology, Z.X.; software, C.G.; validation, J.G., C.G. and L.Y.; formal analysis, L.Y.; investigation, L.Y.; resources, S.H.; data curation, C.G.; writing—original draft preparation, C.G.; writing—review and editing, J.G.; visualization, L.Y.; supervision, Z.X.; project administration, L.Y.; funding acquisition, J.G. and S.H.; All authors have read and agreed to the published version of the manuscript.

Funding: This research was financially supported by the Youth Science Foundation Program, Natural Science Foundation of China, with the project number of NSFC 51805345. And the Natural Science Foundation of Hebei Province, with the project number of E2022203123.

Data Availability Statement: Not applicable.

Conflicts of Interest: The authors declare no conflict of interest.

References

1. Bruce, R.W. *Handbook of Lubrication and Tribology: Theory and Design*, 2nd ed.; CRC Press: Boca Raton, FL, USA, 2012.
2. Klocke, F.; Eisenblatter, G. Dry Cutting. *Ann. CIRP* **1997**, *46*, 519–526. [CrossRef]
3. Mirer, F.E. New Evidence on The Health Hazards and Control of Metalworking Fluids Since Completion of the Osha Advisory Committee Report. *Am. J. Ind. Med.* **2010**, *53*, 792–801. [CrossRef]
4. Choi, S.U.S.; Eastman, J.A. Enhancing Thermal Conductivity of Fluids with Nanoparticles. In Proceedings of the 1995 ASME International Mechanical Engineering Congress & Exposition, San Francisco, CA, USA, 12–17 November 1995; Volume 66, pp. 99–105.
5. Kilincarslan, S.K.; Cetin, M.H. Improvement of the Milling Process Performance by Using Cutting Fluids Prepared with Nano-Silver and Boric Acid. *J. Manuf. Process.* **2020**, *56*, 707–717. [CrossRef]
6. Kumar, A.S.; Paul, S.D.S. Tribological Characteristics and Micro-milling Performance of Nanoparticle Enhanced Water Based Cutting Fluids in Minimum Quantity Lubrication. *J. Manuf. Process.* **2020**, *56*, 766–776. [CrossRef]
7. Lee, P.H.; Nam, T.S.; Li, C.J.; Lee, S.W. Experimental Study on Meso-Scale Milling Process Using Nanofluid Minimum Quantity Lubrication. *Trans. Korean Soc. Mech. Eng. A* **2010**, *34*, 1493–1498. [CrossRef]
8. Rahmati, B.; Sarhan, A.A.D.; Sayuti, M. Morphology of Surface Generated by End Milling AL6061-T6 Using Molybdenum Disulfide (MoS2) Nano-lubrication in End Milling Machining. *J. Clean Prod.* **2013**, *66*, 685–691. [CrossRef]
9. Marcon, A.; Melkote, S.; Kalaitzidou, K.; Debra, D. An Experimental Evaluation of Graphite Nanoplatelet Based Lubricant in Micro-Milling. *CIRP Ann.-Manuf. Technol.* **2010**, *59*, 141–144. [CrossRef]
10. Sayuti, M.; Sarhan, A.A.D. Cutting Force Reduction and Surface Quality Improvement in Machining of Aerospace Duralumin AL-2017-T4 Using Carbon Onion Nano-Lubrication System. *Int. J. Adv. Manuf. Technol.* **2013**, *65*, 1493–1500. [CrossRef]
11. Sayuti, M.; Erh, O.M.; Sarhan, A.A.D.; Hamdi, M. Investigation on the Morphology of The Machined Surface in End Milling of Aerospace AL6061-T6 for Novel Uses of SiO2 Nano Lubrication System. *J. Clean Prod.* **2013**, *66*, 655–663. [CrossRef]
12. Pham, Q.D.; Tran, M.D.; Ngo, M.T.; Tran, T.L.; Nguyen, V.T. Improvement in the Hard Milling of AISI D2 Steel under the MQCL Condition Using Emulsion-Dispersed MoS2 Nanosheets. *Lubricants* **2020**, *8*, 62.
13. Duc, T.M.; Long, T.T.; Tuan, N.M. Novel Uses of Al2O3/MoS2 Hybrid Nanofluid in MQCL Hard Milling of Hardox 500 Steel. *Lubricants* **2021**, *9*, 45. [CrossRef]
14. Iijima, S. Helical Microtubes of Graphitic Carbon. *Nature* **1991**, *354*, 56–58. [CrossRef]
15. Clancy, T.C.; Gates, T.S. Modeling of Interfacial Modification Effects on Thermal Conductivity of Carbon Nanotube Composites. *Polymer* **2006**, *47*, 5990–5996. [CrossRef]
16. Sharma, A.K.; Arun, K.T.A.K.; Dixit, A.R. Effects of Minimum Quantity Lubrication (MQL) in Machining Processes Using Conventional and Nanofluid Based Cutting Fluids: A Comprehensive Review. *J. Clean Prod.* **2016**, *127*, 1–18. [CrossRef]

17. Prabhu, S.; Vinayagam, B.K. AFM Investigation in Grinding Process with Nanofluids Using Taguchi Analysis. *Int. J. Adv. Manuf. Technol.* **2012**, *60*, 149–160. [CrossRef]
18. Sharma, P.; Sidhu, B.S.; Sharma, J. Investigation of Effects of Nanofluids on Turning of AISI D2 Steel Using Minimum Quantity Lubrication. *J. Clean Prod.* **2015**, *108*, 72–79. [CrossRef]
19. Rao, S.N.; Satyanarayana, D.B.; Venkatasubbaiah, D.K. Experimental Estimation of Tool Wear and Cutting Temperatures in MQL Using Cutting Fluids with CNT Inclusion. *Int. J. Eng. Sci.* **2011**, *4*, 2928–2931.
20. Roy, S.; Ghosh, A. High-Speed Turning of AISI 4140 Steel Using Nanofluid Through Twin Jet SQL System. *ASME Int. Manuf. Sci. Eng. Conf.* **2013**, *55461*, 10–14.
21. Wang, P.; Zhang, D. Effect of Molecular Structure on Dispersion of Carbon Nanotubes by Natural Organic Matter Surrogates. *China Environ. Sci.* **2018**, *38*, 3429–3436.
22. Yuba, R.P.; Li, W.Z. Synthesis, Properties, and Applications of Carbon Nanotubes Filled with Foreign Materials: A Review. *Mater. Today Phys.* **2018**, *7*, 7–34.
23. Sinha, A.K.; Hwang, D.W.; Hwang, L.P. A Novel Approach to Bulk Synthesis of Carbon Nanotubes Filled with Metal by A Catalytic Chemical Vapor Deposition Method. *Chem. Phys. Lett.* **2000**, *332*, 455–460. [CrossRef]
24. Wang, G.J.; Qu, Z.H. Modification of Carbon Nanotubes by Chemical Reaction. *Prog. Chem.* **2006**, *18*, 1305–1312.
25. Connell, M.J.; Boul, P.; Ericson, L.M. Reversible Water-Solubilization of Single-Walled Carbon Nanotubes by Polymer Wrapping. *Chem. Phys. Lett.* **2001**, *342*, 265–271. [CrossRef]
26. Ajayan, P.M.; Iijima, S. Capillarity Induced Filling of Carbon Nanotubes. *Nature* **1993**, *361*, 333–335. [CrossRef]
27. Dujardin, E.; Ebbesen, T.W.; Hiura, H. Capillarity of Carbon Nanotubes. *Science* **1994**, *265*, 1850–1852. [CrossRef] [PubMed]
28. Pederson, M.R.; Broughton, J.Q. Nano-Capillarity in Fullerene Tubules. *Phys. Rev. Lett.* **1992**, *69*, 2689–2692. [CrossRef]
29. Kumar, P.G.; Renuka, M.; Velraj, R. Thermal and Electrical Conductivity Enhancement of Solar Glycol-Water Mixture Containing MWCNTs. *Fuller. Nanotub. Car. N.* **2018**, *26*, 871–879.
30. Guan, J.J.; Wang, J.; Lv, T.; Xu, X.F. Dispersion Stability and Enhanced Heat Transfer of Cutting Use Nanofluids Prepared by Composite of Carbon Nanotubes and Dialkyl Pentasulfide. *Mater. Res. Express* **2019**, *8*, 085633.
31. Nurettin, S.; Muammer, K. Stabilization of the Aqueous Dispersion of Carbon Nanotubes Using Different Approaches. *Therm. Sci. Eng. Prog.* **2018**, *8*, 411–417.
32. Dhar, N.; Islam, M.; Islam, S.; Mithu, M. An Experimental Investigation on Effect of Minimum Quantity Lubrication in Machining AISI 1040 Steel. *Int. J. Mach. Tools Manuf.* **2007**, *47*, 748–753. [CrossRef]
33. Bikash, C.B.; Chetana, D.S.; Sudarsan, G.P.; Venkateswara, R. Spread Ability Studies of Metal Working Fluids on Tool Surface and Its Impact on Minimum Amount Cooling and Lubrication Turning. *J. Mater. Process. Tech.* **2017**, *244*, 1–16. [CrossRef]
34. Xu, J.; Yamda, K.; Sekiya, K. Study of Comparing Cutting Force Signal Features for Dry, Air Cooling and Minimum Quantity Lubrication (MQL) Drilling. *J. Adv. Mech. Des. Syst.* **2017**, *11*, JAMDSM0030. [CrossRef]
35. Xia, J.; Li, B.Z.; Zhang, X.P. The Effects of Minimum Quantity Lubrication (MQL) on Machining Force, Temperature, And Residual Stress. *Int. J. Precis. Eng. Man.* **2014**, *15*, 2443–2451.
36. Zhang, Y.B.; Li, C.H.; Jia, D.Z.; Li, B.K. Experimental Study on The Effect of Nanoparticle Concentration on The Lubricating Property of Nanofluids for MQL Grinding of Ni-Based Alloy. *J. Mater. Process. Tech.* **2016**, *232*, 100–115. [CrossRef]
37. Chinchanikar, S.; Kore, S.S.; Hujare, P.A. Review on Nanofluids in Minimum Quantity Lubrication Machining. *J. Manuf. Process.* **2021**, *68*, 56–70. [CrossRef]
38. Abubakr, M.; Hegab, H.; Osman, T.A.; Elharouni, F.; Kishawy, H.A.; Esawi, A.M. Carbon Nanotube–Based Nanofluids: Properties and Applications. In *Handbook of Carbon Nanotubes*; Springer International Publishing: Cham, Switzerland, 2022.
39. Guan, J.J.; Liu, D.L.; Wang, Y.; Feng, B.H.; Xu, X. Tribological Properties of Nanofluid Prepared by Composite of Multi-Walled Carbon Nanotube and Oleic Acid. *Tribology* **2020**, *40*, 290–299.
40. Dosbaeva, G.K.; Hakim, M.A.; Shalaby, M.A. Cutting Temperature Effect on PCBN And CVD Coated Carbide Tools in Hard Turning of D2 Tool Steel. *Int. J. Refract. Met. H* **2015**, *50*, 1–8. [CrossRef]
41. Briggs, D. *Surface Analysis of Polymers by XPS and Static SIMS*, 1st ed.; Cambridge University Press: Cambridge, NY, USA, 1998.
42. Rashi, G.; Amzad, K.; Om, P.K. Fatty Acid-Derived Ionic Liquids as Renewable Lubricant Additives: Effect of Chain Length and Unsaturation. *J. Mol. Liq.* **2020**, *301*, 112322.
43. Hegab, H.; Umer, U.; Esawi, A.; Kishawy, H.A. Tribological Mechanisms of Nano-Cutting Fluid Minimum Quantity Lubrication: A Comparative Performance Analysis Model. *Int. J. Adv. Manuf. Technol.* **2020**, *108*, 3133–3139. [CrossRef]

 *lubricants*

Article

Lubrication Performance and Mechanism of Water-Based TiO$_2$ Nanolubricants in Micro Deep Drawing of Pure Titanium Foils

Muyuan Zhou, Fanghui Jia, Jingru Yan, Hui Wu * and Zhengyi Jiang *

School of Mechanical, Materials, Mechatronic and Biomedical Engineering, University of Wollongong, Wollongong, NSW 2522, Australia
* Correspondence: hwu@uow.edu.au (H.W.); jiang@uow.edu.au (Z.J.)

Abstract: Micro deep drawing (MDD) is a fundamental process in microforming which has wide applications in micro electromechanical system (MEMS) and biological engineering. Titanium possesses excellent mechanical properties and biocompatibility, which makes it a preferred material in micromanufacturing. In this study, eco-friendly and low-cost water-based TiO$_2$ nanolubricants were developed and applied in the MDD with 40 μm-thick pure titanium foils. The lubricants consisting of TiO$_2$ nanoparticles (NPs), 10 wt% glycerol, 0.1 wt% sodium dodecyl-benzene sulfonate (SDBS) and balanced water were synthesised in a facile process. The MDD with 40 μm-thick pure titanium was carried out using the lubricants with varying concentrations of 0.5, 1.0 and 2.0 wt%. The results show that the formability of micro cups could be significantly improved when the nanolubricants are applied. Especially, the use of 1.0 wt% TiO$_2$ nanolubricant demonstrates the best lubrication performance by significantly reducing the final drawing forces, and surface roughness, and the wrinkles by up to 24.2%, 12.55% and 4.82%, respectively. The lubrication mechanisms including the ball bearing and mending effects of NPs on open lubricant pockets (OLPs) and close lubricant pockets (CLPs) areas were then revealed through microstructure observation.

Keywords: water-based TiO$_2$ nanolubricant; micro deep drawing; pure titanium; lubrication mechanism; size effects

Citation: Zhou, M.; Jia, F.; Yan, J.; Wu, H.; Jiang, Z. Lubrication Performance and Mechanism of Water-Based TiO$_2$ Nanolubricants in Micro Deep Drawing of Pure Titanium Foils. *Lubricants* **2022**, *10*, 292. https://doi.org/10.3390/lubricants10110292

Received: 17 October 2022
Accepted: 1 November 2022
Published: 2 November 2022

Publisher's Note: MDPI stays neutral with regard to jurisdictional claims in published maps and institutional affiliations.

1. Introduction

Recently, micro metal parts have been widely utilised due to the miniaturisation trend in industries [1–3]. It has been determined that microforming is a suitable method for producing micro parts. Microforming has excellent economic and environmental benefits compared to other micro manufacturing processes [4]. With the reduction in the size and weight of microforming equipment, the amount of energy consumption, exhaust emissions, noise, and environmental pollution will be decreased. These characters make the microforming process more sustainable and environmentally friendly while also decreasing the expense of environmental protection [5]. The fundamental microforming process includes micro forging, micro extrusion, micro bending and micro deep drawing (MDD). MDD is suitable to produce micro cups, shells and more complex-shaped parts [6], such as the micro cups for drug containers of targeted therapy for cancer. Due to the excellent mechanical properties and biocompatibility, titanium has been frequently employed in recent years, including aerospace, military, and the medical industry. Thus, the MDD products with titanium could have the potential application in the future micro manufacturing industry, which needs to be further investigated. The deformation behaviour of materials in MDD process has been extensively studied. Zhao et al. [1] obtained different microstructures of austenitic stainless-steel foil by conducting heat treatment between 700 and 1100 °C and found that the optimal temperature to decrease the wrinkle of the micro cup was between 900 and 950 °C. Frank et al. [7] found that the deformation behaviour of specimens could be changed with different punching velocities in the MDD process. Luo et al. [8] carried out micro hydromechanical deep drawing (MHDD) with thin stainless steel 304

(SUS 304) foil and found that the wrinkling and the earring of the drawn cup could be limited in a critical hydraulic pressure. Luo et al. [9] also investigated the influences of different blank holder-die gaps in MDD machine and found that the appropriate gap can provide in-process springback before the occurrence of the peak drawing force, which could improve the quality of the micro cups.

The same as macroforming, friction has been recognised as the most important factor in microforming process. Friction has a significant impact on the forming quality in microforming because it changes the forming force of the contact area between the material and the tools and impedes the material flow [10]. Generally, frictional behaviour is related to three main factors: the material property, the surface roughness of material and the lubricant applied [11]. Due to the miniaturisation of materials, friction plays a more important role in micro-scaled forming than that of conventional metal forming [12]. Due to the so-called size effects, the peaks and valleys of the surfaces are a non-negligible parts of sample geometry in microforming [13]. Engel [14] developed the models of open lubricant pockets (OLPs) and close lubricant pockets (CLPs) to explain the tribology in microforming. When the lubricated workpiece is moved, a portion of the lubricant is trapped in the roughness valleys and the others are squeezed out. For the open lubricant pockets (OPLs), the roughness valleys have no connection with the edge of the surface and cannot hold the lubricant. On the contrary, the asperity valleys are connected to the edge of the surface, thus holding the lubricant and the CLPs is formed. Several studies have utilised different kinds of lubricants in MDD and obtained the drawn products with better forming quality. Gong et al. [15] compared the effects of different lubricants on the quality of micro-cup products, and the micro cups were shown to have good drawability under polyethylene (PE) film which is better than that under other lubricants, such as soybean and castor oil. Hu et al. [16] coated diamond-like carbon (DLC) film to the forming tools of MDD and found that the DLC film had great adhesion strength, and a relatively lower drawing force could be achieved. Wang et al. [17] also applied DLC film in MDD and found that the surface quality and shape accuracy of the micro cups could be significantly improved under this lubrication condition.

Following the trend of sustainable development, water-based lubricants with nanoparticles (NPs) are becoming an alternative to traditional lubrication methods [18–20]. He et al. [21] systematically studied the lubrication performance of nanolubricants with various concentrations of Al_2O_3 NPs. Compared with the other conditions, the Al_2O_3 nanolubricant could decrease the friction by 27% through the shaping of thin films and mending effect. Wu et al. [22–26] studied the tribological behaviour of TiO_2 nanolubricant using ball-on-disk tribometer and found that the ball wear was decreased by 97.8% compared to that under dry condition. They subsequently applied water-based TiO_2 nanolubricant in hot steel rolling with different concentrations of TiO_2 and different formulas. The results showed the nanolubricant can effectively reduce the rolling force, surface roughness, oxide scale thickness of the rolled steel. Its surface finish can be also improved due to the rolling and mending effects of the TiO_2 NPs on the surface of materials. In MDD, Kamali et al. [27] applied oil-based TiO_2 nanolubricant in the MDD experiments with magnesium alloy, and it was found that 2.0 wt% nanolubricant showed the best lubrication performance by obtaining the lowest drawing force and best surface quality. Nevertheless, oil-based lubricants usually contain a high concentration of oil. After the use of oil-based lubricants, cleaning and disposal of the used lubricants is an unavoidable issue due to the non-biodegradable nature of oil [24,28]. To improve the lubrication performance of lubricants but reduce the usage of oil, the novel water-based TiO_2 nanolubricants were developed in this study.

Water-based nanolubricants are extensively used in macro metal forming due to their excellent eco-friendly and lubrication performance, which has the potential application in microforming [29–31]. However, the use of water-based nanolubricants in the MDD process has been scarcely reported. In this study, novel water-based TiO_2 nanolubricants were developed and applied in MDD with pure titanium foils. The lubrication performance of water-based TiO_2 nanolubricants in the MDD process was then revealed by analysing the forming quality of micro cups and drawing force. Finally, the lubrication mechanism of the TiO$_2$ nanolubricant in MDD was unveiled.

2. Experimental Procedure

2.1. Material Preparation

The novel nano additive water-based lubricants are made of P25 TiO$_2$ NPs, glycerol, sodium dodecyl-benzene sulfonate (SDBS) and balanced water. All additives are non-toxic materials and environmentally friendly. P25 TiO$_2$ NPs is a mixture which composed of 75% anatase and 25% rutile with approximately 20 nm in diameter [32]. Glycerol is a colourless, odorless, viscous liquid and miscible with water. Glycerol is used to enhance the viscosity of nanolubricants. SDBS is a white to light yellow sand-like solid which could be easily rinsed off the surface of the material. Its solution is used as a dispersing agent to improve the dispersion stability and viscosity of nanolubricants [33]. The water-based TiO$_2$ nanolubricants were prepared according to the flow chart shown in Figure 1. The dropwise glycerol and SDBS solution were first added into the distilled water followed by stirring with a dropper. Next, the TiO$_2$ NPs were gradually added into the solution and then mechanical stirred for 10 min until no obvious agglomeration appears. Afterwards, the remaining agglomeration in the dispersive solution was eliminated by ultrasonication for 10 min. Finally, the water-based TiO$_2$ nanolubricants with excellent dispersion stability were prepared. Four groups of lubrication conditions were designed according to the chemical compositions of the nanolubricants, as shown in Table 1. In this study, the pure titanium foil with the thickness of 40 µm was used and its chemical compositions are listed in Table 2. As a comparison, the dry friction condition was also designed in this study.

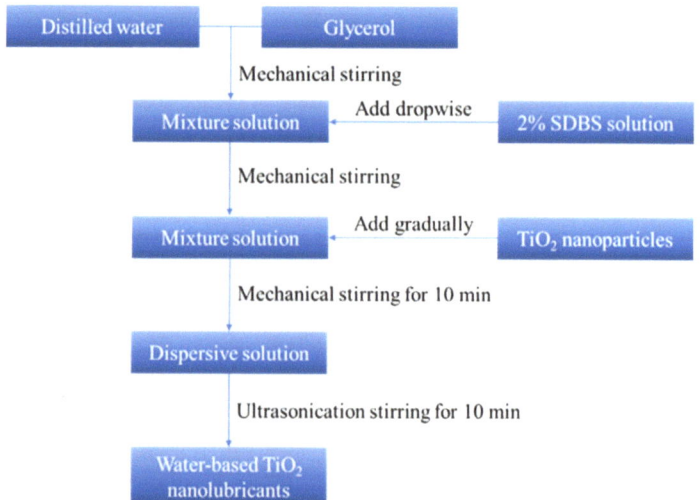

Figure 1. The flow chart of preparation of water-based TiO$_2$ nanolubricants.

Table 1. Chemical compositions of applied lubricants.

Type	Description
Group 1	Dry condition
Group 2	0.5 wt% TiO$_2$ + 10 wt% Glycerol + 0.1 wt% SDBS + Balanced water
Group 3	1.0 wt% TiO$_2$ + 10 wt% Glycerol + 0.1 wt% SDBS + Balanced water
Group 4	2.0 wt% TiO$_2$ + 10 wt% Glycerol + 0.1 wt% SDBS + Balanced water

Table 2. Chemical compositions of the pure titanium foils (wt%).

Materials	Fe	C	N	H	O
Pure titanium	0.20	0.08	0.03	0.015	0.18

2.2. Micro Deep Drawing

In this study, a desktop servo-press machine DT-3AW (Keyence Corporation, Osaka, Japan) was used for MDD process. As shown in Figure 2, the MDD system contains press machine, die sets and control box. The die sets consist of the upper and lower dies where the punch and cavity are placed. The press machine provides the load, and the control box is used to set the operation of the press machine and obtain the position of punch during the process. The real-time drawing force is recorded by a load cell which is located on the top of the upper die. In this study, the MDD process includes two stages: blanking and drawing. In the blanking stage, a round blank with the radius of 0.8 mm was cut from the pure titanium foils, and subsequently the round blank was drawn with a velocity of 0.1 mm/s. The geometrical dimensions of the MDD system are shown in Figure 2e. The friction between the blank and the lower die dominates the total friction force and ultimately affects the drawing force. Therefore, two drops (approx. 0.1 mL) of the water-based TiO$_2$ nanolubricants were dripped into the cavity of the lower die before the MDD process, as shown in Figure 2f. The MDD process was repeated five times for each test to minimise the data scattering of the drawing force.

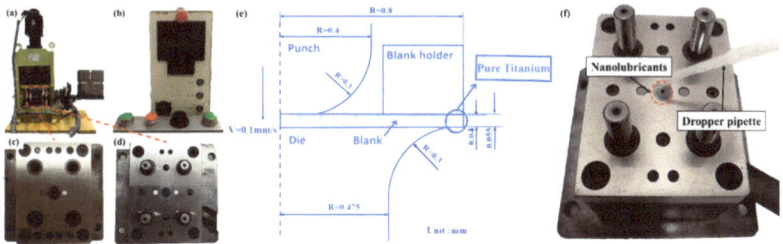

Figure 2. Micro deep drawing system (**a**) press machine, (**b**) control box, (**c**) upper die, (**d**) lower die, (**e**) dimension of the system and (**f**) the supply method of nanolubricants.

2.3. Observations

The sedimentation analysis method was used to measure the dispersion stability of water-based TiO$_2$ nanolubricants. The nanolubricants were photographed every 24 h. The sedimentation phenomenon of TiO$_2$ NPs and the appearance of supernatant were then recorded.

The phase and morphology of the P25 TiO$_2$ NPs were characterised by X-ray diffractometer (XRD, GBC Scientific Equipment Pty Ltd., Melbourne, Australia), JEOL^{TM} JEM-ARM200F transmission electron microscope (TEM, JEOL Ltd., Tokyo, Japan) and JEOL^{TM} JSM-7001F field emission scanning electron microscope (SEM, JEOL Ltd., Tokyo, Japan). The SEM equipped with the energy dispersive spectroscopy (EDS, JEOL Ltd., Tokyo, Japan) was used to analyse the distribution of elements. In addition, the distribution of TiO$_2$ NPs on the micro cups was observed by SEM to explore the lubrication mechanism.

The surface roughness and dimensions of the micro cups were observed by KEYENCE^{TM} VK-X100K 3D laser scanning microscope (Keyence Corporation, Osaka, Japan). To ensure the reliability and accuracy of the data, four regions on each drawn cups were randomly selected to measure the surface roughness values.

3. Results and Discussion

3.1. Material Characterisation

The XRD pattern of the P25 TiO_2 NPs is shown in Figure 3. The phase of the particles can be identified as typical P25 TiO_2 with 75% of anatase and 25% of rutile based on the XRD standard atlas.

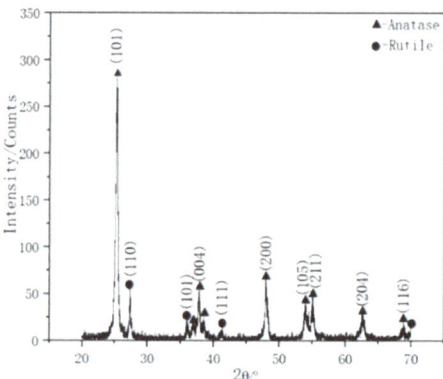

Figure 3. XRD pattern of P25 TiO_2 NPs.

Figure 4 shows the TEM images of the P25 TiO_2 NPs with different concentrations of water-based nanolubricants from 0.5 to 2.0 wt%. It can be observed that the P25 TiO_2 NPs are almost spherical with an average size of about 20 nm. The P25 TiO_2 NPs are uniformly distributed and show excellent dispersion stability with no significant aggregations even at high concentrations such as 2.0 wt%. The nanolubricants with good dispersion stability are expected to provide a stable lubrication performance in the MDD process.

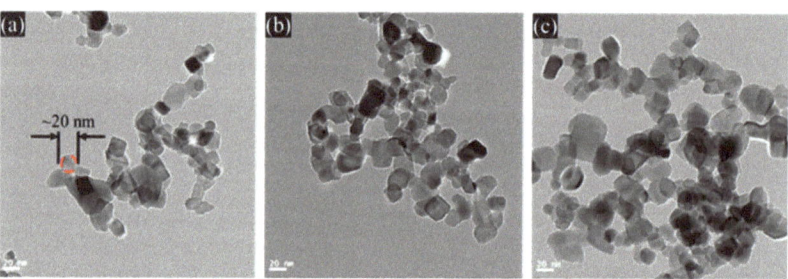

Figure 4. TEM images of P25 TiO_2 NPs with different concentrations of water-based nanolubricants: (**a**) 0.5 wt%, (**b**) 1.0 wt%, and (**c**) 2.0 wt%.

Photographs of the nanolubricants in a period of time are shown in Figure 5. Regardless of the concentration of TiO_2 NPs, all the nanolubricants remain stable for 24 h with insignificant NPs sedimentation at the bottom. After 24 h, the nanolubricants begin to settle slightly and a shallow supernatant appears (Figure 5c,d). This indicates that the water-based TiO_2 nanolubricants have excellent dispersion stability. Stable nanolubricant is usually considered as a prerequisite for the successful application in the MDD process.

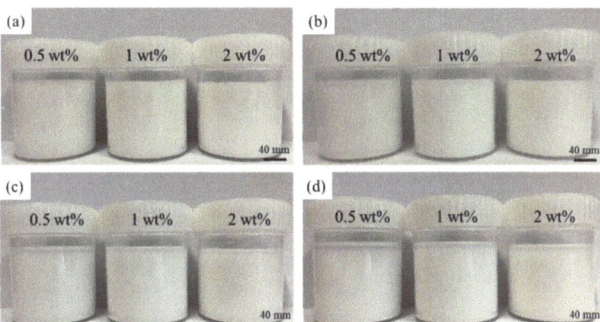

Figure 5. P25 TiO$_2$ NPs dispersed in different water-based nanolubricants at time of (**a**) 0 h, (**b**) 24 h, (**c**) 48 h, and (**d**) 72 h.

3.2. Drawing Force

The average drawing force and displacement curves were obtained, as shown in Figure 6. It indicates that the largest peak drawing force appears in the group when 2.0 wt% nanolubricant was used while the other groups share similar and relatively lower values in the peak drawing force. This is due to the resistance of bending that dominates the maximum drawing force initially [34]. With the increase of drawing depth, the drawing force drops drastically at the end of the MDD process. When the micro cup fully enters the cavity of the lower die [35], there is a significant decrease in the final drawing force after the utilisation of lubricant.

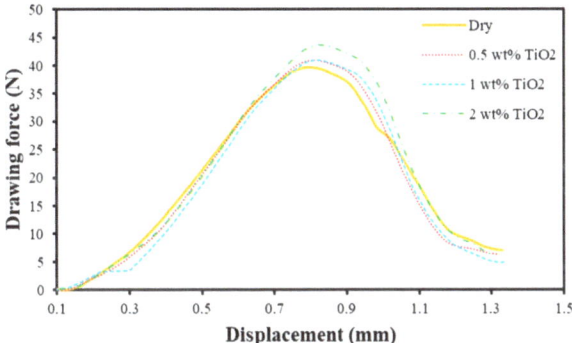

Figure 6. The displacement and drawing force curves obtained under various lubrication conditions.

The maximum drawing force and the final drawing force for each group were summarised in Figure 7. As analysed previous, there is an insignificant change in the peak drawing force as the drawing force is dominated by the resistance of bending before a full cup shape is formed. The effect of the 0.5% and 1.0% nanolubricants on maximum drawing force is insignificant. For the 2.0% nanolubricant, the peak drawing force (42.94 N) increases slightly as the TiO$_2$ NPs agglomeration impedes the material flow. It is evident that the highest final drawing force value (6.97 N) is found in the dry condition among all the lubrication conditions. For the final drawing force, the lubrication effect of the 0.5% and 1.0% nanolubricants is obvious, which achieves reductions of 12.7% and 24.2%, respectively, compared with that in dry condition. In addition, the final drawing force experiences a relatively lower drop to 6.67 N when the 2.0 wt% nanolubricant is employed.

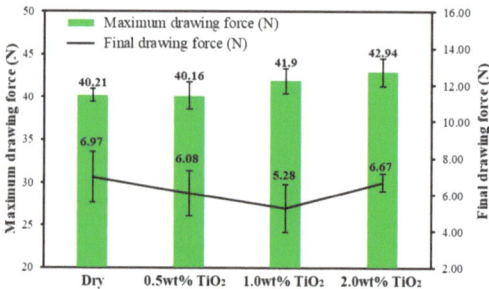

Figure 7. The comparison of maximum and final drawing forces under various lubrication conditions.

3.3. Quality Evaluation of Micro Cups

The surface roughness is an important parameter to evaluate the forming quality of the micro cups. Figure 8 shows the 3D surface morphology of the micro cups from different lubrication conditions. As can be seen in Figure 8a, considerable bumps are observable on the cup surface without the use of nanolubricants and the measured surface is the roughest among four groups. For Groups 2, 3 and 4, the surface quality of the micro cups has been improved from the observation of Figure 8b–d. It is worth noting that the micro cups from Group 3 have the smoothest surface, which displays the least surface peaks and valleys. In addition, the surface quality of the micro cup from Group 4 shows a slight improvement compared with that from dry conditions, but not as evident as those from Groups 2 and 3. Based on the 3D surface morphology, the surface roughness values of the micro cups from four groups were summarised in Figure 9. For MDD process, the nanolubricant with 1.0 wt% TiO_2 NPs shows the best lubrication capability in achieving a good surface quality by reducing the surface roughness value of 12.55% compared with that obtained under dry condition.

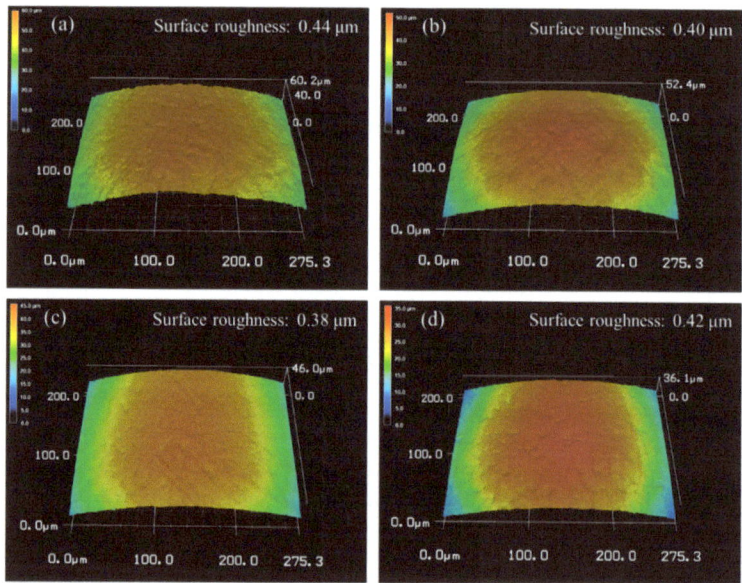

Figure 8. 3D surface morphologies of micro cups from four conditions (**a**) dry condition, (**b**) 0.5 wt% TiO_2, (**c**) 1.0 wt% TiO_2, and (**d**) 2.0 wt% TiO_2.

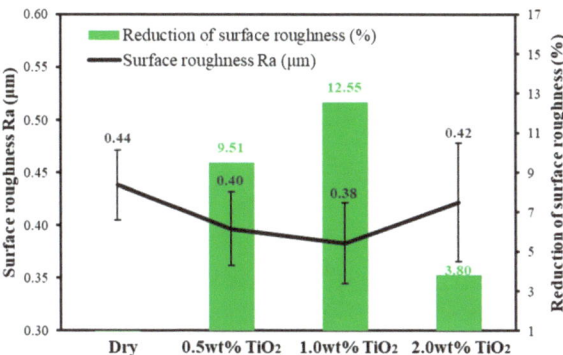

Figure 9. The surface roughness of micro cups from various lubrication conditions.

Wrinkling is an important feature to evaluate the forming quality of the micro cups. Less wrinkles mean better forming quality and higher shape accuracy. During the MDD process, the blank undergoes large radial drawing stress and tangential compressive stress [36]. The surface roughness affects the metal flow behaviour, thereby affecting the forming performance [2]. A large surface roughness will cause the difficulty in material flow and induce more winkles [9]. To clearly exhibit the reduction of wrinkles, the difference between the outer diameter and the minimum inner diameter of the micro cups is defined as the maximum distance, and the difference between the outer and maximum inner diameters of the micro cups is defined as the minimum distance [2], as shown in Figure 10. The amount of wrinkle is calculated by dividing the difference between the maximum and minimum distances by the maximum distance. Figure 11 shows the top views of the micro cups under various lubrication conditions. It can be seen that obvious wrinkling occurs on the flange of the micro cup under dry condition, as shown in Figure 11a. The number of wrinkles is significantly reduced with the use of nanolubricant, especially for the cup from Group 3 (Figure 11c), from which the winkles are hardly seen on the edge of parts. The wrinkles of micro cups obtained using various nanolubricants are summarised in Figure 12. The wrinkles decreased from 15.47% to 14.73% when the 1.0 wt% TiO_2 nanolubricant was used, showing an improvement of 4.82%.

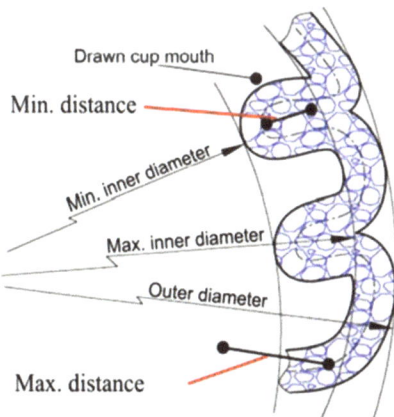

Figure 10. Definitions for parameters of wrinkle.

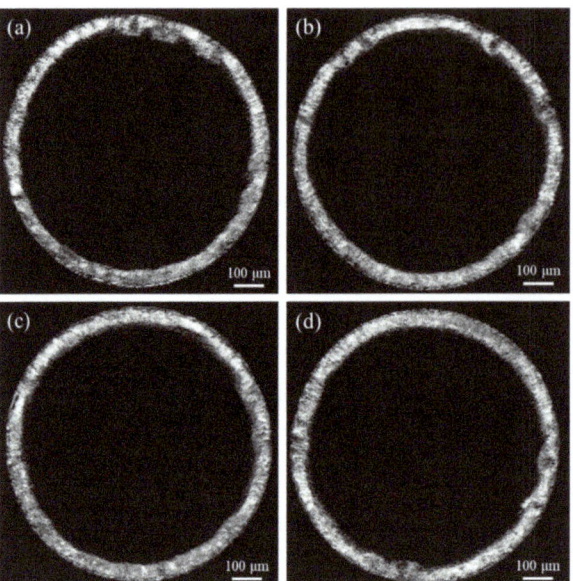

Figure 11. Top views of the micro cups under various nanolubricants (**a**) dry condition, (**b**) 0.5 wt% TiO$_2$, (**c**) 1.0 wt% TiO$_2$, and (**d**) 2.0 wt% TiO$_2$.

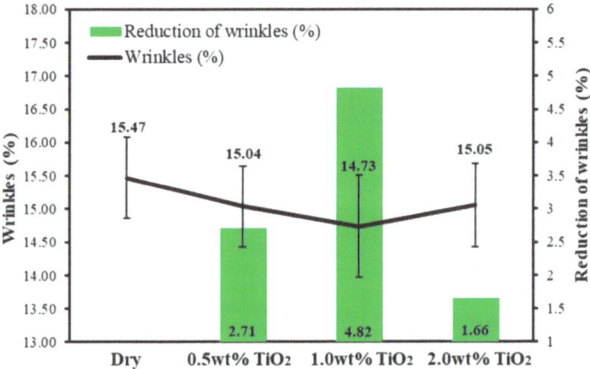

Figure 12. Wrinkles (%) of the micro cups under various lubrication conditions.

3.4. Surface Observation

Figure 13 shows the SEM images and EDS mappings of the side view of the drawn cups under different lubrication conditions. Since both titanium foil and TiO$_2$ NPs contain the element of titanium, the distribution of TiO$_2$ NPs could be determined by the distribution of oxygen element. When the 0.5 wt% nanolubricant is applied, the distribution of TiO$_2$ NPs on the micro cup is observable but not as noticeable as that indicated under other lubrication conditions. As the concentration of nanolubricant increases from 0.5 to 2.0 wt%, there is a significant increase in the number of TiO$_2$ NPs on the surface of the drawn cups according to the EDS mappings. In order to investigate the lubrication mechanism of the TiO$_2$ NPs on the micro cups surface, three areas on the micro cup wall were selected from each micro cup and marked as Areas A, B and C.

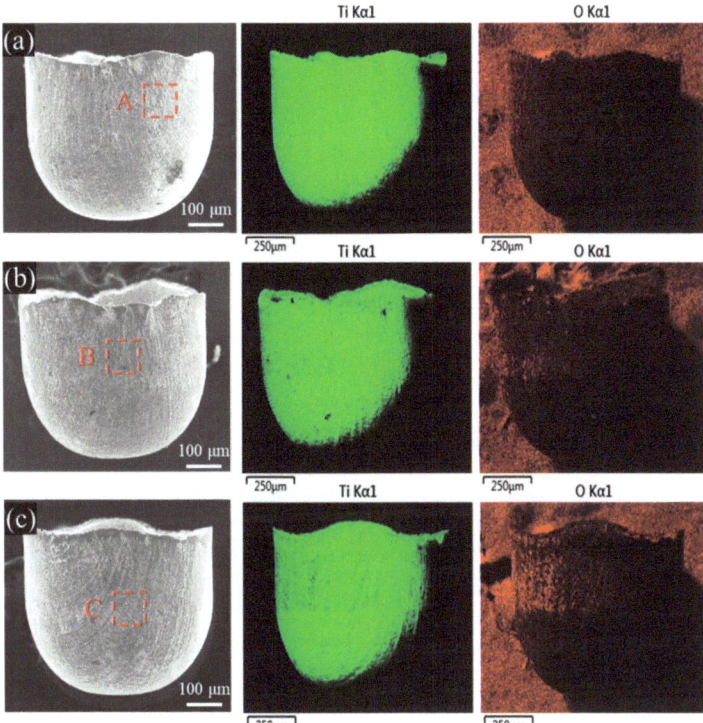

Figure 13. SEM images and EDS mappings of side view of the drawn cups using different lubricants: (**a**) 0.5 wt% TiO$_2$, (**b**) 1.0 wt% TiO$_2$, and (**c**) 2.0 wt% TiO$_2$.

SEM images with higher magnification and EDS spectrum ($\times 10{,}000$) of the Areas A, B and C were selected to explore the lubrication mechanism of TiO$_2$ NPs in MDD process, as shown in Figure 14. The distribution of TiO$_2$ NPs on the surface and the agglomeration in the roughness valleys can be clearly observed according to the SEM images. The corresponding EDS spectrums of the TiO$_2$ NPs were also detected and exhibited along with the SEM observation. As can be seen from Area A (Figure 14a), a few TiO$_2$ NPs are distributed but not fully covered on the surface, especially in the CLPs area, which could result in the incomplete mending effect in this region. A portion of dispersed TiO$_2$ NPs on the contact area can be used to bear the load by converting the sliding friction to the rolling friction, which reduces the rubbing between the tool and the workpiece. This phenomenon is called the rolling effect. As for the Area B shown in Figure 14b, the TiO$_2$ NPs are evenly dispersed on the surface of the micro cup and the CLPs region are well filled with TiO$_2$ NPs. This greatly enhances the synergism action of the rolling and mending effects, and further reduces the friction between the blank and the lower die. However, compared to the 0.5 and 1.0 wt% nanolubricants, excessive TiO$_2$ NPs are provided to fill the CLPs region when the 2.0 wt% nanolubricant is applied, and the redundant TiO$_2$ NPs are agglomerated on the surface of the material, as presented in Area C (Figure 14c). The agglomeration impedes the material flow during the MDD process and affects the further supply of the TiO$_2$ NPs into the contact area, which results in a relatively poor lubrication performance.

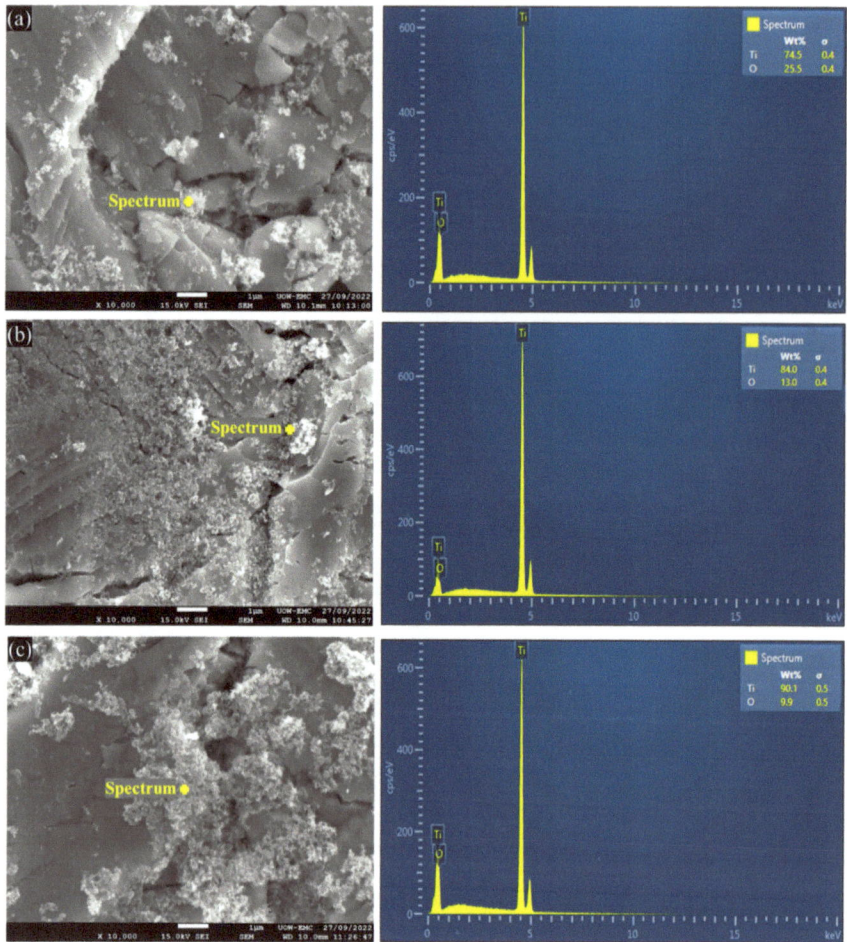

Figure 14. SEM images and EDS spectrums of the drawn cups with different lubricants (**a**) 0.5 wt% TiO_2, (**b**) 1.0 wt% TiO_2, and (**c**) 2.0 wt% TiO_2.

The OLPs area is located in the mouth area of the drawn micro cup. Generally, the liquid lubricants will be squeezed out and make it hardly bear the applied load in this region. However, due to the high viscosity of the nanolubricant, the TiO_2 NPs of the nanolubricants could remain in the OLPs area and thus provide a part of load bearing capacity. This thus leads to a reduction of the friction between the blank and the lower die. Figure 15 shows the SEM images and EDS mappings of the cup mouth area using different lubricants. Compared to the 0.5 and 2.0 wt% nanolubricants, the TiO_2 NPs in the 1.0 wt% nanolubricant were more uniformly distributed at the OLPs area. For the 0.5 wt% nanolubricant, there are insufficient TiO_2 NPs that provide load bearing capacity at the OLPs area. For the 2.0 wt% nanolubricant, more aggregations of TiO_2 NPs were observed at the OLPs area. By comparison, it can be observed that there are a few TiO_2 NPs that exist along the cup mouth area. Therefore, the friction near the mouth area reduces, and the winkles decrease significantly, which meets well with the observations in Figure 9. This phenomenon illustrates that the nanolubricants can effectively enhance the lubrication performance at the OLPs area.

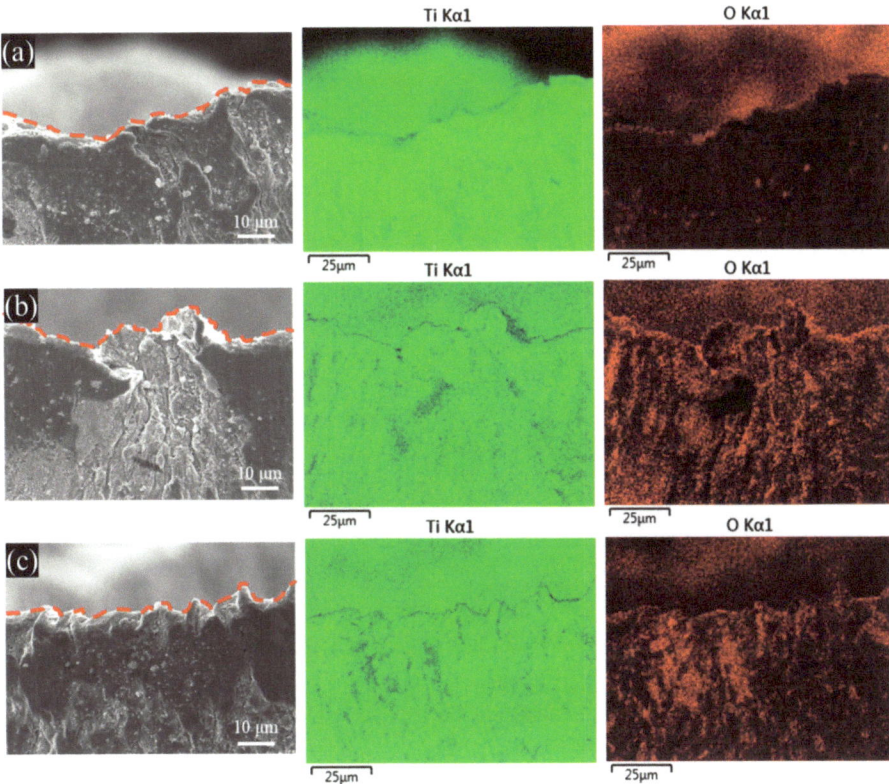

Figure 15. SEM images and EDS mappings of mouth area of the micro cups with different lubricants (**a**) 0.5 wt% TiO_2, (**b**) 1.0 wt% TiO_2, and (**c**) 2.0 wt% TiO_2.

3.5. Lubrication Mechanism

Figure 16 illustrates the lubrication mechanism of water-based TiO_2 nanolubricants in the MDD process. The nanolubricants with excellent dispersion stability provide the necessary prerequisites for obtaining excellent lubrication performance between the blank and the lower die. Some TiO_2 NPs worked as ball bearings to reduce the friction of real contact areas between the blank and the lower die. These TiO_2 NPs rotate in the contact area during the MDD process, which results in a lower coefficient of friction than that obtained under dry conditions where the two surfaces slide directly against each other, as shown in Figure 16a,b. In addition, TiO_2 NPs aggregate in the CLPs area, and thus the pits and valleys on the cup surface can be compensated by the mending effect of TiO_2 NPs, resulting in a relatively lower surface roughness, as shown in Figure 16c,d. In this study, as the concentration of TiO_2 NPs increases from 0.5 to 1.0 wt%, more TiO_2 NPs effectively participate in the MDD process in the contact area and CLPs areas, which enables the lubrication performance to gradually improve. However, further increasing the TiO_2 NPs concentration to 2.0 wt% will cause excessive aggregation of TiO_2 NPs, which leads to the increase in friction and decrease in forming quality. At the mouth area of the drawn cups, the TiO_2 NPs can be kept in the OLPs area due to high viscosity of the nanolubricant. Hence, the remained nanolubricant could provide load bearing capacity and therefore result in a reduction in friction. In addition, the aggregation of the appropriate concentration of TiO_2 NPs can reduce the shear stress in the contact area during MDD process, which also results in the reduction of friction on the material surface [37].

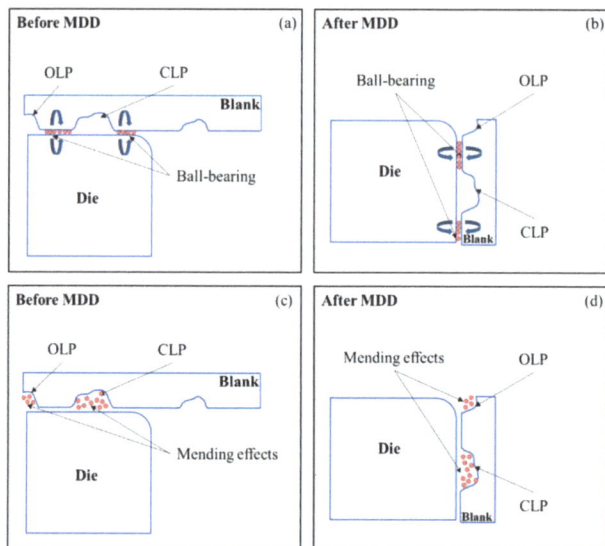

Figure 16. The schematic view of the nanolubricant mechanism during MDD process (**a,b**) rolling/ball bearing mechanism, (**c,d**) mending effects.

4. Conclusions

In this study, the MDD of pure titanium were carried out using 0.5, 1.0 and 2.0 wt% water-based nanolubricants to examine their lubrication performance. The forming quality of the micro drawn cups in terms of drawing force, surface roughness and shape accuracy was systematically evaluated. Finally, the lubrication mechanism of the nanolubricant in the MDD process was unveiled. The main conclusions can be drawn as follows.

(1) The TEM images and sedimentation observations indicate the novel water-based TiO_2 nanolubricants possess excellent dispersion stability.

(2) As for the final drawing force, the 1.0 wt% nanolubricants exhibit the best lubrication performance, resulting in a 24.2% reduction in the final drawing force compared to the dry condition.

(3) The 1.0 wt% nanolubricants show a significant improvement in the quality of the micro cups. There is a 12.55% reduction in the surface roughness compared to the dry condition. The wrinkles are decreased from 15.46% (dry conditions) to 14.73%.

(4) The TiO_2 NPs can fill in the OLPs area and therefore provide load bearing capacity, resulting in a reduction in friction during MDD. At the CLPs area, the lubrication mechanism can be contributed by the synergistic effect of ball bearing and mending effect of the TiO_2 NPs.

Author Contributions: M.Z. performed the experiments in addition to analysing the data and writing the paper; F.J. contributed to the SEM operation and editing; J.Y. contributed to proofreading; H.W. contributed to the supervision, experimental design, result analysis and revise the paper; Z.J. contributed to the conceptualization, supervision, funding acquisition, result analysis and editing. All authors have read and agreed to the published version of the manuscript.

Funding: This research was funded by the Australian Research Council (ARC, Grant Nos. DP190100738 and DP190100408).

Data Availability Statement: Not applicable.

Acknowledgments: The authors acknowledge the financial support from Australian Research Council (ARC, Grant Nos. DP190100738 and DP190100408). The authors are grateful to Matthew Franklin in the workshop of SMART Infrastructure Facility at the University of Wollongong for his great support on sample machining and preparation.

Conflicts of Interest: The authors declare no conflict of interest.

References

1. Zhao, J.; Wang, T.; Jia, F.; Li, Z.; Zhou, C.; Huang, Q.; Jiang, Z. Experimental Investigation on Micro Deep Drawing of Stainless Steel Foils with Different Microstructural Characteristics. *Chin. J. Mech. Eng.* **2021**, *34*, 11. [CrossRef]
2. Jiang, Z.; Zhao, J.; Xie, H. *Microforming Technology Theory, Simulation and Practice*, 1st ed.; Academic Press: London, UK, 2017.
3. Hasan, M.; Zhao, J.; Jiang, Z. Micromanufacturing of composite materials: A review. *Int. J. Extreme Manuf.* **2019**, *1*, 26. [CrossRef]
4. Luo, L.; Jiang, Z.; Wei, D.; Manabe, K.-I.; Zhao, X.; Wu, D.; Furushima, T. Effects of surface roughness on micro deep drawing of circular cups with consideration of size effects. *Finite Elements Anal. Des.* **2016**, *111*, 46–55. [CrossRef]
5. Chinchanikar, S.; Kolte, Y. A review on experimental and numerical studies on micro deep drawing considering size effects and key process parameters. *Aust. J. Mech. Eng.* **2022**, 14. [CrossRef]
6. Gong, F.; Guo, B.; Wang, C.; Shan, D. Micro deep drawing of micro cups by using DLC film coated blank holders and dies. *Diam. Relat. Mater.* **2011**, *20*, 196–200. [CrossRef]
7. Vollertsen, F.; Hu, Z. Analysis of punch velocity dependent process window in micro deep drawing. *Prod. Eng.* **2010**, *4*, 553–559. [CrossRef]
8. Luo, L.; Wei, D.; Wang, X.; Zhou, C.; Huang, Q.; Xu, J.; Wu, D.; Jiang, Z. Effects of hydraulic pressure on wrinkling and earing in micro hydro deep drawing of SUS304 circular cups. *Int. J. Adv. Manuf. Technol.* **2016**, *90*, 189–197. [CrossRef]
9. Luo, L.; Wei, D.; Zu, G.; Jiang, Z. Influence of blank holder-die gap on micro-deep drawing of SUS304 cups. *Int. J. Mech. Sci.* **2020**, *191*, 106065. [CrossRef]
10. Tan, X. Comparisons of friction models in bulk metal forming. *Tribol. Int.* **2002**, *35*, 385–393. [CrossRef]
11. Peng, L.; Lai, X.; Lee, H.-J.; Song, J.-H.; Ni, J. Friction behavior modeling and analysis in micro/meso scale metal forming process. *Mater. Des.* **2010**, *31*, 1953–1961. [CrossRef]
12. Vollertsen, F.; Hu, Z.; Niehoff, H.; Theiler, C. State of the art in micro forming and investigations into micro deep drawing. *J. Mater. Process. Technol.* **2004**, *151*, 70–79. [CrossRef]
13. Han, J.; Zheng, W.; Wang, G.; Yu, M. Experimental study on size effect of dry friction in meso/micro-upsetting process. *Int. J. Adv. Manuf. Technol.* **2017**, *95*, 1127–1133. [CrossRef]
14. Engel, U. Tribology in microforming. *Wear* **2006**, *260*, 265–273. [CrossRef]
15. Chan, W.; Fu, M.; Lu, J. The size effect on micro deformation behaviour in micro-scale plastic deformation. *Mater. Des.* **2011**, *32*, 198–206. [CrossRef]
16. Hu, Z.; Schubnov, A.; Vollertsen, F. Tribological behaviour of DLC-films and their application in micro deep drawing. *J. Mater. Process. Technol.* **2012**, *212*, 647–652. [CrossRef]
17. Wang, C.; Guo, B.; Shan, D.; Bai, X. Experimental research on micro-deep drawing processes of pure gold thin sheet using DLC-coated female die. *Int. J. Adv. Manuf. Technol.* **2012**, *67*, 2477–2487. [CrossRef]
18. Rahman, M.H.; Warneke, H.; Webbert, H.; Rodriguez, J.; Austin, E.; Tokunaga, K.; Rajak, D.K.; Menezes, P.L. Water-Based Lubricants: Development, Properties, and Performances. *Lubricants* **2021**, *9*, 73. [CrossRef]
19. Cui, X.; Li, C.; Ding, W.; Chen, Y.; Mao, C.; Xu, X.; Liu, B.; Wang, D.; Li, H.N.; Zhang, Y.; et al. Minimum quantity lubrication machining of aeronautical materials using carbon group nanolubricant: From mechanisms to application. *Chin. J. Aeronaut.* **2021**, *35*, 85–112. [CrossRef]
20. He, Y.; Wang, L.; Wu, T.; Wu, Z.; Chen, Y.; Yin, K. Facile fabrication of hierarchical textures for substrate-independent and durable superhydrophobic surfaces. *Nanoscale* **2022**, *14*, 9392–9400. [CrossRef] [PubMed]
21. He, A.; Huang, S.; Yun, J.-H.; Wu, H.; Jiang, Z.; Stokes, J.; Jiao, S.; Wang, L.; Huang, H. Tribological Performance and Lubrication Mechanism of Alumina Nanoparticle Water-Based Suspensions in Ball-on-Three-Plate Testing. *Tribol. Lett.* **2017**, *65*, 40. [CrossRef]
22. Wu, H.; Zhao, J.; Xia, W.; Cheng, X.; He, A.; Yun, J.H.; Wang, L.; Huang, H.; Jiao, S.; Huang, L.; et al. A study of the tribological behaviour of TiO2 nano-additive water-based lubricants. *Tribol. Int.* **2017**, *109*, 398–408. [CrossRef]
23. Wu, H.; Jia, F.; Li, Z.; Lin, F.; Huo, M.; Huang, S.; Sayyar, S.; Jiao, S.; Huang, H.; Jiang, Z. Novel water-based nanolubricant with superior tribological performance in hot steel rolling. *Int. J. Extreme Manuf.* **2020**, *2*, 025002. [CrossRef]
24. Wu, H.; Wei, D.; Hee, A.C.; Huang, S.; Xing, Z.; Jiao, S.; Huang, H.; Jiang, Z. The influence of water-based nanolubrication on mill load and friction during hot rolling of 304 stainless steel. *Int. J. Adv. Manuf. Technol.* **2022**, *121*, 7779–7792. [CrossRef]
25. Wu, H.; Zhao, J.; Cheng, X.; Xia, W.; He, A.; Yun, J.-H.; Huang, S.; Wang, L.; Huang, H.; Jiao, S.; et al. Friction and wear characteristics of TiO$_2$ nano-additive water-based lubricant on ferritic stainless steel. *Tribol. Int.* **2018**, *117*, 24–38.
26. Wu, H.; Zhao, J.; Luo, L.; Huang, S.; Wang, L.; Zhang, S.; Jiao, S.; Huang, H.; Jiang, Z. Performance Evaluation and Lubrication Mechanism of Water-Based Nanolubricants Containing Nano-TiO$_2$ in Hot Steel Rolling. *Lubricants* **2018**, *6*, 57. [CrossRef]
27. Kamali, H.; Xie, H.; Jia, F.; Wu, H.; Zhao, H.; Zhang, H.; Li, N.; Jiang, Z. Effects of nano-particle lubrication on micro deep drawing of Mg-Li alloy. *Int. J. Adv. Manuf. Technol.* **2019**, *104*, 4409–4419. [CrossRef]

28. Haus, F.; German, J.; Junter, G.-A. Primary biodegradability of mineral base oils in relation to their chemical and physical characteristics. *Chemosphere* **2001**, *45*, 983–990. [CrossRef]

29. Kamali, H.; Xie, H.; Zhao, H.; Jia, F.; Wu, H.; Jiang, Z. Frictional Size Effect of Light-Weight Mg–Li Alloy in Micro Deep Drawing under Nano-Particle Lubrication Condition. *Mater. Trans.* **2020**, *61*, 239–243. [CrossRef]

30. Huo, M.; Wu, H.; Xie, H.; Zhao, J.; Su, G.; Jia, F.; Li, Z.; Lin, F.; Li, S.; Zhang, H.; et al. Understanding the role of water-based nanolubricants in micro flexible rolling of aluminium. *Tribol. Int.* **2020**, *151*, 106378. [CrossRef]

31. Sun, J.L.; Zhu, Z.X.; Xu, P.F. Study on the Lubricating Performance of Nano-TiO$_2$ in Water-Based Cold Rolling Fluid. *Mater. Sci. Forum* **2015**, *817*, 219–224. [CrossRef]

32. Wu, H.; Kamali, H.; Huo, M.; Lin, F.; Huang, S.; Huang, H.; Jiao, S.; Xing, Z.; Jiang, Z. Eco-Friendly Water-Based Nanolubricants for Industrial-Scale Hot Steel Rolling. *Lubricants* **2020**, *8*, 96. [CrossRef]

33. Wu, K.; Zhang, Y.; Ge, F.; Huang, X.; Ge, H.H.; Zhao, Y.Z.; Meng, X.J.; Dou, B.L. Corrosion behavior of brass in SCW-SDBS-TiO$_2$ nanofluid. *J. Alloy. Compd.* **2021**, *855*, 9. [CrossRef]

34. Luo, L.; Jiang, Z.; Wei, D.; Manabe, K.-I.; Sato, H.; He, X.; Li, P. An experimental and numerical study of micro deep drawing of SUS304 circular cups. *Manuf. Rev.* **2015**, *2*, 27. [CrossRef]

35. Luo, L.; Jiang, Z.; Wei, D. Influences of micro-friction on surface finish in micro deep drawing of SUS304 cups. *Wear* **2017**, *374*, 36–45. [CrossRef]

36. Wei, D.; Luo, L.; Sato, H.; Jiang, Z.; Manabe, K. Simulations of hydro-mechanical deep drawing using Voronoi model and real microstructure model. *Procedia Eng.* **2017**, *207*, 1033–1038. [CrossRef]

37. Wu, Y.; Tsui, W.; Liu, T. Experimental analysis of tribological properties of lubricating oils with nanoparticle additives. *Wear* **2007**, *262*, 819–825. [CrossRef]

 *lubricants*

Article

Lubrication Performance and Mechanism of Electrostatically Charged Alcohol Aqueous Solvents with Aluminum–Steel Contact

Xiaodong Hu [1,2], Ying Wang [1,2], Hongmei Tang [1,2], Yu Xia [1,2], Shuiquan Huang [3], Xuefeng Xu [1,2,*] and Ruochong Zhang [1,2,*]

1   College of Mechanical Engineering, Zhejiang University of Technology, Hangzhou 310023, China
2   Key Laboratory of Special Purpose Equipment and Advanced Processing Technology, Ministry of Education and Zhejiang Province, Zhejiang University of Technology, Hangzhou 310023, China
3   School of Mechanical Engineering, Yanshan University, Qinhuangdao 066004, China
*   Correspondence: xuxuefeng@zjut.edu.cn (X.X.); zhangruochong@zjut.edu.cn (R.Z.)

Abstract: Alcohol aqueous solvents were prepared by individually adding n-propanol, isopropanol, 1,2-propanediol, and glycerol to deionized water for use as lubricants for the electrostatic minimum quantity lubrication (EMQL) machining of aluminum alloys. The tribological characteristics of those formulated alcohol solvents under EMQL were assessed using a four-ball configuration with an aluminum–steel contact, and their static chemisorption on the aluminum surfaces was investigated. It was found that the negatively charged alcohol lubricants (with charging voltages of −5 kV) resulted in 31% and 15% reductions in the coefficient of friction (COF) and wear scar diameter (WSD), respectively, in comparison with those generated using neutral alcohol lubricants. During the EMQL, static charges could help dissociate the alcohol molecules, generating more negative ions, which accelerated the chemisorption of those alcohol molecules on the aluminum surfaces and thereby yielded a relatively homogeneous-reacted film consisting of more carbon and oxygen. This lubricating film improved the interfacial lubrication, thus producing a better tribological performance for the aluminum alloys. The results achieved from this study will offer a new way to develop high-performance lubrication technologies for machining aluminum alloys.

Keywords: alcohols; electrostatic minimum quantity lubrication; aluminum–steel contact; friction and wear

Citation: Hu, X.; Wang, Y.; Tang, H.; Xia, Y.; Huang, S.; Xu, X.; Zhang, R. Lubrication Performance and Mechanism of Electrostatically Charged Alcohol Aqueous Solvents with Aluminum–Steel Contact. *Lubricants* **2022**, *10*, 322. https://doi.org/10.3390/lubricants10110322

Received: 24 October 2022
Accepted: 15 November 2022
Published: 21 November 2022

Publisher's Note: MDPI stays neutral with regard to jurisdictional claims in published maps and institutional affiliations.

1. Introduction

Owing to their low density, high specific strength, atmospheric corrosion resistance, and good electrical and thermal conductivity [1,2], aluminum alloys have been widely used in aerospace and automobile manufacturing [3,4]. However, problems such as the sticking phenomenon and built-up edges are prone to occur in the cutting process of the aluminum alloy due to its low hardness and high plasticity, which may negatively affect the machining quality of the workpiece [5]; the transfer of aluminum to steel in the aluminum–steel system is easy to occur in the aluminum–steel system, resulting in various degrees of damage on the surface of the aluminum parts, and even serious failure of the friction pair [6,7]. Therefore, in order to meet the requirements of the machining precision of aluminum parts and address the problems of quick-wear and the difficult lubrication of the aluminum–steel contact, it is crucial to realize the high-efficiency lubrication of aluminum alloys.

In the study of aluminum alloy lubrication, alcohols are well-known as highly effective anti-friction and anti-wear additives [8,9]. Montgomery [10] and Hironaka [11] studied the wear behavior of an aluminum–steel system under boundary lubrication conditions. It was discovered that functional groups, such as the hydroxyl groups in the alcohol compounds, can react with the aluminum surface to produce amorphous substances (aluminum complexes or salts), which may form sufficient adsorption films. Hotten [12] and Wan et al. [13] found that diol compounds are efficient boundary lubricants, which may react with the aluminum to form five- or six-element complexes for lubrication, and they also pointed out that the molecular structure has an important influence on the lubrication. Hu et al. [9,14] investigated the lubrication performance of a series of alcohols. The results showed that alcohols can significantly improve the lubrication state of the aluminum surface, whether it is a long chain or a short chain. The generation of an organic aluminum alkoxide on the aluminum alloy surface to produce continuous boundary films serves as the lubrication mechanism, thus lessening or avoiding wear. Its anti-wear ability and load-carrying capacity were affected by the chain length, hydroxyl number, and concentration.

Kajdas proposed the negative-ion-radical mechanism (NIRAM) based on the theory of exoelectron emission to explain the lubrication mechanism of alcohols during friction [15,16], and this hypothesis was subsequently supported by much research [17–19]. They believed that the tribochemical reaction between the alcohols and aluminum was primarily initiated by electrons. During the friction, the electrons emitted from the aluminum surface cause the dissociation of the alcohol molecules to form negative ions and free radicals and, finally, form an anti-friction and anti-wear protective layer on the metal surface. Inspired by the NIRAM, we assumed that when the lubricating fluids are charged by an external electric field, a large number of electrons will play a facilitating role in generating more negative ions or free radicals and enhancing the film-forming chemical reaction at the friction interface.

Electrostatic minimum quantity lubrication (EMQL) uses a high-voltage electrostatic electrode to contact-charge lubricants and produces charged mists under compressed air that are sprayed into the machining area for lubrication and cooling. The research showed that [20,21] the lubricant droplets carry electrostatic charges as the atomization of the charged lubricant, and their charge-to-mass ratio [22] sharply rises with the increase of the charging voltage's absolute value; the particle size of the lubricating oil droplets decreased and was more uniformly distributed after the charging. The wettability, penetration, and deposition of the lubricant were all optimized. This technology has demonstrated excellent lubrication performance in the cutting of difficult-to-machine materials, such as stainless steel and titanium alloy [23,24], compared to the conventional minimum quantity lubrication (MQL). Previous research mainly focused on the analysis of the changes in the physical properties of the droplets after charging, but the impacts of the electrostatic interaction on the interfacial chemical reactions were rarely involved.

In this study, four short-chain alcohols were selected as lubricants, and the lubricants were contact-charged by high-voltage electrodes (EMQL was used as the charging method). Firstly, the tribological properties of the charged alcohols in the aluminum–steel system were investigated. Secondly, the impacts of the various molecular structures on the friction-reduction and wear-resistance properties were explored by evaluating the lubricating performance of the four alcohols and the corrosion tendency of the aluminum in different alcohol solutions. Finally, the lubrication mechanism of the charged alcohols was analyzed through the static reaction experiment of the alcohol with the aluminum, in combination with the surface morphology and element distribution of the aluminum ball-wear scar after the friction. To improve the machining performance and lubrication efficiency of aluminum alloys, the study findings may be applied in these fields.

2. Materials and Methods

2.1. Preparation of the Lubricants

Four analytically pure alcohols were chosen: n-propanol, isopropanol, 1,2-propanediol, and glycerol (purchased from Sinopharm Chemical Reagent Co., Ltd., Shanghai, China). The properties of the alcohols are shown in Table 1. The effect of the hydroxyl position on the friction and wear of the aluminum was studied by n-propanol and isopropanol, and the effect of the hydroxyl number was studied by n-propanol/isopropanol, 1,2-propanediol, and glycerol. Alcohols must be diluted with neutral solvents to lessen their reactivity since they easily corrode aluminum substrates [25,26]. The final alcohol lubricants were formulated at a volume concentration of 25% (25% alcohol and 75% deionized water).

Table 1. Properties of alcohols.

Alcohol	Formula	Molecular Mass	Boiling Point (°C)	Viscosity at 20 °C (mPa·s)
n-Propanol		60.1	97	2.256
Isopropanol		60.1	83	2.038
1,2-Propanediol		76.1	195	60.5
Glycerol		92.1	290	1412

2.2. Tribological Testing

The four-ball friction tests were conducted by an MMW-1 multi-specimen test system (Jinan Shijin Group Co., Ltd., Jinan, China). The upper friction pair consisted of a GCr15 steel ball, and the lower friction pair were three 1060 aluminum balls (the diameters were all 12.7 mm). The friction tester was grounded (as shown in Figure 1). The test conditions were as follows: loads of 12, 24, and 49 N; rotational speeds of 50, 150, and 250 r/min; the EMQL system was used for drip lubrication (air pressure: 0 MPa, flow rate: 20 mL/h, and the charging voltage, −5 kV); the test time was 20 min. The friction coefficient was recorded by the computer in real-time, and three parallel tests were conducted for each group. After the test, the test balls were removed and ultrasonically cleaned with petroleum ether for 10 min, and the wear scar diameters of the aluminum balls under various lubrication conditions were measured with a VW-6000 optical microscope (Keyence, Osaka, Japan). SEM and EDS (EVO18, Zeiss, Oberkochen, Germany) were used to observe and analyze the morphology and elemental composition of the worn surfaces of the aluminum balls.

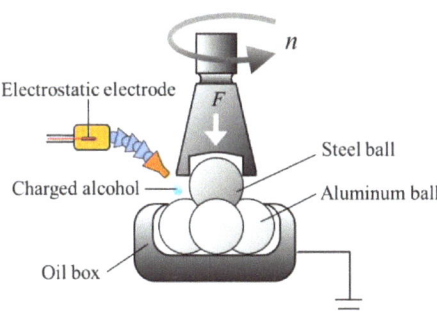

Figure 1. Schematic of the four-ball test.

2.3. Static Reaction Experiment

The static reaction tests of the aluminum plates and alcohols were carried out. The test device is shown in Figure 2. We selected 1060 aluminum plates (10 mm × 10 mm × 1 mm), and the alcohol solutions were n-propanol, isopropanol, 1,2-propanediol, and glycerol aqueous solutions, with a volume concentration of 25%. The aluminum plate was heated by an oil bath at 25 °C, 75 °C, or 140 °C for 30 min. The alcohol solutions were dropped on the surface of the aluminum plate by EMQL, with a flow rate of 20 mL/h and charging voltages of 0 kV and −5 kV. The surfaces of the aluminum plates were polished to mirrors before the experiment. After the experiment, the aluminum plates were ultrasonically cleaned in anhydrous ethanol for 10 min to remove the residual of the physical adsorption solution. SEM/EDS (EVO18, Zeiss, Oberkochen, Germany) was used to observe and analyze the morphology and element content of the reaction film on the aluminum plate surface.

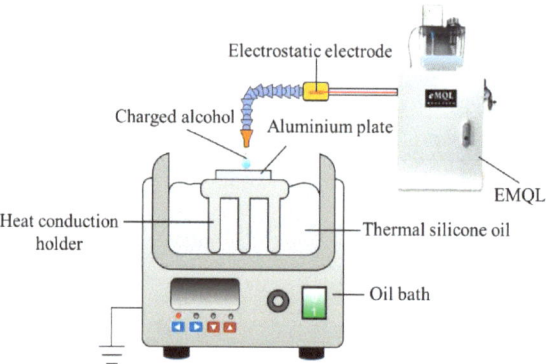

Figure 2. Diagram of static reaction device.

2.4. Electrochemical Polarization Test

A CH1760E electrochemical workstation (Shanghai Chenhua Instrument Co., Ltd., Shanghai, China) was used for the electrochemical polarization test (Figure 3). It consisted of a three-electrodes system; the working electrode was a 1060 pure aluminum plate (Φ 14 mm × 2 mm), the reference electrode was a calomel electrode, and the auxiliary electrode was a platinum electrode. The working electrode was immersed in four kinds of a 25% alcohol solution (n-propanol, isopropanol, 1,2-propanediol, and glycerol), and the self-corrosion potential of the aluminum plate was measured by the Tafel curve. Each test group was divided into three groups, and the average value was taken as the corrosion potential.

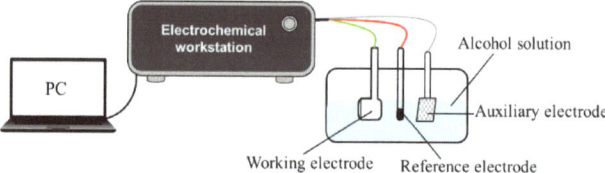

Figure 3. Diagram of electrochemical polarization test.

3. Results and Discussion

3.1. Tribological Performance

3.1.1. Lubrication Performance of Charged Alcohol Lubricants

Figure 4 shows the average COF and WSD as a function of the load under the lubrication conditions of the charged and uncharged four alcohol solutions. It can be seen that the COF and WSD lubricated by the charged four alcohol solutions are lower than those by the uncharged solutions at all loads. The largest reduction among them in the COF is around 31% (glycerol, 49 N), and the highest reduction in the WSD is about 12%, i.e., the wear volume is reduced by nearly 30% (n-propanol, 49 N). According to the findings, charging may effectively enhance the anti-friction and anti-wear properties of alcohol solutions. Since the adsorption capacity of alcohol molecules is enhanced after charging, it is easier to form protective lubricating films on the friction surface, thus avoiding the direct contact of the aluminum–steel friction pair.

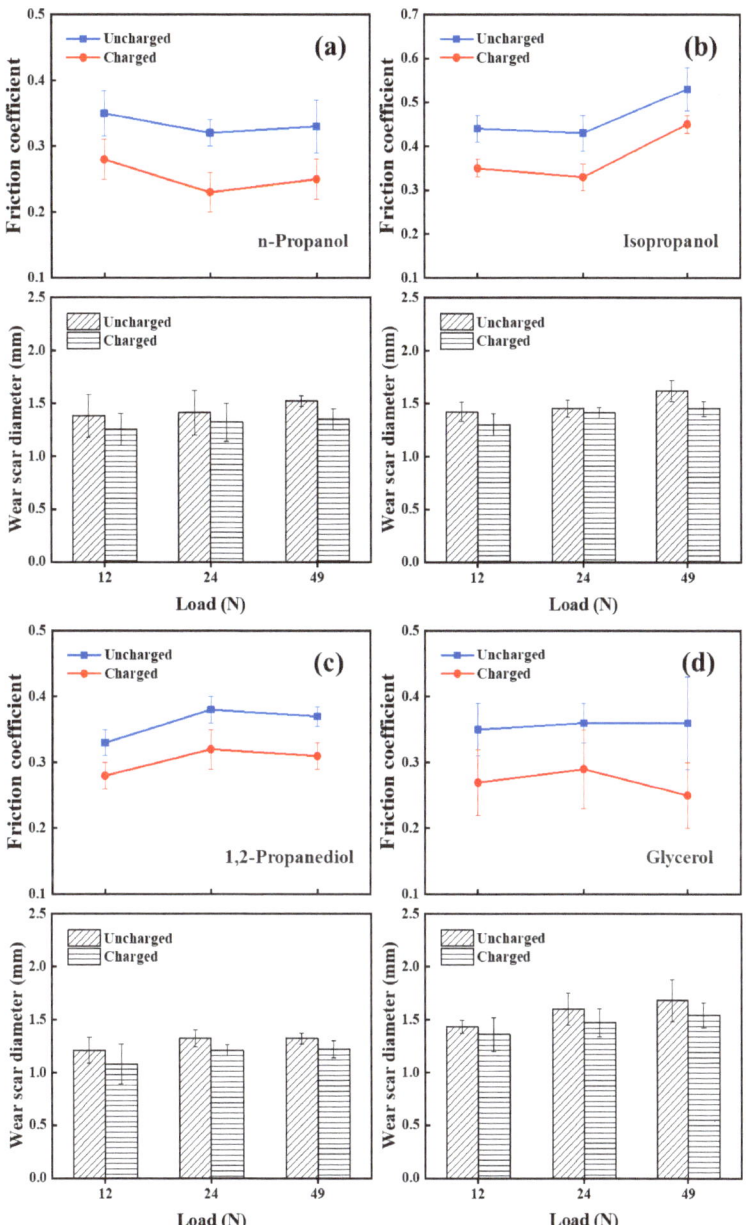

Figure 4. Comparison of the COF and WSD lubricated by charged and uncharged (**a**) n-propanol, (**b**) isopropanol, (**c**) 1,2-propanediol, and (**d**) glycerol under different loads (rotational speed: 150 r/min).

The average COF and WSD lubricated by the four alcohols, both charged and uncharged, at different rotational speeds are shown in Figure 5. The charged alcohol solutions exhibit good tribological properties at various rotational speeds. Among them, when the rotational speed is 150 r/min, the COF of the charged glycerol decreases by approximately 30%, and when the rotational speed is 50 r/min, the WSD of 1,2-propanediol reduces by about 15%, i.e., the wear volume decreases by nearly 36%. This demonstrates that when four kinds of alcohol solutions are charged, the lubrication performances of the aluminum–steel friction pair are improved.

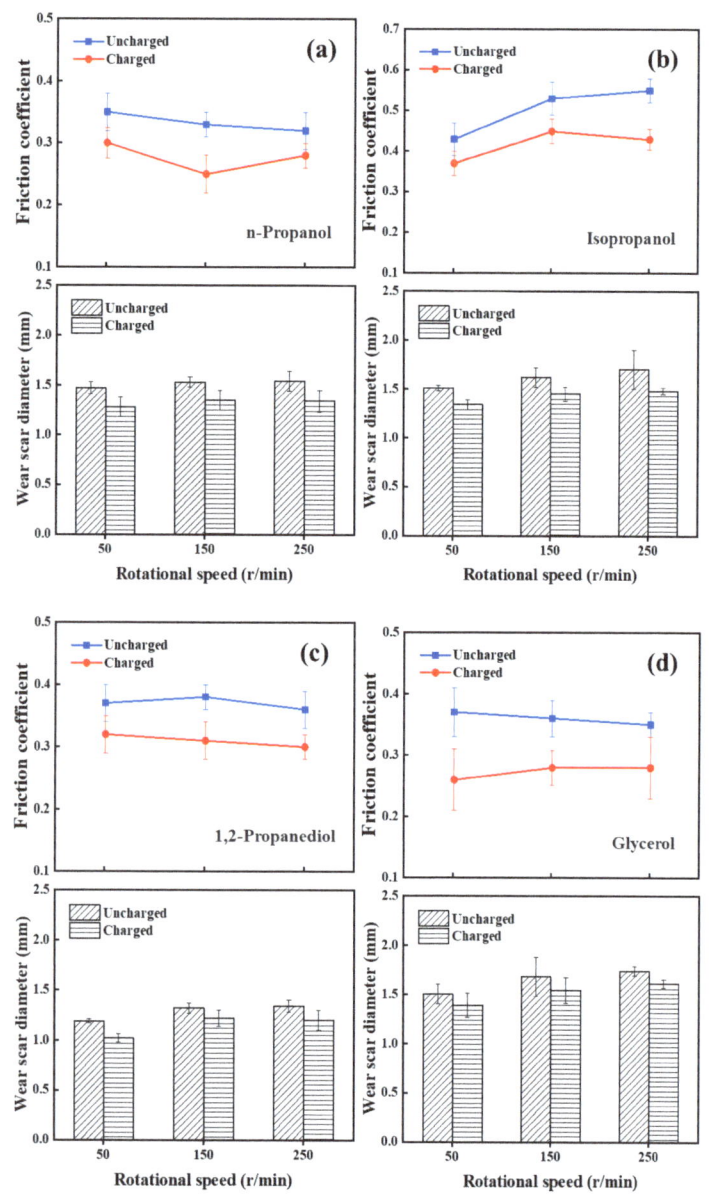

Figure 5. Comparison of the COF and WSD lubricated by charged and uncharged (**a**) n-propanol, (**b**) isopropanol, (**c**) 1,2-propanediol, and (**d**) glycerol under different rotational speeds (load: 49 N).

3.1.2. Comparison of Four Alcohols' Lubrication Performance

Figure 6 shows the COF of the aluminum–steel contact lubricated by the four alcohol solutions under different loads and speeds. As can be observed, the order of the COF is:

n-propanol < glycerol < 1,2-propanediol < isopropanol. Among them, isopropanol has the largest COF, with a large gap compared to the other three alcohols. Although the average COF of the glycerol solution is small, the real-time COF curve fluctuates considerably (Figure 7), indicating that its lubrication stability is poor.

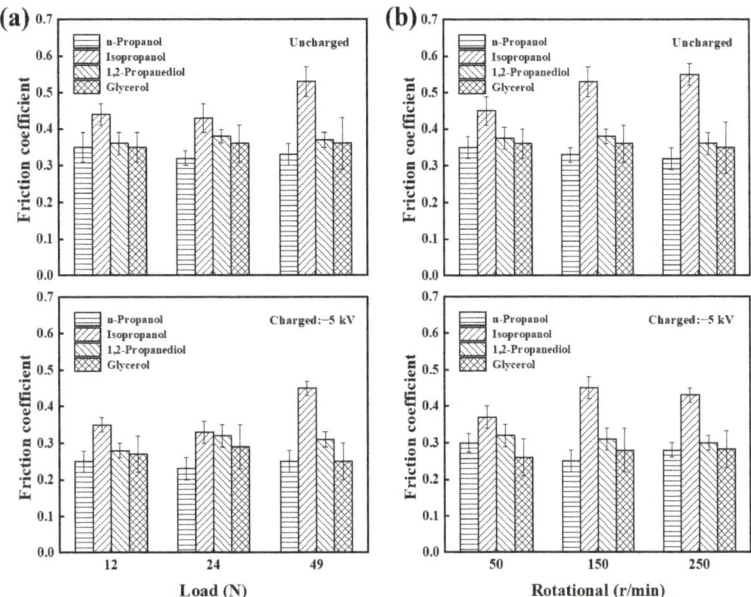

Figure 6. COF lubricated by four alcohol solutions under different (**a**) loads (speed is 150 r/min) and (**b**) rotational speeds (load is 49 N).

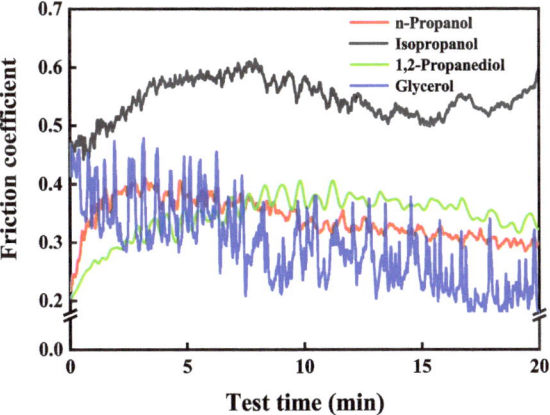

Figure 7. COF curves of aluminum–steel friction pairs with time (uncharged; load: 49 N; rotational speed: 250 r/min).

Figure 8 compares the WSD on the surface of the aluminum balls lubricated by the four alcohol solutions under different loads and rotational speeds. The order of the wear scar diameter is: 1,2-propanediol < n-propanol < isopropanol < glycerol. 1,2-Propanediol has the smallest WSD, which is significantly lower than that of the other alcohols and shows better anti-wear performance. The most severe wear is seen with glycerol, which corresponds to its poor stability. A large amount of flaking debris is generated as a result of the severe wear, causing remarkable fluctuations in the COF [27,28] and worse lubrication performance.

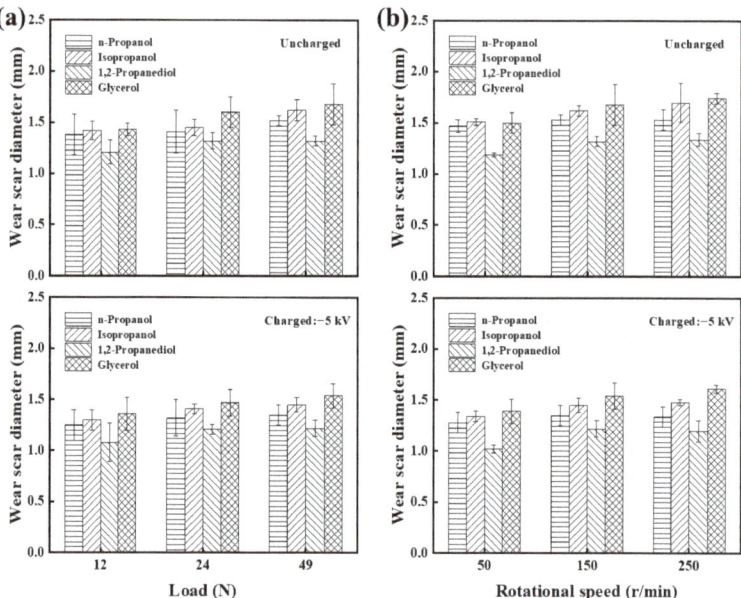

Figure 8. The WSD lubricated by four alcohol solutions under different (**a**) loads (speed is 150 r/min) and (**b**) rotational speeds (load is 49 N).

Glycerol, which has a unique polyhydroxy structure, possesses the strongest adsorption capability among the four alcohols [29]. Theoretically, it can form a more durable chemisorption film with the aluminum surface, which can effectively prevent the transfer and adhesion of the aluminum and shows high anti-wear and bearing performance. Glycerol did not, however, exert excellent lubrication performance in our study. It is speculated that the aluminum matrix may have experienced over-corrosion owing to the higher concentration of the glycerol solution. The self-corrosion potential of the aluminum in the four alcohol solutions was measured to judge the corrosion tendency of the aluminum in different alcohol solutions. As can be seen in Figure 9, glycerol has the largest negative potential values, suggesting that aluminum loses electrons easier, i.e., it is more easily corroded.

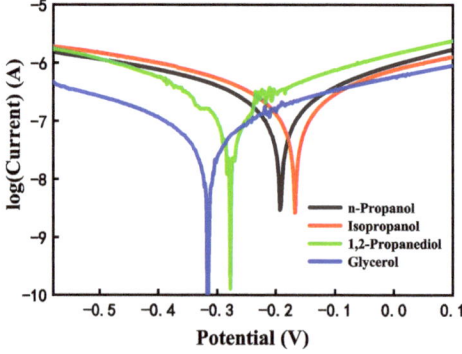

Figure 9. Polarization characteristic curve of aluminum plate in 25% of four alcohol solutions.

The self-corrosion potential of the aluminum plate is −0.251 V when the concentration of glycerol is lowered to 10% (as shown in Figure 10). Compared to a 25% concentration, the corrosion resistance has increased. Additionally, the COF curve fluctuates slightly, and the aluminum ball's WSD decreases by about 19% (Figure 11). The lubrication stability and anti-wear capacity of glycerol increase as its concentration decreases. It is concluded that the over-corrosion of glycerol on the aluminum matrix at a 25% concentration affects its lubrication performance and results in inadequate anti-friction and anti-wear performance.

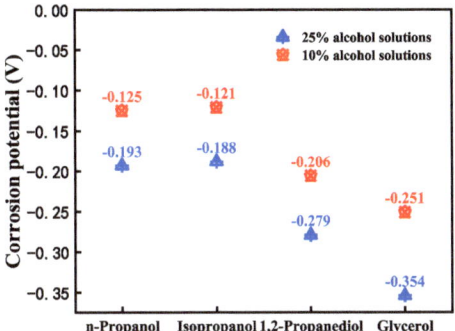

Figure 10. Corrosion potential of aluminum plate in four alcohol solutions.

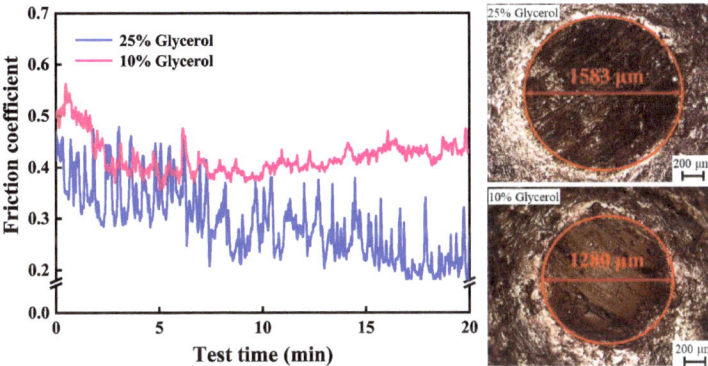

Figure 11. The COF and wear scar morphology under 10% and 25% glycerol lubrication (uncharged; load: 49 N; rotational speed: 250 r/min).

3.2. Static Reaction Experiment Results

Table 2 shows the image of the reaction films on the aluminum plate surface following the static reaction experiment. It can be seen that, at ambient temperature, there is no adsorption film on the aluminum plate surface when the solution is uncharged; the white adsorption films are visible on the local area of the aluminum plate surface after the treatment of the four alcohol solutions with a high voltage electrostatic charge. The SEM morphology and EDS energy spectrum of the reaction film area on the aluminum plate surface after the charged isopropanol treatment were analyzed (as shown in Figure 12). Some deposits may be observed on the aluminum plate surface. C and O elements are present in this substance. So, obviously, charging stimulates the adsorption of the alcohol molecules, which form relatively stable compounds on the aluminum surface.

Table 2. Static reaction films on aluminum plate surface.

	25 °C		75 °C		140 °C	
	Uncharged	**Charged**	**Uncharged**	**Charged**	**Uncharged**	**Charged**
n-Propanol						
Isopropanol						
1,2-Propanediol						
Glycerol						

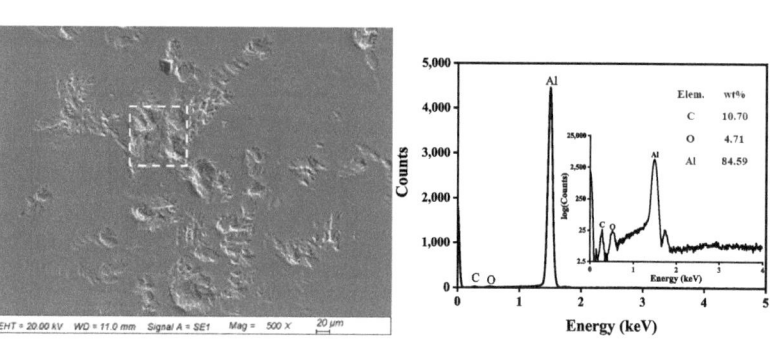

Figure 12. SEM and EDS of the aluminum plate surface after charged isopropanol static reaction (at room temperature: 25 °C).

The surface of the aluminum plate is covered with a film at 75 °C, whether charged or not. However, the surface film is not obvious when glycerol is used, indicating that the chemical interaction between glycerol and aluminum is negligible at 75 °C. When the temperature is raised to 140 °C, a reaction film appears on the aluminum plate surface (Table 2), indicating that glycerol can react with aluminum.

Figure 13 depicts the aluminum plate surface's SEM picture and EDS energy spectrum, detected within the dashed boxed areas after three different types of treatment. Compared with the untreated aluminum plate, the surface of the aluminum plate is covered with a reaction film after the n-propanol static reaction treatment, and the content of the C and O elements rises, suggesting that the reaction film might be an organic film or an oxide film. Meanwhile, the surface film is dense, and the surface scan reveals fewer black holes following the charged n-propanol treatment than in the uncharged, demonstrating a uniform distribution of the C and O elements. The percentages of C and O increase by around 9.8% and 61.9%, respectively, according to energy spectrum analysis. This indicates that the chemical activity of the alcohols can be effectively boosted by charging.

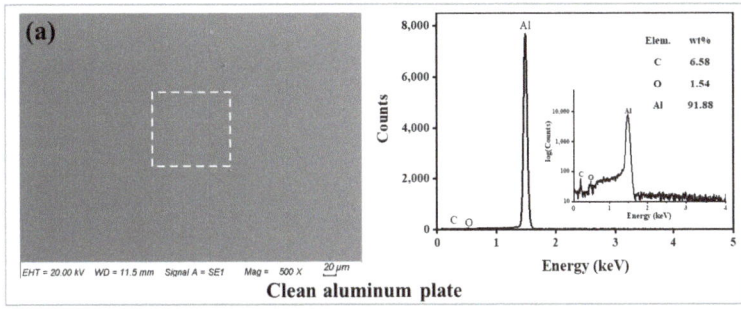

Clean aluminum plate

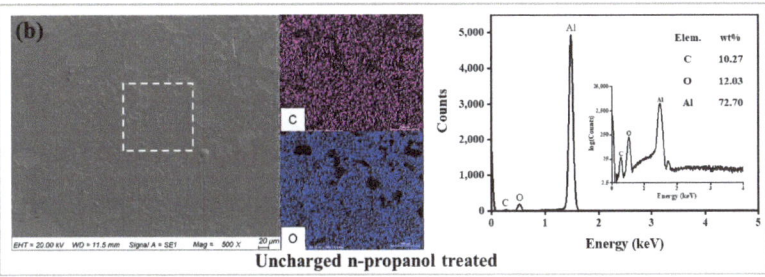

Uncharged n-propanol treated

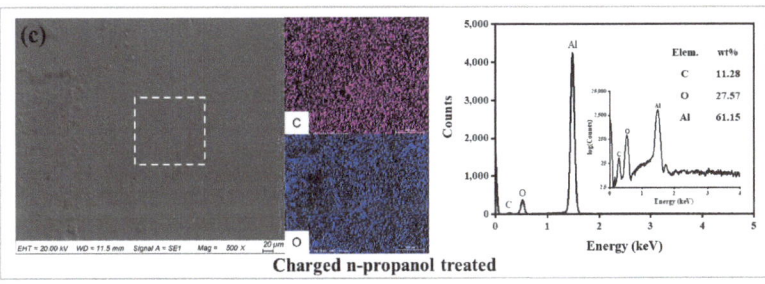

Charged n-propanol treated

Figure 13. SEM and EDS of aluminum plate surface after (**a**) untreated, (**b**) uncharged n-propanol-treated, and (**c**) charged n-propanol-treated (75 °C heating).

It can be seen from Table 3 that the C and O content all increased after the four alcohol solutions were charged, compared to uncharged, which proves that charging promotes the reactivity between alcohol and aluminum, and thus promotes the formation of a reaction film. In contrast to other alcohols, n-propanol and 1,2-propanediol have higher C and O contents, and the rise in the C and O content is more pronounced after charging, which is consistent with their better anti-friction and anti-wear performance.

Table 3. Element content of aluminum surface after four alcohols' static reaction (75 °C heating).

Alcohols Content %	n-Propanol		Isopropanol		1,2-Propanediol		Glycerol	
	Uncharged	Charged	Uncharged	Charged	Uncharged	Charged	Uncharged	Charged
C	10.27	11.28	6.13	12.88	11.07	16.28	7.89	15.02
O	17.03	127.57	10.82	9.67	4.28	17.91	2.86	5.89
Al	72.70	61.15	83.05	77.45	84.65	65.81	89.25	79.09

3.3. Lubrication Mechanism of Alcohol with Different Molecular Structures

The four alcohols with different molecular structures exerted different effects on the lubrication of the aluminum, according to the tribological testing and static reaction experiment.

The effect of the molecular structure on the anti-friction properties was analyzed. It showed that a three-layer carbon chain structure is formed when n-propanol adsorbs to the friction surface (Figure 14a). This multi-layer structure may better separate the steel–aluminum friction pair and achieves lower COFs in the four alcohol solutions. 1,2-propanediol possesses a two-layer carbon chain structure, while a five-membered ring structure is stably adsorbed, and the COF is likewise low. Isopropanol has only one hydroxyl group and performs poor adsorption despite forming a two-layer carbon chain structure on the friction surface (Figure 14b). Since the adsorption film is thin, the friction increases.

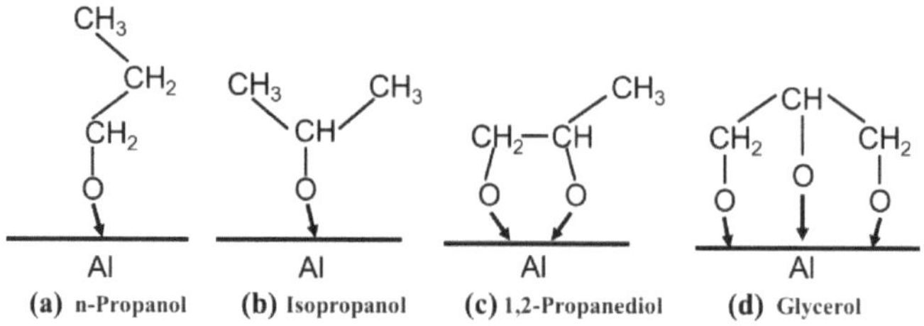

Figure 14. Chemisorption model of alcohols on aluminum surface during friction. (**a**) n-propanol, (**b**) isopropanol, (**c**) 1,2-propanediol, (**d**) glycerol.

By analyzing the influence of the molecular structure on the anti-wear properties, it can be concluded that 1,2-propanediol containing two hydroxyl groups forms a bidentate bond with the aluminum atom on the surface (Figure 14c). The stable five-membered ring structure [12,13] efficiently prevents the transfer of aluminum to steel and provides good anti-wear properties. Although glycerol with three hydroxyl groups might produce stronger adsorption (Figure 14d), Igari et al. [30] found that alcohol with this structure has a great impact on the aluminum wear when hydroxyl groups exist at both ends of the lubricant molecule. Glycerol exhibits the highest amount of wear among the four alcohols, which is attributed to its corrosive wear on the aluminum matrix at high concentrations.

3.4. Lubrication Mechanism of Charged Alcohols

In order to investigate the lubrication mechanism of the charged alcohols, n-propanol and 1,2-propanediol, which exhibit preferable anti-friction and anti-wear properties, were selected for further analysis. The wear scar morphology and the EDS energy spectrum of the aluminum ball surface after tribological tests with n-propanol and 1,2-propanediol lubrication are depicted in Figures 15 and 16. Deposits can be seen on the surface of the wear scar (the darker regions in the figures), demonstrating that the alcohol molecules build a protective film on the aluminum surface through physical and chemical adsorption or a reaction during the friction and have effective lubrication. At the same time, energy spectrum analysis revealed the presence of C and O components in the wear scar, with a very high concentration of O, which indicates that the surface film is dominated by the oxide film. This may be because the organic aluminum compounds are soft, shearable, and generally sacrificial layers [31] produced by the reaction of the alcohol and aluminum. The internal C-O chain is eventually broken by the action of the high shear force, leaving just alumina and its hydrate [32]. This viewpoint can be confirmed by the results of the static reaction experiment (Section 3.2). The surface films produced by the static interaction between the alcohol and aluminum are mostly organic films and oxide films without high shear forces, as shown in Table 2 by the slight differences in the C and O contents of the surface films.

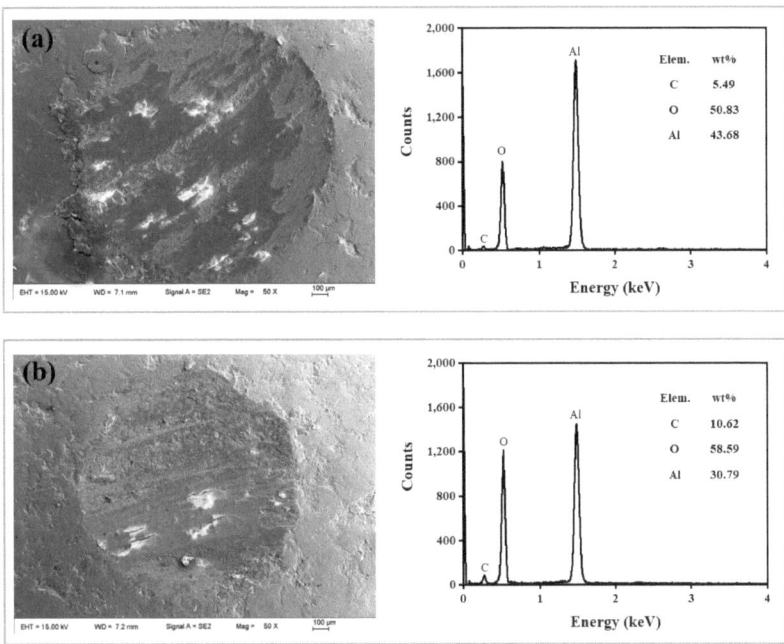

Figure 15. SEM and EDS of aluminum ball surface under (**a**) uncharged n-propanol lubrication and (**b**) charged n-propanol lubrication (load: 49 N; rotational speed: 150 r/min).

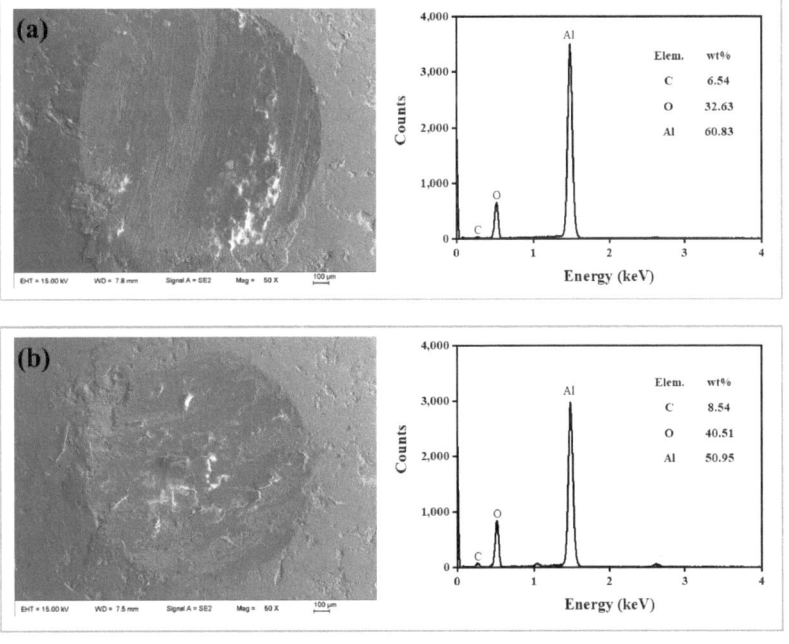

Figure 16. SEM and EDS of aluminum ball surface under (**a**) uncharged 1,2-propanediol lubrication and (**b**) charged 1,2-propanediol lubrication (load: 49 N; rotational speed: 150 r/min).

Comparing the impacts of uncharged and charged on alcohol lubrications, it is discovered that the wear scar is larger, with visible flaking and adhesion, surrounded under uncharged conditions (as illustrated in Figures 15a and 16a), which is mainly adhesive wear. This may be due to the low adsorption strength of the lubricants on the friction surface, which desorbs at a specific temperature [33] and cannot lubricate the surface properly, resulting in direct contact with some surfaces. Furthermore, because aluminum material is relatively soft, friction can cause plastic deformation on its surface. Adhesion points will be formed, accompanied by an instantaneously high temperature. When the alcohol is charged, the wear scar decreases, and its surrounding area becomes smooth and flat, with nearly no adhesion (as shown in Figures 15b and 16b). The energy spectrum shows that the percentages of C and O on the surface rise in comparison to the uncharged (n-propanol: 93.4%, 15.3%; 1,2-propanediol: 30.6%, 24.1%). It indicates that charging improves the lubrication performance of the alcohols. This might be because the reactivity between the alcohol and aluminum is enhanced after charging, and a high-strength chemical reaction film is easier to form on the friction surface. This film plays a sacrificial protective role, effectively preventing direct contact between the metal surfaces and reducing or avoiding the adhesion phenomenon, thereby reducing surface wear.

According to the Hard and Soft Acids and Bases (HSAB) theory [34] and the NIRAM, the reason why alcohols effectively lubricate the aluminum surface is that compounds with polar functional groups (e.g., alcohols) are hard bases, which are simple to adsorb to fresh hard acid surfaces (e.g., aluminum). There are several tribochemical reactions that occur during the friction process [35,36], including surface oxidation, high-energy nascent surface or abrasive particle catalysis [37], exoelectron emission, the oxidation and degradation of lubricating oil molecules, and local temperature rise. Among them, the interaction between the emitted electrons and alcohol molecules induces the dissociation of the alcohol molecules to form negative ions and radicals, which adsorb on the positive charge spots on the surface to produce an organometallic chemisorption film. The dehydrogenation reaction may take place since the C-H chain in alcohol breaks under a variety of catalytic, high-temperature, and oxidation conditions. This may cause intermolecular crosslinking and the formation of a network polymer film to protect the friction surface. If the shear strength is too high, the chemical bonds of organometallic compounds will break, and inorganic films (such as Al_2O_3) will be produced. The protective films will ultimately be destroyed by friction and wear, and then new organic and inorganic films will be produced. It is a dynamic process.

The above process demonstrates that the electron is very important for the reaction between alcohol and aluminum. However, the electron only originated from the friction process when general alcohols were employed as lubricants. The alcohols' lubrication effectiveness may not be completely utilized for aluminum—a material with weak exoelectron emission intensity [38,39]. By introducing electrostatic technology, the alcohol solutions are charged with negative high-voltage static electricity before entering the friction area, and a large number of electrons trigger the dissociation of alcohol molecules to generate negative ions and radicals. Alcohol molecules are, once more, ionized when they come into contact with friction-induced exoelectrons. Then, more negative ions on the aluminum surface will participate in chemical adsorption, thus forming a thicker lubricating film.

The lubrication mechanism of the charged alcohols has been analyzed using n-propanol as an example (Figure 17). Firstly, the alcohol molecules are dissociated to produce negative ions under the action of the electrons induced by charging and friction. Subsequently, the negative ions are chemisorbed to positive charge sites on the aluminum surface, forming an adsorption film. Molecular cross-linking may also take place simultaneously, generating an easily sheared network polymer film that effectively prevents direct contact between the aluminum–steel and lowers the friction. Finally, the C-O bonds break and generate free radicals under the influence of the high shear. The residual O reacts with the Al to create Al_2O_3—a thin, hard, and dense oxide film that provides an anti-wear effect.

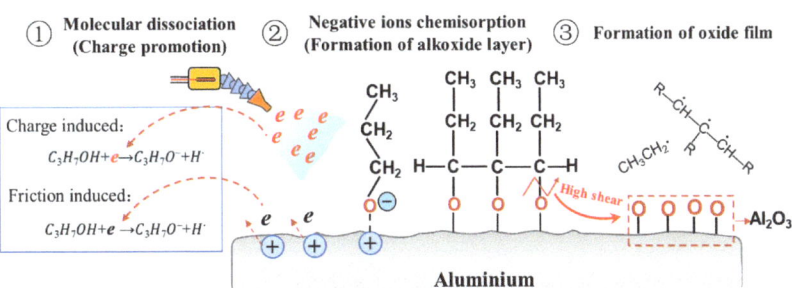

Figure 17. Lubrication model of charged n-propanol.

4. Conclusions

In this paper, the effects of charged alcohol lubricants on the lubrication performance of aluminum and its mechanism were analyzed. Based on the NIRAM, the lubrication characteristics of EMQL technology were studied from the perspective of the chemical reaction film. The following conclusions can be drawn:

(1) Compared with the uncharged case, the COF of the charged four alcohols decreased by about 31% at the highest, and the WSD was reduced by up to 15%. The lubrication performance of n-propanol and 1,2-propanediol was superior to that of the other alcohols;

(2) The static reaction experiment demonstrated that, following alcohol charging, the aluminum plate's surface film had a uniform morphology, the content of C and O grew noticeably, and the molecular structure had a certain impact on the content of C and O;

(3) The tribological properties of aluminum–steel friction pairs under alcohol lubrication were affected by the alcohols' molecular structures. The highest anti-friction performance was provided by n-propanol, which formed a three-layer carbon chain structure after its adsorption, and the best anti-wear performance was offered by 1,2-propanediol, which formed a stable five-membered ring structure on the aluminum surface;

(4) The lubrication mechanism of the charged alcohols was that the introduction of electrostatic technology enhanced the dissociation of the alcohol molecules to produce more negative ions. Numerous negative ions were chemisorbed onto the positive charge sites on the aluminum surface to form an easy shear protective film (the organic aluminum compounds), and then some chemical bonds of the organic aluminum compounds could be broken by the high shear force. Finally, organic aluminum compounds and inorganic aluminum oxide films were formed on the aluminum surface.

The effect of charged alcohol aqueous solvents on the lubrication performance of aluminum was investigated, in this work, using a particular alcohol concentration. In order to improve the practical application value of this research, subsequent work will focus on adding alcohol to the cutting fluid and charging the cutting fluid to evaluate its cutting performance and find the optimum additive concentration.

Author Contributions: Conceptualization, X.H., Y.W. and X.X.; methodology, X.H. and H.T.; validation, X.H., Y.W. and R.Z.; formal analysis, Y.X. and S.H.; investigation, X.H., Y.W. and H.T.; resources, Y.X., H.T. and S.H.; data curation, S.H. and R.Z.; writing—original draft preparation, Y.W.; writing—review and editing, X.H., Y.W., S.H. and X.X.; visualization, H.T. and Y.X.; supervision, X.X. and R.Z.; project administration, S.H., X.X., and R.Z.; funding acquisition, X.H., S.H. and X.X. All authors have read and agreed to the published version of the manuscript.

Funding: This research was financially supported by the National Natural Science Foundation of China (Grant No. 52275468), the National Key Research and Development Program of China (Grant No. 2020YFB2010604), the Natural Science Foundation of Hebei Province (Grant No. E2022203123), and the Hebei Provincial Research Foundation for Returned Scholars (Grant No. C20220335).

Data Availability Statement: The datasets generated during and/or analyzed during the current study are available on request from the corresponding author.

Conflicts of Interest: The authors declare no conflict of interest.

References

1. Li, M.; Wang, J.; Yang, H.; Shi, J. Research on influence of milling process parameters on residual stress of 7055 aluminum alloy. *J. Phys. Conf. Ser.* **2022**, *2268*, 012002. [CrossRef]
2. Dahnel, A.N.; Fauzi, M.H.; Raof, N.A.; Mokhtar, S.; Khairussaleh, N.K.M. Tool wear and burr formation during drilling of aluminum alloy 7075 in dry and with cutting fluid. *Mater. Today Proc.* **2022**, *59*, 808–813. [CrossRef]
3. Sharmila, B.; Selvakumar, G.; Prakash, S.R. Investigations on the effect of dielectric medium and wedm parameters on surface characteristics of al 7068 (ordnance aluminium) alloy. *Surf. Topogr.-Metrol.* **2022**, *10*, 035031.
4. Secgin, O.; Sogut, M.Z. Surface roughness optimization in milling operation for aluminum alloy (al 6061-t6) in aviation manufacturing elements. *Aircraft Eng. Aerosp. Technol.* **2021**, *93*, 1367–1374. [CrossRef]
5. Wei, C.; Shao, T.; Dong, Y.; Yang, C. Preparation and application of aluminum alloy semi-synthetic fluids with high performance. *Lubr. Eng.* **2013**, *38*, 102–107.
6. Nautiyal, P.C.; Schey, J.A. Transfer of aluminum to steel in sliding contact: Effects of lubricant. *J. Tribol.* **1990**, *112*, 282–287. [CrossRef]
7. Decrozant-Triquenaux, J.; Pelcastre, L.; Courbon, C.; Prakash, B.; Hardell, J. Effect of surface engineered tool steel and lubrication on aluminium transfer at high temperature. *Wear* **2021**, *477*, 203879. [CrossRef]
8. Udupa, A.; Sugihara, T.; Viswanathan, K.; Chandrasekar, S. Altering the stability of surface plastic flow via mechanochemical effects. *Phys. Rev. Appl.* **2019**, *11*, 014021. [CrossRef]
9. Hu, Y.; Liu, W. Investigation of aliphatic aacohols as lubricating additives for aluminum on steel contact. *Tribology* **2000**, *20*, 75–77.
10. Montgomery, R.S. The effect of alcohols and ethers on the wear behavior of aluminum. *Wear* **1965**, *8*, 466–473. [CrossRef]
11. Hironaka, S.; Sakurai, T. The effect of pentaerythritol partial ester on the wear of aluminum. *Wear* **1978**, *50*, 105–114. [CrossRef]
12. Hotten, B.W. Bidentate organic compounds as boundary lubricants for aluminum. *Lubr. Eng.* **1974**, *30*, 398–402.
13. Wan, Y.; Liu, W.; Xue, Q. Effects of diol compounds on the friction and wear of aluminum alloy in a lubricated aluminum-on-steel contact. *Wear* **1996**, *193*, 99–104. [CrossRef]
14. Hu, Y.; Liu, W. Tribological properties of alcohols as lubricating additives for aluminum-on-steel contact. *Wear* **1998**, *218*, 244–249.
15. Kajdas, C.; Liu, W. Tribochemistry of aluminium and aluminium alloy systems lubricated with liquids containing alcohol or amine additive types and some other lubricants—A review. *Lubr. Sci.* **2004**, *16*, 267–292. [CrossRef]
16. Kajdas, C. About an anionic-radical concept of the lubrication mechanism of alcohols. *Wear* **1987**, *116*, 167–180. [CrossRef]
17. Kajdas, C. On a negative-ion concept of ep action of organo-sulfur compounds. *ASLE Trans.* **1985**, *28*, 21–30. [CrossRef]
18. Kajdas, C. Importance of anionic reactive intermediates for lubricant component reactions with friction surfaces. *Lubr. Sci.* **1994**, *6*, 203–228. [CrossRef]
19. Hiratsuka, K.; Kajdas, C. Mechanochemistry as a key to understand the mechanisms of boundary lubrication, mechanolysis and gas evolution during friction. *Proc. Inst. Mech. Eng. Part J-J. Eng. Tribol.* **2013**, *227*, 1191–1203. [CrossRef]
20. Lv, T.; Huang, S.Q.; Hu, X.D.; Feng, B.H.; Xu, X.F. Study on aerosol characteristics of electrostatic minimum quantity lubrication and its turning performance. *J. Mech. Eng. Sci.* **2019**, *55*, 129–138. [CrossRef]
21. Xu, X.; Huang, S.; Wang, M.; Yao, W. A study on process parameters in end milling of aisi-304 stainless steel under electrostatic minimum quantity lubrication conditions. *Int. J. Adv. Manuf. Technol.* **2017**, *90*, 979–989. [CrossRef]
22. Huang, S.; Yao, W.; Hu, J.; Xu, X. Tribological performance and lubrication mechanism of contact-charged electrostatic spray lubrication technique. *Tribol. Lett.* **2015**, *59*, 28. [CrossRef]
23. Lv, T.; Xu, X.; Yu, A.; Hu, X. Oil mist concentration and machining characteristics of sio2 water-based nano-lubricants in electrostatic minimum quantity lubrication-emql milling. *J. Mater. Process. Technol.* **2021**, *290*, 116964. [CrossRef]
24. Huang, S.; Lv, T.; Wang, M.; Xu, X. Enhanced machining performance and lubrication mechanism of electrostatic minimum quantity lubrication-emql milling process. *Int. J. Adv. Manuf. Techol.* **2018**, *94*, 655–666. [CrossRef]
25. Huang, S.; Wu, H.; Jiang, Z.; Huang, H. Water-based nanosuspensions: Formulation, tribological property, lubrication mechanism, and applications. *J. Manuf. Process.* **2021**, *71*, 625–644. [CrossRef]
26. Minami, I.; Sugibuchi, A. Surface chemistry of aluminium alloy slid against steel lubricated by organic friction modifier in hydrocarbon oil. *Adv. Tribol.* **2012**, *2012*, 926870. [CrossRef]
27. Huang, G.; Fan, S.; Li, T.; Sun, T.; Fan, H.; Su, Y.; Song, J. Construction of graphite-copper 3-d composite lubricating layer on cu663 alloy surface and its tribological performances. *Tribology* **2021**, *41*, 304–315.
28. Wang, L.; Ge, X.; Liu, Z.; He, M. Effect of particles on oil film characteristics of journal bearing. *Lubr. Eng.* **2021**, *46*, 51–57.
29. Liang, X.; Yang, X.; Wu, J.; Tu, Z.; Wang, Z.; Yuan, Z. Effect of alcohol additives on the plate-out oil film in cold rolling and its molecular dynamics simulations. *Tribol. Trans.* **2019**, *62*, 504–511. [CrossRef]
30. Igari, S.; Mori, S.; Takikawa, Y. Effects of molecular structure of aliphatic diols and polyalkylene glycol as lubricants on the wear of aluminum. *Wear* **2000**, *244*, 180–184. [CrossRef]
31. Hsu, S.M.; Gates, R.S. Boundary lubricating films: Formation and lubrication mechanism. *Tribol. Int.* **2005**, *38*, 305–312. [CrossRef]
32. Li, J.; Zhang, C.; Luo, J. Superlubricity achieved with mixtures of polyhydroxy alcohols and acids. *Langmuir* **2013**, *29*, 5239–5245. [CrossRef] [PubMed]
33. Li, G.; Li, C.; Yu, Y.; Ma, X. Study on the effects of additives on the lubricity for 7075. *Lubr. Eng.* **2010**, *35*, 52–57.

34. Pearson, R.G. Hard and soft acids and bases—The evolution of a chemical concept. *Coord. Chem. Rev.* **1990**, *100*, 403–425. [CrossRef]
35. Carlton, H.; Huitink, D.; Liang, H. Tribochemistry as an alternative synthesis pathway. *Lubricants* **2020**, *8*, 87. [CrossRef]
36. Podrabinnik, P.; Gershman, I.; Mironov, A.; Kuznetsova, E.; Peretyagin, P. Tribochemical interaction of multicomponent aluminum alloys during sliding friction with steel. *Lubricants* **2020**, *8*, 24. [CrossRef]
37. Kajdas, C.; Hiratsuka, K. Tribochemistry, tribocatalysis, and the negative-ion-radical action mechanism. *Proc. Inst. Mech. Eng. Part J-J. Eng. Tribol.* **2009**, *223*, 827–848. [CrossRef]
38. Nakayama, K. Triboemission of charged particles and resistivity of solids. *Tribol. Lett.* **1999**, *6*, 37–40. [CrossRef]
39. Molina, G.J.; Furey, M.J.; Ritter, A.L.; Kajdas, C. Triboemission from alumina, single crystal sapphire, and aluminum. *Wear* **2001**, *249*, 214–219. [CrossRef]

lubricants

Article

Thickening Properties of Carboxymethyl Cellulose in Aqueous Lubrication

Jan Ulrich Michaelis [1,2], Sandra Kiese [2,*], Tobias Amann [3], Christopher Folland [4], Tobias Asam [4] and Peter Eisner [2,5,6]

[1] TUM School of Life Sciences, Technical University of Munich (TUM), D-85354 Freising, Germany
[2] Fraunhofer Institute for Process Engineering and Packaging (IVV), Giggenhauser Straße 35, D-85354 Freising, Germany
[3] Fraunhofer Institute for Mechanics of Materials (IWM), D-79108 Freiburg, Germany
[4] Carl Bechem GmbH, D-58089 Hagen, Germany
[5] ZIEL-Institute for Food & Health, TUM School of Life Sciences Weihenstephan, Technical University of Munich, Weihenstephaner Berg 1, D-85354 Freising, Germany
[6] Faculty of Technology and Engineering, Steinbeis-Hochschule, George-Bähr-Str. 8, D-01069 Dresden, Germany
* Correspondence: sandra.kiese@ivv.fraunhofer.de; Tel.: +49-8161-491-525

Abstract: Increasingly restricted availability and environmental impact of mineral oils have boosted the interest in sustainable lubrication. In this study, the thickening properties of sodium carboxymethyl celluloses (CMCs) were investigated in order to assess their potential as viscosity modifiers in aqueous gear and bearing fluids. The pressure, temperature and shear dependence of viscosity was studied at different concentrations and molecular weights M_W. The tribological properties were investigated at different viscosity grades in both sliding and rolling contact, and compared to rapeseed oil and polyethylene glycol 400. The viscosity of the CMC solutions was adjustable to all application-relevant viscosity grades. Viscosity indices were similar or higher compared to the reference fluids and mineral oil. Temporary and permanent viscosity losses increased with M_W. Permanent viscosity loss was highest for high M_W derivatives, up to 70%. The pressure-viscosity coefficients α were low and showed a high dependency on shear and concentration. In rolling contact, low M_W CMC showed up to 35% lower friction values compared to high M_W, whereas no improvement of lubricating properties was observed in sliding contact. The results suggest that low M_W CMC has great potential as bio-based thickener in aqueous lubrication.

Keywords: cellulose derivative; carboxymethyl cellulose; viscosity modifier; water-containing lubricant; water-based lubricant; aqueous lubrication; biolubrication; biopolymer

Citation: Michaelis, J.U.; Kiese, S.; Amann, T.; Folland, C.; Asam, T.; Eisner, P. Thickening Properties of Carboxymethyl Cellulose in Aqueous Lubrication. *Lubricants* **2023**, *11*, 112. https://doi.org/10.3390/lubricants11030112

Received: 1 February 2023
Revised: 28 February 2023
Accepted: 2 March 2023
Published: 4 March 2023

1. Introduction

The idea of aqueous lubrication is highly tempting, especially given the increasing environmental and political challenges. Water is not only environmentally friendly, locally and globally available, fire resistant and easily disposable, but furthermore, high thermal conductivity and low friction coefficients of water-containing fluids improve the efficiency of tribological systems [1–4]. Unfortunately, the lubricating properties of plain water are rather poor and the insufficient viscosity strongly limits the minimum lubricating film thickness, a decisive factor in risk assessment of sliding wear, particularly in gear transmission [1,4–6].

Low viscosity is generally compensated by the addition of high molecular polymeric additives, commonly referred to as viscosity modifiers (VM) or thickeners. The functionality of water soluble VM depends in particular on thickening performance, viscosity–temperature relationship, shear stability, viscosity-pressure relationship and film-forming capability [6–8]. In comparison to current applications of aqueous lubricants, such as metal working, requirements in gear transmission are significantly higher. Especially in the contacting area of sliding tooth flanks, thickeners are subjected to high temperatures and extreme shear strain rates [1,9]. Thermosensitive thickeners, such as polyvinyl alcohols [10], polyacrylic acid [11] and polyvinyl pyrrolidone [12], are therefore unsuitable for transmission applications [13]. Polyalkylene glycols provide, in principle, excellent lubricating properties and good biodegradability, but are relatively expensive, susceptible to oxidation and most importantly they are of fossil origin [2,14–16]. Chen et al. [17,18] tribologically investigated lyotropic sugar-based liquid crystals (alkylglucopyranosides) on the macro- and nanoscale. 40% aqueous solution of octyl β-D-glucopyranoside showed minimum coefficients of friction (COF) of 0.02, which is lower than a standard lubricating oil. One of the disadvantages for large scale applications, however, are the high costs of glucopyranosides.

Polymers of biological origin, often referred to as biopolymers, are mostly non-toxic, biodegradable, eco-friendly and reduce the dependence on non-renewable resources [19]. In combination with water, biological thickeners enable the formulation of sustainable, aqueous lubricants. Especially water-soluble biopolymer derivatives with adjustable, multifunctional properties, such as cellulose or chitosan ether, are a promising alternative to fossil-based polymeric thickeners [20,21]. However, most cellulose derivatives are also thermosensitive, form gels or precipitate when heated. The only exceptions are hydroxyethyl and sodium carboxymethyl cellulose (CMC) [13,22]. Naik et al. [13] studied the potential of hydroxyethyl celluloses as viscosity modifiers in aqueous hydraulic fluids. Benchmarked against standard hydraulic fluids, the resulting wear scar diameters were at least twice as high. In order to improve the lubricating performance, the authors promoted the utilization of lower molecular weight derivatives at higher concentrations. In a more recent study, Gelinski et al. [23] developed and evaluated a hydraulic medium, including glycerol and carboxymethyl chitosan. Although, tribological tests showed improved wear performances at increasing chitosan concentrations, the influence of the polymers on elastohydrodynamic film formation has not been investigated. Sagraloff et al. [4,9] experimentally investigated the influence of different cellulose ethers on the scuffing and wear performance of aqueous gear fluids. Depending on the polymer used, the scuffing load carrying capacity was increased by up to 3 failure load stages. However, due to low elastohydrodynamic film thicknesses of the polymer solutions, all samples showed a high risk of sliding wear.

CMC is a water soluble, anionic, non-toxic, biodegradable, linear polysaccharide of anhydro-glucose, covalently linked by β-1,4-glycosidic bonds [24]. Carboxymethylation of wood- or cotton-based cellulose with sodium hydroxide-chloroacetic acid results in a partial substitution of protons by carboxymethyl groups ($-CH_2COOH$) [25]. Current application fields of CMC include e.g., the food, textile, cosmetic, pharmaceutical, biomedical and paper industry [24,26]. The rheological behavior of CMC solutions is primarily characterized by their average molecular weight M_η and the degree of substitution (DS), the average number of substituted hydroxyl groups per monomer unit [27]. Additionally, uniformity or distribution of substitution is an influencing factor, as less substituted, crystalline segments promote aggregation and thus, thixotropy [28]. For DS above 1.0, crystallinity is close to zero and solubility in water reaches its maximum [29,30]. In the literature, the DS varies widely, going from 0.2 to 2.84 [20,31]; commercially available grades, however, are limited to the range between 0.38 and 1.4 [32]. Polymers with DS below 0.4 are considered as water insoluble [33]. Figure 1 shows the idealized molecular structure of CMC with a DS of 1.0.

In contrast to most cellulose derivatives, CMCs show no thermosensitive behavior and are thermally stable up to 140 °C [34,35]. However, the anionic character of CMC polymers promotes interaction with polyvalent cations and other polyelectrolytes [27]. Guan et al. [36] investigated the anti-wear properties of a fluid containing 2% triethanolamine, 0.5% of a zinc alkoxyphosphate and varying amounts of CMC. Experiments were carried out by means of an MQ-800 four-ball tester at a speed of 1450 rpm and a load of 392 N at 20 °C. The results showed a concentration-dependent improvement of the anti-wear capacity of the fluid, with an optimum at 0.7 wt.% CMC. To the best knowledge of the authors, the hydrodynamic lubricating properties of aqueous CMC solutions have not been investigated yet. Our aim was to study the rheological and tribological properties of aqueous CMC solutions and investigate the influence of molecular weight, concentration and degree of substitution on elastohydrodynamic film formation. In the first part, shear, temperature and pressure dependence of solution viscosity was investigated. In the second part, the lubricating performance of the CMC solutions was studied at different viscosity grades under sliding and rolling conditions. The results were benchmarked against rapeseed oil and polyethylene glycol 400, two common biodegradable lubricating fluids.

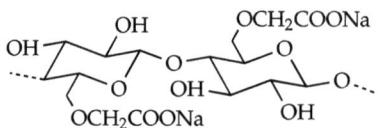

Figure 1. Molecular structure of carboxymethyl cellulose (CMC) with a degree of substitution (DS) of 1.0.

2. Materials and Methods

2.1. Materials

Five types of commercially available CMC were investigated. Table 1 shows the trade name, abbreviation, viscosity range and degree of substitution of all cellulose derivatives. Blanose™ 7ULC and Ambergum™ 1221 were provided by Ashland Industries Europe GmbH (Schaffhausen, Switzerland) and Walocel™ CRT 30, Walocel™ CRT 1000 and Walocel™ CRT 10000 by DuPont de Nemours (Wilmington, NC, USA). Blanose™ 9LCF was provided by Ashland Industries Europe GmbH (Schaffhausen, Switzerland) and used as reference for molecular weight determination.

Disodium hydrogen phosphate, sodium dihydrogen phosphate, ethanolamine, triethanolamine, 2-amino-2-methylpropanol, azelaic acid and glycerol were purchased from Sigma Aldrich (Steinheim, Germany). Polyethylene glycol 400 (PEG) was purchased from Carl Roth (Karlsruhe, Germany). 2-propanol and ethanol were purchased from Th. Geyer (Renningen, Germany). The commercially available rapeseed oil vitaDor (RSO) was supplied by Lidl, Germany. Acticide MBS was provided by Thor (Speyer, Germany). Hordaphos® 145 was provided by Clariant (Pratteln, Switzerland). Gray cast iron chips were purchased from profluid® (Ulm, Germany).

Table 1. Trade names, abbreviations, viscosity ranges (at solution concentration) and DS of the CMC derivatives used in this study.

Trade Name	Abbr.	Viscosity at 25 °C [mPas]	DS
Blanose 7ULC	C7XS	10–20 (6%)	0.65–0.90
Ambergum 1221	C2XS	10–20 (5%)	1.15–1.45
Walocel CRT 30 G	C9S	20–40 (2%)	0.82–0.95
Blanose 9CLF	B9S	25–50 (2%)	0.80–0.95
Walocel CRT 1000 PA	C9M	550–800 (2%)	0.82–0.95
Walocel CRT 10000 PA	C9L	900–1500 (1%)	0.82–0.95

2.2. Flow Behavior and Critical Concentration

Sodium phosphate buffer (0.1 M, pH = 7), including 0.1 wt% of Acticide MBS, was mixed with the desired amount of cellulose derivative. To ensure complete homogenization, the solutions were gently stirred using a magnetic stirrer at room temperature for at least 16 h. Prior to all rheological measurements, the CMC solutions were preheated in a water bath to 40 °C for at least 20 min. Ultrasonification, as documented in other publications, was not applied to avoid possible degradation of the cellulose polymers [37].

The rheological investigations were performed on a Physica MCR 301 rotational rheometer (Anton Paar GmbH, Graz, Austria) using a concentric cylinder measurement system (CC27-SN24807, Anton Paar GmbH) over the shear rate $\dot{\gamma}$ range of 0.1 to 1000 s^{-1}. The solutions were prepared and measured at least in duplicate. During all measurements, the temperature T was kept constant at 40.0 ± 0.1 °C and a solvent trap was used to minimize water evaporation. Depending on the rheological behavior, zero-shear viscosities were either calculated by fitting the measurement points to the Herschel-Bulkley or Cross model using OriginPro2018 software. Dilute polymer solutions generally show Newtonian behavior [38]. However, arising turbulences at higher shear rates effected alleged dilatant behavior, found to be depicted more accurately by the Herschel-Bulkley model [39]:

$$\eta = \eta_0 + K\dot{\gamma}^{n-1}, \tag{1}$$

where η is dynamic viscosity, η_0 is zero-shear viscosity, $\dot{\gamma}$ is shear rate, K is flow consistency index and n is flow behavior index. The Cross model (Equation (2)) is used to describe pseudoplastic behavior:

$$\eta_{\dot{\gamma}} = \eta_\infty + [\eta_0 - \eta_\infty]/[1 + (C_{\dot{\gamma}})^P], \tag{2}$$

where $\eta_{\dot{\gamma}}$ is viscosity as a function of shear rate, η_∞ is viscosity at infinite-shear rate, $C_{\dot{\gamma}}$ is the cross time-constant or consistency of a solution and P is the (cross-)rate constant [40]. Salt-free solutions of flexible polyelectrolytes can be classified into dilute, semidilute (non-entangled and entangled), and concentrated conditions. The transition concentration c_e from the semidilute entangled to the concentrated region, was determined by plotting zero-shear specific viscosity η_{sp} [28],

$$\eta_{sp} = \frac{\eta_0 - \eta_s}{\eta_s}, \tag{3}$$

where η_s is the solvent viscosity, as a function of concentration [wt%] on a double logarithmic scale and separately fitting the diluted and concentrated region to the power law equation using OriginPro2018 software [41]:

$$\eta = K\dot{\gamma}^{n-1}. \tag{4}$$

2.3. Influence of Pressure

To investigate the effect of pressure on the viscosity and determine the pressure-viscosity coefficient α, 10 and 15% of C7XS were mixed with water including 0.1% of Acticide MBS. Measurements were performed on a rotational rheometer (Physica MCR 501, Anton Paar GmbH, Graz, Austria) using a high pressure cell in double gap configuration (DG, Anton Paar GmbH). The inner gap of the measurement system was 0.4 mm and the outer gap 0.44 mm. The amount of liquid used was 8 mL. The viscosity was measured at ambient pressure η_a and 200 bar η_p at shear rates of 50, 100, 500 and 1000 s^{-1}. The temperature T was kept constant at 20 °C during all measurements and the pressure-viscosity coefficient α was calculated using the Barus equation [42]:

$$\alpha = \frac{ln\left(\frac{\eta_p}{\eta_a}\right)}{p - p_a}, \tag{5}$$

where p and p_a are the high (200 bar) and low (ambient) pressure.

2.4. Influence of Temperature

To measure the influence of temperature on the solution viscosity at two specific viscosity grades, zero-shear viscosities were set to 46.0 ± 2.3 and 220.0 ± 11.0 mPas by adjusting the CMC concentrations. Additivation consisted of 1.0 wt% triethanolamine, 0.5 wt% Hordaphos 145 and 0.1 wt% of Acticide MBS. The viscosity as a function of temperature was determined using a concentric cylinder measurement system (CC27-SN24807, Anton Paar GmbH). Measurements were performed at least in duplicate over a temperature range of 10 to 90 °C. The shear rate was kept constant at 50 s^{-1} during all measurements. The effect of temperature on the dynamic viscosity of CMC solutions was determined by fitting the measurement points to the Arrhenius equation, which is given in its logarithmic form by [43],

$$ln\, \eta = ln\, A - \frac{E_a}{R_G T}, \tag{6}$$

where A is a pre-exponential constant, E_a is the Arrhenius activation energy and R_G is the gas constant. The viscosity-temperature relationship was quantified by adjusting the factor Q to adequate temperatures for aqueous lubrication. The adjusted factor Q was calculated by

$$Q = \frac{\eta_{90\,°C}}{\eta_{30\,°C}}, \tag{7}$$

where $\eta_{90\,°C}$ and $\eta_{30\,°C}$ are the dynamic viscosities at 90 °C and 30 °C.

2.5. Tribological Investigations

The tribological performance of the CMC solutions under pure sliding conditions was evaluated by means of a rotational rheometer (Physica MCR 301, Anton Paar GmbH, Graz, Austria) equipped with a ball-on-three-plates tribological cell (Anton Paar GmbH). Figure 2a shows the schematics of the tribological cell. Three plates (15 × 6 × 3 mm) are arranged at a 45° angle to center one ball ($r = 0.00635$ m) and evenly distribute the normal load F_L. While the plates remain stationary, the ball rotates at a given rotation speed n, resulting in a sliding speed u_s in each contact point of $2\pi r n / cos(45°)$ and mean entrainment speed u_e of $u_s/2$ The balls were produced from non-stainless steel 100Cr6 (1.3505) and the plates from stainless steel X5CrNi18-10 (1.4301). Young's modulus E and poisson's ratio v were 2.1 × 10^{11} Nm^{-2} and 0.30 for 100Cr6 and 1.8 × 10^{11} Nm^{-2} and 0.24 for X5CrNi18-10, respectively. The initial surface roughness R_a of ball and plates was 30 nm. The solution viscosities were set to 46.0 ± 2.3 and 220.0 ± 11.0 mPas for each derivative by adjusting the polymer concentrations. In order to prevent excessive wear on the specimen and microbiological growth, all solutions were additivated with 1.0 wt% triethanolamine, 0.5 wt% Hordaphos 145 and 0.1 wt% Acticide MBS. Rheological investigations were performed on a Physica MCR 301 rotational rheometer (Anton Paar GmbH, Graz, Austria) using a concentric cylinder measurement system (CC27-SN24807, Anton Paar GmbH) over the shear rate $\dot{\gamma}$ range of 0.1 to 1000 s^{-1}. During all measurements, the temperature T was kept constant at 40.0 ± 0.1 °C. Dynamic viscosity at zero and infinite high shear rate were determined by fitting the measurement values to the Cross' equation (Equation (2)). The maximum temporary viscosity loss (TVL) was calculated by

$$TVL = (1 - \frac{\eta_\infty}{\eta_0}) \cdot 100. \tag{8}$$

Before each tribological test, the cell and specimen were thoroughly cleaned with isopropanol in an ultrasonic bath. To ensure even temperature distribution, the solutions were preheated to 40 °C for at least 15 min before each measurement. The normal load F_L was set to 10 N, equaling a maximum Hertz contact pressure of 0.63 GPa. Each tribological system was first run-in for 10 min at a constant sliding speed u_S of 500 mms^{-1}. Afterwards, the sliding speed u_S was increased and decreased from 0.1 to 1400 mm/s for 8 consecutive runs. The temperature T was kept constant at 40.0 ± 0.1 °C. After completion of the measurement, the wear scar diameters (WSD) on the plates were measured by means of an MZ16 stereo microscope from Leica microsystems (Wetzlar, Germany). Each diameter was measured at least four times. Friction regimes were classified according to the dimension-less lambda ratio λ, representing the ratio of minimum lubricating film thickness h_{min} to the composite roughness of the contacting surfaces $R_{a,1}$ and $R_{a,2}$ [44],

$$\lambda = \frac{h_{min}}{\sqrt{R_{a,1}^2 + R_{a,2}^2}}. \tag{9}$$

According to Hamrock and Dowson [45], the minimum film thickness h_{min} in point contact at initial Hertzian pressure is calculated as follows:

$$h_{min} = 3.63 \cdot U^{0.68} \cdot G^{0.49} \cdot R \cdot W^{-0.073}(1 - e^{-0.68k}), \tag{10}$$

$$U = \frac{\eta u_E}{2E'R'}, \tag{11}$$

$$G = \alpha E', \tag{12}$$

$$W = \frac{F_N}{E'R^2'}, \tag{13}$$

$$F_N = \frac{F_L\sqrt{2}}{3}, \tag{14}$$

where U is the dimensionless speed parameter, G the geometry parameter, W the load parameter, E' the effective modulus of elasticity ($2.11 \times 10^{11}\,\mathrm{Nm}^{-2}$), R is the effective Radius and k the ellipticity of the ball ($\sim$1).

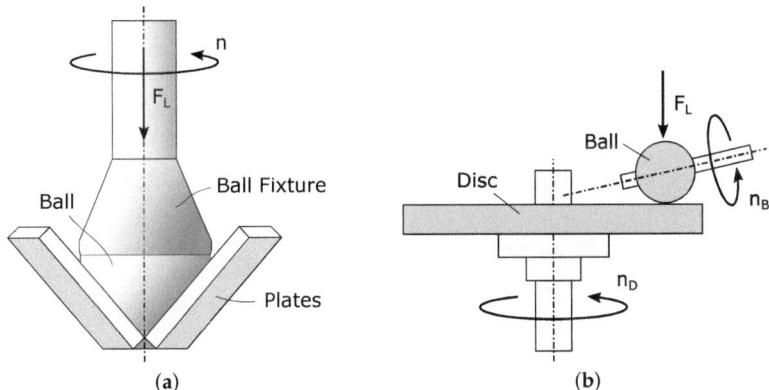

Figure 2. Schematic of (**a**) the ball-on-three-plates and (**b**) ball-on-disc tribological setup.

Further testing in rolling contact was performed on a disc-on-ball tribological system (Mini Traction Machine, PCS Instruments, London, UK). Figure 2b shows the structure and operating principle of the measurement system. The slide to roll ratio (SRR) is defined as the ratio of absolute sliding speed $|u_B - u_D|$ to the entrainment speed $u_E = (u_B - u_D)/2$, where u_B and u_D are the surface speeds of the ball and disk, respectively. The coefficient of friction μ was measured in the velocity range between 1 to 3500 mm/s at a constant temperature of 40.0 $\pm 0.1\,^\circ$C and normal load of 30 N, equaling a maximum Hertz contact pressure of 0.95 GPa. The SRR was set to 30% during all measurements. The specimen, ball and disc were produced from steel grade 100Cr6, both with a surface roughness R_a of 20 nm. The diameter of the ball was 9.5 mm.

2.6. Shear Stability—Permanent Viscosity Loss

The shear stability test evaluated the permanent viscosity loss (PVL) under high shear stress. The test setup and implementation are based on the tapered roller bearing test, a methodology commonly applied in gear oil applications [46]. As shown in Figure 3, the cell consists of a tapered roller bearing (FAG 32008-XDY), case (bearing seat), base plate, top plate and shaft. The cell was sealed by two O-rings and a rotary shaft seal to avoid evaporation and spillage. The case, shaft and both plates were made from steel and manufactured by means of a CNC milling machine in the internal manufacturing. The stress and load, specified in DIN 51350-6 (1450 rpm; 60 °C and 20 h), were adjusted to aqueous lubrication. The lubricant volume was 22 mL. The rotational speed was set to 1000 rpm by a stepper motor engine. The cell temperature was kept at constant 40 ± 0.1 °C using a water bath and the axial load was set to 50 N. Solution viscosities were set to 46 ± 4.6 mPas for each derivative and all samples were additivated with 1.0 wt% ethanolamine, 0.5 wt% triethanolamine, 0.5 wt% Hordaphos 145 and 0.1 wt% of Acticide MBS. Zero viscosity of the fresh and sheared solutions was determined by means of a parallel plate measurement system (PP50, Anton Paar GmbH) at least in duplicate over a shear rate range of 0.1 to $10,000\,\mathrm{s}^{-1}$. The PVL is defined as the decrease in zero-shear viscosity due to shear and was calculated by

$$\mathrm{PVL} = \frac{\eta_{0,f} - \eta_{0,s}}{\eta_{0,s}}, \tag{15}$$

where $\eta_{0,f}$ is the zero viscosity of the fresh (before) and $\eta_{0,s}$ of the sheared (after) solution.

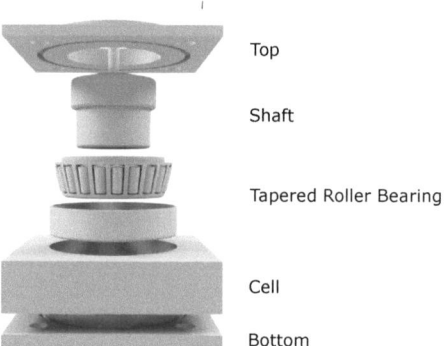

Figure 3. Design of the tapered roller bearing shearing cell.

3. Results and Discussion

3.1. Viscosity-Concentration Relationship

The flow curves were measured in a series of at least eight concentrations ranging from 0.05 to 17.5 wt%, depending on the molecular weight of the CMC. The respective weight percentages were selected to cover the viscosity range between 1 and 220 mPas, corresponding to the relevant viscosity grades of available lubricants for many target applications. Figure 4 exemplarily shows the resulting flow curves for C9L on the left and C2XS on the right. The flow curves of the other derivatives are available in Supplementary Materials. At low concentrations, the flow characteristics were nearly Newtonian (colored in dark gray) for all derivatives. More concentrated solutions (colored in light gray) showed shear-thinning behavior, in which case the curves can be divided into a "Newtonian viscosity plateau" at low, and a shear-thinning part at high shear rates. The beginning of shear thinning behavior, also referred to as critical shear rate $\dot{\gamma}_c$, shifted to lower shear rates with increasing concentration and molecular weight M_W, which is in agreement with literature [47].

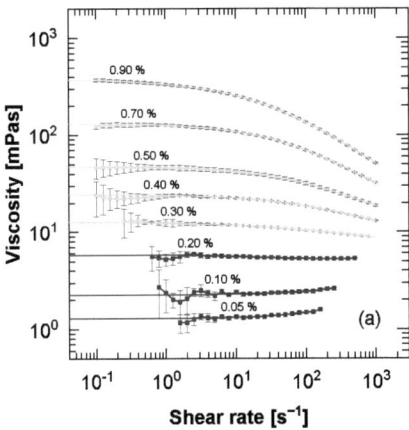
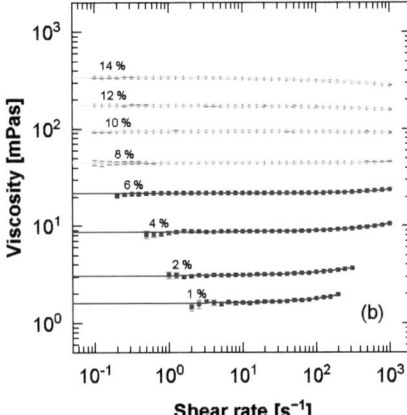

Figure 4. Viscosity as a function of shear rate at different concentrations, exemplary shown for (a) the high-molecular weight C9L and (b) the low-molecular weight derivative C2XS. The plotted points display the measurement results and the lines the fitted curves according to Cross' (colored in light gray) or Herschel's model (colored in dark grey).

The zero-shear viscosities η_0 were determined by fitting the data of dilute solutions to the Herschel-Bulkley equation (Equation (1)) and of the concentrated, shear-thinning solutions to Cross' equation (Equation (2)). At lower concentrations, inertial effects or turbulences within the measurement gap resulted in a viscosity increase with shear rate and thus, erroneous Newtonian fittings. The effect was successfully compensated by using the Herschel-Bulkley model. All derivatives showed the expected increase in zero-shear viscosity with increasing CMC concentration [47]. At maximum concentration, C7XS was slightly thixotropic in behavior, which is usually effected by low DS [28]. All other solutions showed no time-dependent viscosity changes.

3.2. Critical Concentration

Figure 5a shows the specific viscosity at zero-shear rate η_{sp} as a function of solution concentration on double logarithmic scale. The lines represent the best fit power laws in the semidilute unentangled and entangled regimes. At c_e, polymer chains begin to entangle, which corresponds to the transition from Newtonian to shear thinning behavior [48,49]. Table 2 lists the critical concentrations c_e and the respective zero-shear viscosities η_e for all derivatives. Despite the lower average molecular weight M_W, the critical concentration of C7XS was lower compared to C2XS, which is attributed to partially unsubstituted microcrystalline parts. Higher DS increases the amount of soluble molecules and thus the solution viscosity [50].

Table 2. Average molecular weight M_W, degree of substitution (DS), critical concentration c_e, critical viscosity η_e, slope values in the entangled m_e and concentrated regime m_c of the investigated derivatives.

Derivative	M_W [kg/mol]	DS	c_e [wt%]	η_e [mPas]	m_e [1]	m_c [1]
C7XS	24 *	0.7	8.25	13.95	1.66	4.59
C2XS	35 *	1.2	6.60	21.67	1.74	3.64
C9S	88 *	0.9	2.10	14.60	1.71	3.44
B9S	100	0.9	1.41	7.71	1.45	3.64
C9M	240 *	0.9	0.73	11.27	1.62	3.30
C9L	520 *	0.9	0.27	8.17	1.54	3.10

* estimated according to Miehle et al. [48].

In comparison to the literature data, a critical concentration of C9M fairly corresponds to the results of Wagoner et al. [51], who reported a critical concentration of 0.67 wt% for a CMC of 250 kDa (DS 0.7). The resulting zero-shear viscosity of 22 mPas, however, is significantly higher and more comparable to the results for C2XS or C7XS. The same applies to the value of 1.5 g L^{-1} reported by Charpentier et al. [52] for a CMC of 300 kDa (DS 0.9). In contrast, values published by Miehle et al. [48], 1.3 wt% and 6.4 mPas for a CMC of 100 kDa (DS 0.9), are clearly below the results for C9S. Overall, c_e is strongly dependent on molecular weight, uniformity of substitution and degree of substitution [48]. As intrinsic viscosity is strongly influenced by ionic strength and thus molarity of the buffer, the same might apply to the level of critical zero-shear viscosity [53,54]. Moreover, the molecular weights provided by manufacturers are sometimes inaccurate, complicating the direct comparison of the results [55].

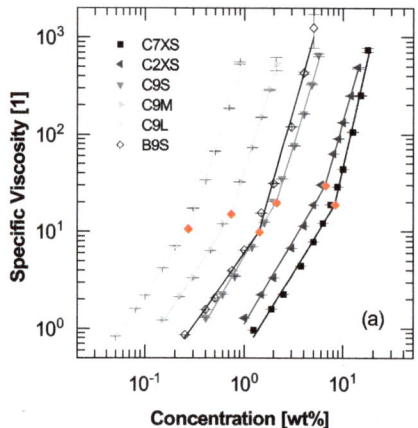 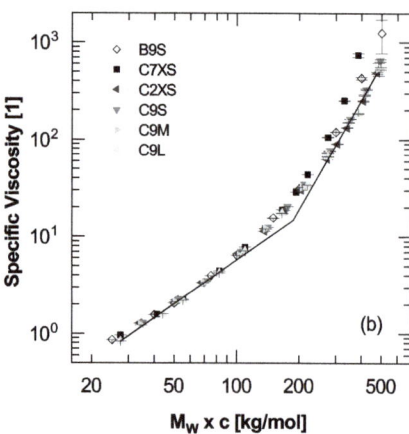

Figure 5. (**a**) Double logarithmic plot of the specific viscosity as a function of concentration. Plotted lines correspond to the best fit power laws and red dots to the critical concentrations c_e (**b**) Double logarithmic plot of the specific viscosity as a function of product of viscosity average molecular weight and concentration. Plotted lines show the scaling predictions for CMC solutions of 1.5 and 3.75, published by Lopez et al. [28].

According to Miehle et al. [48], multiplying solution concentration with average molecular weight results in near superimposition of the viscosity graphs. Using B9S, with an average molecular weight of 100 kDa, as given by the supplier, allowed for an estimation of the remaining average molecular weights. Figure 5b shows the resulting superimposition. The estimated average molecular weights are listed in Table 2.

Noticeable in both figures, specific zero viscosity of B9S and C7XS increased faster with concentration compared to the other investigated derivatives. Similar behavior was observed by Barba et al. [56] in relation with interchain interactions of less substituted molecules. All curves showed no sharp transition or kink at the calculated critical concentration, which is similar to literature results [28,48,49,57]. In all publications, smooth transitions are attributed to polydispersity of the polymers. The slope values of all power law fittings for the entangled semidilute m_e and concentrated m_c regions are given in Table 2. The lines in Figure 5b show the scaling predictions for CMC of 1.5 and 3.75, published by Lopez et al. [28]. At high and low solution viscosities, the measured values are in close agreement with the predictions. The slope values of C9L in the semidilute and C2XS in the concentrated region showed the highest correlation. Overall, deviations from the predicted slope values increased with the number of fitted measurement points, as smooth transition and curvature resulted in higher gradients in the entangled and lower gradients in the concentrated region.

With the exception of C7XS and B9S, the critical concentration c_e and the respective zero-shear viscosity clearly decreased with molecular weight. Plotted on a double logarithmic scale, the measurement points of both parameters were in close agreement with linear fits. Equation (16) describes the resulting correlation between c_e and M_η and Equation (17) between η_e and M_η.

$$log(c_e) = 2.62087 - 1.17071 \cdot log(M_\eta) \tag{16}$$

$$log(\eta_e) = 1.86543 - 0.34915 \cdot log(M_\eta) \tag{17}$$

The theoretical critical concentration c_e of 50 wt% therefore equals a molecular weight M_W of 6.13 kDa and a critical zero-shear viscosity η_e of 38.95 mPas. In conclusion, all solutions with zero-shear viscosities η_0 above 21.67 mPas and thus fluids of viscosity classes above ISO VG 22 are expected to show shear-thinning behavior. Nevertheless, low molecular weight derivatives showed less shear thinning behavior and are therefore preferable for lubricating applications.

3.3. Viscosity-Pressure-Relationship

The pressure-viscosity coefficients of aqueous 10 and 15% C7XS solutions were determined as a function of the shear rate at 20 °C (Figure 6). In each case, dynamic viscosities were measured at 1 and 200 bar and the pressure-viscosity coefficient was calculated using Equation (5).

At the lower concentration of 10%, the pressure-viscosity coefficients α were clearly smaller than at 15% C7XS. However, at 15% α showed a greater dependence on shear rate $\dot{\gamma}$, which corresponds to the flow behavior of the solutions, determined in Section 3.1. Whereas the 10% solution showed near Newtonian flow behavior at ambient pressure and shear rates below $1000\,s^{-1}$, the 15% solution behaved in a pseudoplastic way. According to Cook et al. [58], the effect of pressure on solution viscosity is comparable to an increase in polymer concentration, indicating that the critical shear rate $\dot{\gamma}_c$ decreases with increasing pressure and the shear-thinning properties become even more prominent. In contrast to most organic solvents, water does not show an exponential increase in viscosity with pressure [58]. The pressure-dependence observed for the CMC solutions is therefore assumed to be entirely effected by the polymers. Addition of CMC increased the pressure-viscosity coefficient α of pure water ($\sim$0.8 GPa^{-1} [59]) to a maximum of about 3.5 GPa^{-1} at 20 °C. However, in comparison to rapeseed oil with a coefficient of 14.1 GPa^{-1} at 25 °C, polyethylene 400 with $\sim$11.8 GPa^{-1} at 20 °C (calculated from the viscosity data in [60]) and glycerol with 5.9 GPa^{-1} at 30 °C [61] coefficients of the CMC solutions are clearly lower. Pressure-viscosity coefficients of water-containing polyalkylene glycol-based gear lubricants with water contents of up to 70% are between 5.61 and 6.26 GPa^{-1} at 40 °C [2]. Wang et al. [62] calculated a coefficient of 6.08 GPa^{-1} for a 40% polyalkylene glycol solution and Wan et al. [63,64] measured pressure-viscosity coefficients around $\sim$2 GPa^{-1} for 15% polyalkylene glycol solutions and differing concentrations of propylene glycol at 22 °C, which are slightly higher than the viscosity-pressure coefficients of the C7XS solutions at high shear. In general, highly stressed machine elements, such as gears and rolling bearings, are governed by elastohydrodynamic lubrication and the minimum lubricating film thickness is an important factor in risk assessment [4,65]. As the minimum film thickness in line contact is proportional to $\alpha^{0.54}$ [66], shear-dependence results in a decrease of film thickness with increasing shear stress, counteracting the sliding speed related increase in film thickness. Since the pressure-viscosity coefficient α of aqueous glycerol and polyalkylene glycol solutions decreases with increasing temperature and water content, similar is to be expected for CMC solutions [59,62]. Accordingly, in order to increase the pressure viscosity coefficient α, an increase in CMC concentration is required.

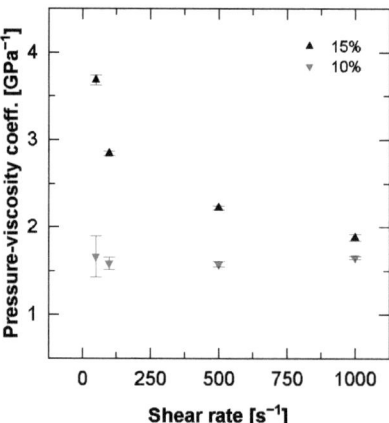

Figure 6. Viscosity -pressure coefficient as a function of shear rate for aqueous 10 and 15% C7XS solutions at 20 °C.

3.4. Viscosity-Temperature Relationship

Figure 7a shows the viscosity as a function of temperature in a semi-logarithmic plot for C7XS, C9S and C9L, at zero-shear viscosity grade (ZSVG) 46 and 220 mPas and a shear rate of 50^{-1}. Due to the similarity of the measurement results with C9S, the curves for C2XS and C9M are not shown. Figure 7b shows dynamic viscosity as a function of temperature for PEG and RSO. Table 3 lists the temperature related viscosity losses in terms of Q for all tested solutions.

At temperatures below 20 °C, all samples showed running-in behavior, caused by uneven heat distribution inside the measuring system. Overall, the dynamic viscosities of all fluids, CMC solutions and references, decreased with increasing temperature and the Arrhenius equation (Equation (6)) could be adapted to the measurement points above 30 °C. Lower viscosities of C9L at the starting temperature of 10 °C can be explained by shear-thinning flow behavior. As can be seen in Figure 4a, the viscosities of aqueous C9L solutions at shear rates of $50\,\text{s}^{-1}$ are clearly lower than the zero-shear viscosities. The solutions of the highest molecular weight derivative C9L were the least affected by changes in temperature, with a maximum Q factor of 0.252. The Q factors of C2XS, C9S and C9M at zero solution viscosity of 46 mPas were between 0.160 and 0.178, which corresponds well to the Q factor of RSO. The viscosity of vegetable oils generally exhibits low temperature dependency and the viscosity indices are more than twice those of mineral oils [67–69]. Thus, the comparability of Q indicates excellent temperature behavior. Especially in applications with fluctuating temperatures, fluids with higher Q values are preferable in order to ensure proper operation [7,46]. By using the Arrhenius fit functions and estimating the densities of the aqueous solutions to 0.99 g cm^{-3} at 40 °C and 0.96 g cm^{-3} at 100 °C [70], the kinematic viscosities at 40 and 100 °C were calculated, leading to viscosity indices between 160 and 305. In comparison, viscosity indices of concentrated hydroxyethyl cellulose solutions, measured by Naik et al. [13], were clearly smaller between 114 and 126. Yilmaz et al. [2] determined viscosity indices between 135 and 189 for water-containing gear fluids, based on polyalkylene glycols.

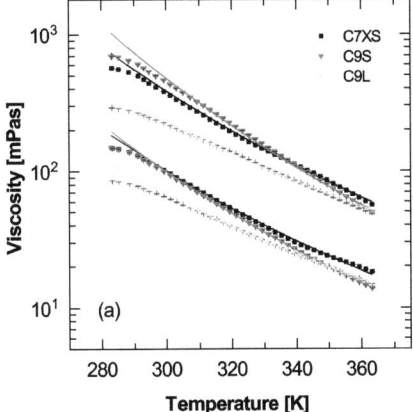 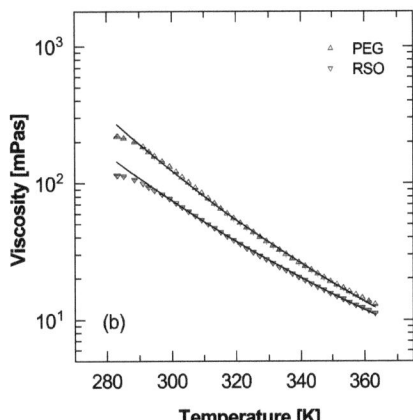

Figure 7. Semi-logarithmic plot of the dynamic viscosity as a function of temperature at shear rate 50 $^{-1}$ for (**a**) the cellulose derivatives C7XS, C9S and C9L at zero-shear viscosity grade 46 and 220 and (**b**) reference fluids PEG and RSO. The plotted points display the measurement results and the lines the fitted curves according to the Arrhenius equation.

Table 3. Viscosity-temperature relationship, quantified by the Q factor, for the all tested CMC solutions and reference fluids.

	C7XS	C2XS	C9S	C9M	C9L	PEG	RSO
Q at ZSVG 46	0.203	0.170	0.160	0.178	0.252	0.123	0.168
Q at ZSVG 220	0.170	0.141	0.124	0.139	0.234		

Courses of C7XS showed a slight deviation from Arrhenius behavior at higher temperatures, above 70 °C at ZSVG 46 and 50 °C at ZSVG 220, generally indicating changes in structure due to gelation or even precipitation [71]. Small deviations of the C7XS solutions are most likely caused by lower and less homogeneous substitution compared to the other derivatives [50]. Accordingly, in order to avoid temperature-dependent changes in structure and solution viscosity, DS above 0.95 are preferable.

3.5. Temporary Viscosity Loss

In order to determine the TVL and analyze the influence of flow behavior, concentration and viscosity on the tribological properties of CMC solutions, dynamic zero-shear solution viscosities were set to the ZSVG 46 and 220.

Figure 8a,b shows the dynamic viscosity η as a function of shear rate $\dot{\gamma}$ at 40 °C for ZSVG 46 and 220, respectively, both displayed on a double logarithmic scale. The zero-shear η_0 and infinite-shear η_∞ viscosities were determined by fitting the measurement values to the Cross' equation (Equation (2)). Table 4 shows concentration c_{CMC}, zero-shear η_0 and infinite-shear η_∞ viscosity, maximum temporary shear loss (TVL) and wear scar diameter (WSD) for all tested fluids. Adjustment of zero-shear solution viscosity, allows for a direct comparison of flow behavior, infinite viscosity and shearing loss.

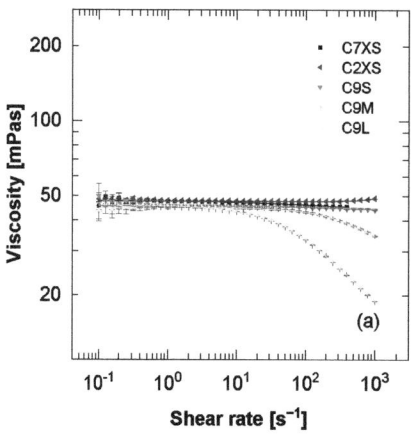 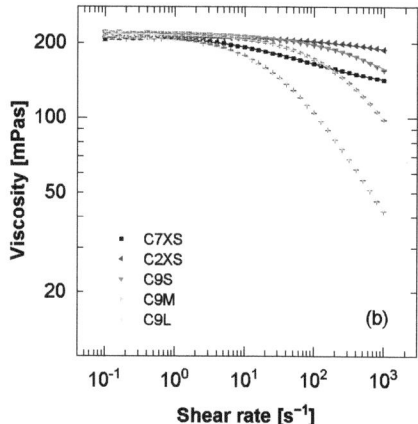

Figure 8. Viscosity as a function of shear rate at (**a**) ZSVG 46 and (**b**) ZSVG 220. The plotted points display the measurement results and lines the fitted curves according to Cross' model ($T = 40\,^\circ$C).

Noticeably, the flow behavior of C7XS is clearly different compared to the higher substituted derivatives. Especially in direct comparison to the similar sized C2XS, shear-thinning of C7XS starts at lower shear rates and TVL is significantly higher. Similar behavior has been observed with other, lower substituted derivatives, in preliminary tests. The most likely reason is the aforementioned lower and less homogeneous substitution. Investigations by Kulicke et al. [30,50] showed that CMC molecules with DS below $\sim$0.95 are only partially solvated, which results in a lower zero viscosity at an equal concentration and molecular weight compared to higher substituted derivatives. Moreover, solutions of C7XS were rather turbid suspensions in appearance than clear solutions, typical for CMC with DS below $\sim$0.85 [30]. In conclusion, DS around 0.95 are preferable to exploit the highest thickening capacity at equal concentration and average molecular weight.

Table 4. Concentration c_{CMC}, viscosity at zero η_0 and infinite η_∞ shear rate, temporary viscosity loss TVL, wear scar diameter WSD, minimum lubricating film thickness h_{min} and lambda ratio λ for all CMC solutions at ZSVG 46 and 220, the reference fluids and additivated water.

Derivative	c_{CMC} [wt%]	η_0 [mPas]	η_∞ [mPas]	TVL [%]	WSD [µm]	h_{min} [nm]	λ [1]
C7XS	11.670	47.90 ± 0.06	41.49 ± 1.20	13.38	482	32.10	0.76
	15.740	213.60 ± 0.75	129.80 ± 1.69	39.23	1593	74.32	1.75
C2XS	8.210	47.63 ± 0.02	-	0	497	35.26	0.83
	12.940	210.06 ± 0.04	181.30 ± 6.42	13.69	1592	93.28	2.20
C9S	3.060	45.70 ± 0.02	43.08 ± 0.03	5.73	328	32.93	0.78
	4.990	219.80 ± 0.20	97.89 ± 0.83	55.46	357	61.35	1.45
C9M	1.090	44.89 ± 0.02	27.70 ± 0.56	38.29	372	24.39	0.57
	1.870	217.19 ± 0.20	25.98 ± 1.92	88.04	547	24.89	0.59
C9L	0.415	46.07 ± 0.07	12.30 ± 0.19	73.30	301	14.04	0.33
	0.750	213.73 ± 0.43	14.18 ± 0.69	93.37	332	16.49	0.39
AW		0.77 ± 0.00			347		
RSO		32.64 ± 0.05			352		
PEG		45.42 ± 0.01			233		

Excluding C7XS from the evaluation, shear losses increased with average molecular weight and ZSVG. Unexpectedly, the flow behavior of C2XS at 46 mPas was near-

Newtonian, if not slightly dilatant, which is in contradiction to the maximum Newtonian zero-shear viscosity of 21.67 mPas, determined in Section 3.2. The upward deviation is attributed to lower salt contents of the solutions, resulting in an expansion of the polyelectrolyte coils and hence a viscosity increase [54]. For the upper ZSVG 220, all solutions showed shear thinning flow behavior. As expected, the highest molecular weight derivative CL9 showed the greatest losses under shear, with a maximum TVL of 93.37%. Differences in shear loss, between C2XS and C9L, were around 80%, regardless of the viscosity grade. Comparable polyalkylene glycol (970 Da) based lubricating oils with water contents of up to 80% showed Newtonian flow behavior at shear rates below $1000 \, \text{s}^{-1}$ [62]. As expected, the reference lubricating fluids, PEG and RSO showed Newtonian flow behavior. The measured dynamic viscosities are in good correlation with the literature [60,72] and listed in Table 4. As maximum shear rates in highly loaded machine elements, such as gears or bearings, reach values of up to $10^7 \, \text{s}^{-1}$ [73], calculated infinite-shear viscosities η_∞ correspond approximately to the dynamic viscosity under application conditions. Infinite-shear viscosities could be increased by increasing the CMC concentration, however, high zero-shear viscosities would considerably reduce the energy efficiency of the lubricated contact and moreover complicate production and handling of the fluids [74]. In order to achieve Newtonian flow behavior at ZSVG 220, a further reduction in molecular weight and an increase in CMC concentration would be required.

3.6. Lubricating Performance under Sliding Conditions

The effect of flow behavior on the tribological properties was first studied in pure sliding motion in steel/steel tribological contacts at 10 N normal load and 40 °C. Figure 9a,b show the coefficient of friction (COF) as a function of sliding speed v at ZSVG 46 and 220 respectively, both plotted on a semi-logarithmic scale. The values shown, are the averages and standard deviations of the last three test runs at decreasing velocity. The measured curves are compared to the reference curve of additivated water (AW), including 1.0 wt% triethanolamine, 0.5 wt% Hordaphos 145 and 0.1 wt% Acticide MBS without the addition of CMC, with a Newtonian dynamic viscosity of 0.77 mPas at 40 °C (Figure 9a), and the reference lubricating fluids PEG and RSO (Figure 9b). The resulting wear scar diameter (WSD) of all tested solutions is listed in Table 4. Figure 10 exemplarily shows the optical images of the WSD at ZSVG 46 for C9L, C2XS and C7XS.

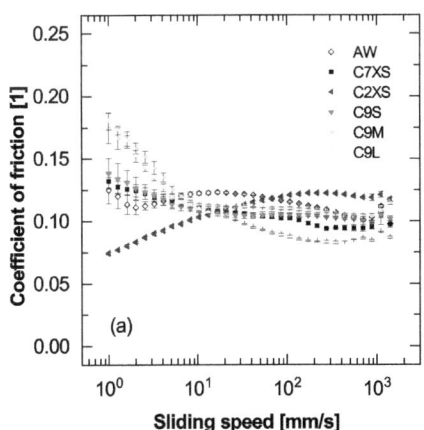 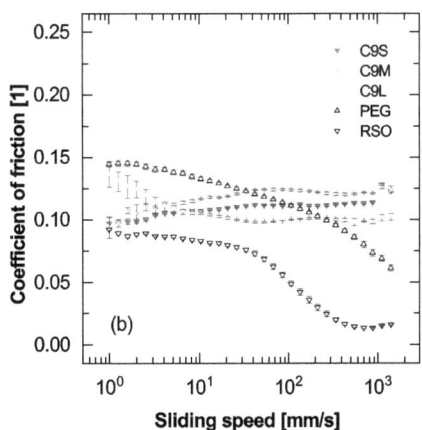

Figure 9. COF as a function of sliding speed at (**a**) ZSVG 46 and (**b**) ZSVG 220 (ball-on-three-plates, $F = 10 \, \text{N}$, $T = 40 \, °\text{C}$).

For AW, the COF was between 0.12 and 0.10. Due to the low viscosity, the contact operates solely in the boundary lubrication regime. At ZSVG 46, the friction curves of the aqueous CMC solutions varied with concentration and molecular weight. Whereas the COF for C9L decreased with the logarithm of sliding speed from 0.17 to 0.08, the friction values for C2XS increased steadily from 0.08 to 0.12. The curves for C7XS, C9S and C9M showed a decrease in friction at low sliding speeds, but COF stayed almost constant around 0.1 above $10 \, mm \, s^{-1}$. At ZSVG 220, the measurements for C2XS and C7XS resulted in scuffing. The corresponding curves are not displayed, due high fluctuations of COF. COF for C9M increased initially and reached a constant value of 0.13. The same applies to C9S, though constant friction values were slightly lower, around 0.11. C9L showed a slight decrease at low sliding speeds and stayed nearly constant, around 0.10 at velocities above $10 \, mm \, s^{-1}$. The COF of RSO was highest at low sliding speeds and showed a sharp decrease in friction at speeds above $50 \, mm \, s^{-1}$, indicating a hydrodynamic lift. RSO reached the minimum measured COF of 0.01. The friction values of PEG started around 0.15 and decreased to 0.08. In both cases, measurement deviations between the test runs were very low.

The WSD are in agreement with the measured friction values. The lowest diameter of 233 µm was measured for PEG. Although its COF is higher compared to rapeseed oil, lower wear indicates an overall better separation of the tribological surfaces. Surprisingly, the wear scar diameters of rapeseed oil (352 µm) and additivated water (347 µm) were in close agreement. The low molecular weight derivatives C2XS and C7XS resulted in elevated wear scar diameters, irrespective of the viscosity grade. Addition of C9M and C9S had no clear effect on wear. Relatively low friction and wear of water is due to the addition of the antiwear additive. Comparative measurements without additivation showed substantially higher wear and friction.

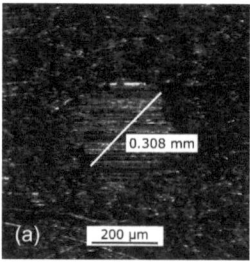 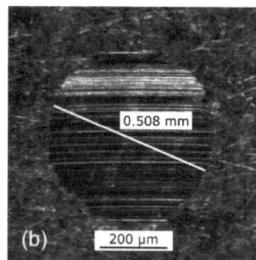 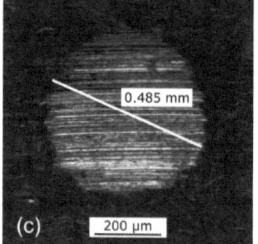

Figure 10. Optical images of the wear scar diameter (WSD) at ZSVG 46 for (**a**) C9L (**b**) C2XS and (**c**) C7XS (ball-on-three-plates, $F = 10 \, N$, $T = 40 \, °C$).

Minimum friction values of 0.08 for the CMC solutions indicate that boundary or mixed lubrication is observed for all biopolymer samples. A clear decline in friction due to elastohydrodynamic film formation was not detected in any case. On the contrary, zigzag deviations or leaps at sliding velocities above $1000 \, mm \, s^{-1}$. point to unstable lubricating films and contact of the steel surface asperities. Preliminary experiments, without the addition of antiwear additive, support this assumption. In all cases, high sliding speeds resulted in severe scuffing.

Assuming initial Hertzian pressure, the minimum lubricating film thickness h_{min} was calculated at maximum sliding velocity u_s of $1.4 \, ms^{-1}$ and infinite dynamic viscosity η_∞, using Equation (10). According to Section 3.3, the pressure-viscosity coefficients α were $1.65 \, GPa^{-1}$ at ZSVG 46 and $1.88 \, GPa^{-1}$ at ZSVG 220. Due to the lower solution concentrations of C9S, C9M and C9L compared to C7XS, the pressure-viscosity coefficients α and thus minimum film thicknesses are likely to be lower than estimated. Wear of the tribological surfaces during the measurements would decrease the initial Hertzian pressure and thus increase the lubricating film thickness. The corresponding lambda ratios λ were calculated at initial surface roughness R_a using Equation (9). The minimum lubricating film thickness h_{min} and lambda ratio λ of all tested solutions are listed in Table 4.

All the relevant parameters and procedure for calculating h_{min} and λ are provided in Supplementary Materials.

The maximum lambda ratio λ of 2.03 for C2XS confirms the assumption of boundary and mixed lubrication ($\lambda < 3$), even at high sliding speeds [44]. Lambda ratios λ above 3, assigned to elastohydrodynamic lubrication have not been achieved. Declining friction values for C9L and C9M are attributed to an increasing alignment of the polymers in sliding direction and the associated reduction in flow resistance. The lower COF of CL9 at high sliding speeds and the decrease in WSD by up to 15% compared to AW indicate absorbed boundary layers. Measurements by Guan et al. [36] showed a similar improvement in WSD of up to 18% at a concentration of 0.7 wt% CMC. Concentrations above 0.7 wt% also led to an increase in WSD. Gelinski et al. [23] were able to reduce the wear ellipsis from 27.5 to 15.5 mm^{-2}, by the addition of up to 0.8% carboxymethyl chitosan. Experimental studies by Naik et al. [13] showed decreasing WSD with increasing hydroxyethyl cellulose concentration and decreasing molecular weight M_W. Increasing friction values in boundary lubrication have previously been observed by Campen et al. [75] for closed packed films of organic friction modifiers. According to the authors, increasing friction with speed is the consequence of sliding between ordered densely packed monolayers. Furthermore, investigations of Liu et al. [76] suggest that the adverse effects of C2XS are due to high interactions and entanglement of the CMC polymers. According to their results, polymers with radii larger than half the height of the lubricating gap are influenced by the confinement, and viscous layers of adsorbed polymers hinder the access of subsequent molecules. With an estimated gap height of 10 nm for aqueous solutions, the maximum unaffected polymer radius is around 5 nm [76]. The polymer radius of a 90 kDa CMC is around 16 nm in 0.1 M NaCl solution and $\sim$40 nm in pure water, thus considerably higher than 5 nm [77]. Increasing COF for C2XS at ZSVG 46 and scuffing for C2XS and C7XS at ZSVG 220 indicate that the chain lengths of both molecules are too large for them to enter the lubricating gap.

Further increase of the zero-shear viscosity by increasing the concentration did not lead to an improvement of elastohydrodynamic film formation. In conclusion, achieving elastohydrodynamic lubrication under sliding conditions is not possible with the commercially available CMC derivatives studied and an increase in minimum lubricating film thickness, would require smaller molecules, with polymer radii below 5 nm.

3.7. Lubricating Performance under Rolling Conditions

Figure 11a,b show the coefficient of friction as a function of speed at 30 N normal load and a SRR of 30% at 40 and 60 °C, respectively. In all cases, friction decreased with sliding speed.

Independently of the measurement temperature, the COF of C7XS was about 35% lower, compared to the other tested derivatives, reaching a minimum value of 0.01, at a maximum speed of 3500 mm s^{-1}. As measurements were conducted at decreasing velocity, slight fluctuations at high speeds might be explained by running-in behavior and polymer orientation. In comparison to pure sliding, rolling seems to hinder polymer entanglement, improve the polymer intake into the lubricating gap and thus enhance film formation. Additionally, the higher entertainment speed u_E of maximum 3500 mm s^{-1} results in a minimum lubricating film thickness of around 230 nm, corresponding to a lambda ratio λ above 5. Therefore, the limit value of 6 kDa, reported by Liu et al. [76] is primarily applicable to sliding conditions. In rolling motion, polymers with average molecular weights around 25 kDa clearly decrease friction and improve the lubricating performance. Slightly lower friction values at 60 °C compared to 40 °C are attributable to the lower solution viscosity at higher temperature. In comparison, COF measured by Sagraloff et al. [9] for aqueous biopolymer solutions at 40 and 60 °C and Stribeck curves published by Vengudusamy et al. [78] for synthetic and mineral oils gear oils at 100 °C were on a comparable level.

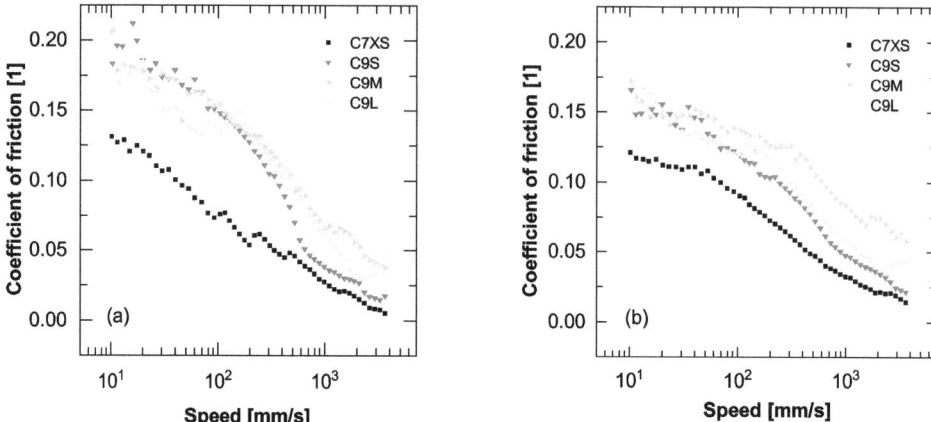

Figure 11. Coefficient of friction as a function of speed for C7XS, C9S, C9M and C9L at (**a**) 40 °C (**b**) 60 °C (ball-on-disc, SRR = 30%).

3.8. Permanent Viscosity Loss

Permanent viscosity loss (PVL) of the fully additivated CMC solutions, already applied in Section 3.7, were measured in a tapered roller bearing shearing cell. The required loading duration was determined in preliminary tests with a high molecular derivative and set to 24 h (data not shown). Figure 12a,b shows the flow curves for the fresh and sheared polymer solutions, respectively. Table 5 lists zero viscosities of the fresh ($\eta_{0,0}$) and sheared ($\eta_{0,24}$) fluids, calculated PVL and concentrations of all investigated polymer solutions.

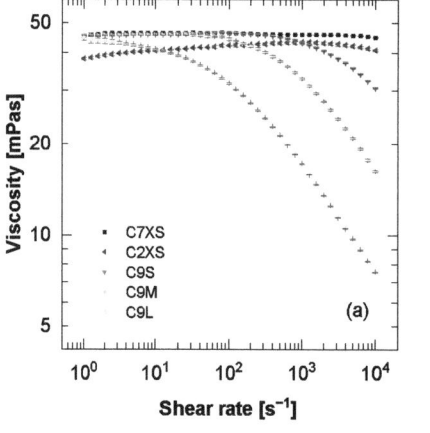
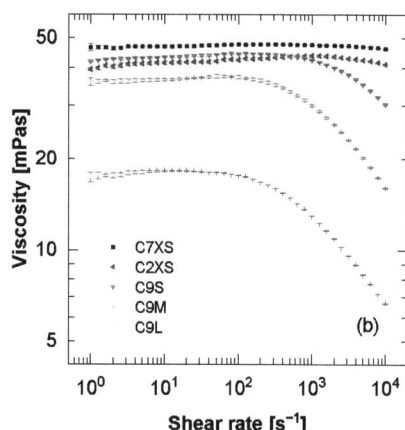

Figure 12. Double logarithmic plot of the viscosity as a function of shear rate for the (**a**) fresh and (**b**) sheared CMC solutions.

While viscosities for C7XS and C2XS show no detectable difference, C9M und C9L clearly deviate downwards. Overall, permanent viscosity losses increased with average molecular weight M_W and decreased with concentration, which corresponds to the results of Antti et al. [79] and Almeida and Dias [80]. Measurements by Schweiger [81] showed similar viscosity losses of up to 70% for high molecular cellulose derivatives. In theory, shear forces are highest in the middle region of the polymers and molecules break into two roughly equal parts. Below a limiting molecular weight, shear forces are insufficient to further fragment the molecules. The smallest fragments are hence a little bigger than half of the limiting molecular weight [7,46] Constant zero-shear viscosities and flow curves of C7XS and C2XS are indicative of shear stability. Molecular weights of both derivatives are below the limiting molecular weight. All other CMC solutions were affected by shear and showed PVL between 6.21 and 58.75%. Consequently, the limiting molecular weight was between 35 (C2XS) and 88 kDa (C9S) under the given shearing conditions. Higher loads and shearing energies, as stated in the current standard, would probably further reduce the limiting value. To conclude, usage of low molecular weight derivatives below the limiting molecular weight are preferable in high shear gear and bearing applications applications. Naik et al. loss of 19.4% in Kurt Orbahn rig test, HEC average molecular weight of 45 KDa. stabalized at a molar mass of ~80 kDa temperatures of 25 to 40 °C [23].

Table 5. Zero-shear viscosity of the fresh $\eta_{0,0}$ and sheared $\eta_{0,24}$ solutions, permanent viscosity loss PVL and polymer concentration c_{CMC}.

Derivative	M_W [kg/mol]	$\eta_{0,0}$ [mPas]	$\eta_{0,24}$ [mPas]	PVL [%]	c_{CMC} [wt%]
C7XS	24 *	46.44	47.68	-	10.68
C2XS	35 *	43.21	43.75	-	7.94
C9S	88 *	45.70	42.86	6.21	3.30
C9M	240 *	45.64	37.10	18.71	1.17
C9L	520 *	45.81	18.89	58.76	0.34

* estimated according to Miehle et al. [48].

4. Conclusions

In this study, the rheological and tribological properties of aqueous CMC solutions and the effect of concentration and molecular weight were studied to assess the potential of sodium carboxymethyl celluloses as thickening agent in sustainable, aqueous lubricants. The results were compared to rapeseed oil, polyethylene glycol 400 and additivated water, serving as biodegradable reference lubricants. The solution viscosities were adjustable to all viscosity grades relevant in gear and bearing applications. Viscosity indices are similar or better compared to the reference fluids. The rheological investigations showed a clear increase in temporary and permanent viscosity loss with molecular weight. Low DS promoted thixotropic behavior and lowered temperature stability. Tribological measurements in a ball-on-three-plates system displayed no improvement in elastohydrodynamic film formation under sliding conditions by the addition of carboxymethyl cellulose. The measurement results indicate that the average molecular weight of all derivatives used was too high for the molecules to enter the lubricating gap. Under rolling conditions, low molecular weight derivatives reduced friction by up to 35%. In order to achieve Newtonian flow behavior, increase the pressure-viscosity coefficient and improve elastohydrodynamic film formation future investigations should focus on lower molecular weight derivatives and higher concentrations.

Supplementary Materials: The following supporting information can be downloaded at: https://www.mdpi.com/article/10.3390/lubricants11030112/s1.

Author Contributions: Conceptualization, J.U.M. and S.K.; methodology, J.U.M.; formal analysis, J.U.M. and S.K.; investigation, J.U.M.; resources, T.A. (Tobias Amann), C.F. and T.A. (Tobias Asam); writing—original draft preparation, J.U.M. and T.A. (Tobias Amann); writing—review and editing, S.K., T.A. (Tobias Asam), C.F. and P.E.; visualization, J.U.M., S.K. and T.A. (Tobias Amann); supervision, P.E.; project administration, J.U.M., S.K. and P.E. All authors have read and agreed to the published version of the manuscript.

Funding: This research was partly funded by the Bavarian Research Foundation grant number Az-1314-17.

Data Availability Statement: The data presented in this study are available on request from the corresponding author. The data are not publicly available due to privacy reason.

Acknowledgments: The authors thank Carl Bechem GmbH, Torqeedo GmbH, Leistritz Pumpen GmbH, Renk AG, Wittenstein alpha GmbH and Forschungsstelle für Zahnräder und Getriebesysteme (FZG) for their participation in the research project.

Conflicts of Interest: The authors declare no conflict of interest.

Abbreviations

The following abbreviations are used in this manuscript:

AW	Additivated Water
COF	Coefficient of Friction
CMC	Sodium Carboxymethyl Cellulose
DS	Degreee of Substitution
MTM	Mini Traction Machine
PEG	Polyethylene Glycol 400
PVL	Permanent Viscosity Loss
RSO	Rapeseed Oil
SRR	Slide-Roll Ratio
TVL	Temporary Viscosity Loss
VM	Viscosity Modifier
WSD	Wear Scar Diameter
ZSVG	Zero-Shear Viscosity Grade

References

1. Ma, L.; Zhang, C.; Liu, S. Progress in experimental study of aqueous lubrication. *Chin. Sci. Bull.* **2012**, *57*, 2062–2069. [CrossRef]
2. Yilmaz, M.; Mirza, M.; Lohner, T.; Stahl, K. Superlubricity in EHL contacts with water-containing gear fluids. *Lubricants* **2019**, *7*, 46. [CrossRef]
3. Yilmaz, M.; Lohner, T.; Michaelis, K.; Stahl, K. Minimizing gear friction with water-containing gear fluids. *Forsch. Ing.* **2019**, *83*, 327–337. [CrossRef]
4. Sagraloff, N.; Dobler, A.; Tobie, T.; Stahl, K.; Ostrowski, J. Development of an oil free water-based lubricant for gear applications. *Lubricants* **2019**, *7*, 33. [CrossRef]
5. Rahman, M.H.; Warneke, H.; Webbert, H.; Rodriguez, J.; Austin, E.; Tokunaga, K.; Rajak, D.K.; Menezes, P.L. Water-based lubricants: Development, properties, and performances. *Lubricants* **2021**, *9*, 73. [CrossRef]
6. Martin, J.M.; De Barros-Bouchet, M.I. Water-like lubrication of hard contacts by polyhydric alcohols. In *Aqueous Lubrication*; IISc Research Monographs Series; Co-Published with Indian Institute of Science (IISc): Bangalore, India, 2011; Volume 3, pp. 219–235.
7. Martini, A.; Ramasamy, U.S.; Len, M. Review of viscosity modifier lubricant additives. *Tribol. Lett.* **2018**, *66*, 58. [CrossRef]
8. Spencer, N.D. Aqueous lubrication with poly(ethylene glycol) brushes. *Tribol. Online* **2014**, *9*, 143–153. [CrossRef]
9. Sagraloff, N.; Winkler, K.J.; Tobie, T.; Stahl, K.; Folland, C.; Asam, T. Investigations on the scuffing and wear characteristic performance of an oil free water-based lubricant for gear applications. *Lubricants* **2021**, *9*, 24. [CrossRef]
10. Zhang, Z.; Ye, Z.; Hu, F.; Wang, W.; Zhang, S.; Gao, L.; Lu, H. Double-network polyvinyl alcohol composite hydrogel with self-healing and low friction. *J. Appl. Polym. Sci.* **2022**, *139*, 51563. [CrossRef]
11. Ma, L.; Zhao, J.; Zhang, M.; Jiang, Z.; Zhou, C.; Ma, X. Study on the tribological behaviour of nanolubricants during micro rolling of copper foils. *Materials* **2022**, *15*, 2600. [CrossRef]
12. Dhakal, N.; Shi, Y.; Emami, N. Tribological behaviour of UHMWPE composites lubricated by polyvinylpyrrolidone-modified water. *Lubr. Sci.* **2022**, *34*, 42–53. [CrossRef]
13. Naik, S.C.; Pittman, J.F.T.; Richardson, J.F.; Lansdown, A.R. Evaluation of hydroxyethyl cellulose ether as a thickener for aqueous lubricants or hydraulic fluids. *Wear* **1978**, *50*, 155–168. [CrossRef]
14. Han, T.; Yi, S.; Zhang, C.; Li, J.; Chen, X.; Luo, J.; Banquy, X. Superlubrication obtained with mixtures of hydrated ions and polyethylene glycol solutions in the mixed and hydrodynamic lubrication regimes. *J. Colloid Interface Sci.* **2020**, *579*, 479–488. [CrossRef]
15. Branch, D.W.; Wheeler, B.C.; Brewer, G.J.; Leckband, D.E. Long-term stability of grafted polyethylene glycol surfaces for use with microstamped substrates in neuronal cell culture. *Biomaterials* **2001**, *22*, 1035–1047. [CrossRef] [PubMed]
16. Han, S.; Kim, C.; Kwon, D. Thermal degradation of poly(ethyleneglycol). *Polym. Degrad. Stab.* **1995**, *47*, 203–208. [CrossRef]
17. Chen, W.; Amann, T.; Kailer, A.; Rühe, J. Thin-film lubrication in the water/octyl β-D-glucopyranoside system: Macroscopic and nanoscopic tribological behavior. *Langmuir* **2019**, *35*, 7136–7145. [CrossRef]
18. Chen, W.; Amann, T.; Kailer, A.; Rühe, J. Macroscopic friction studies of alkylglucopyranosides as additives for water-based lubricants. *Lubricants* **2020**, *8*, 11. [CrossRef]
19. George, A.; Sanjay, M.R.; Srisuk, R.; Parameswaranpillai, J.; Siengchin, S. A comprehensive review on chemical properties and applications of biopolymers and their composites. *Int. J. Biol. Macromol.* **2020**, *154*, 329–338. [CrossRef]

20. Wüstenberg, T. *Cellulose und Cellulosederivate: Grundlagen, Wirkungen und Applikationen*; Behr's GmbH: Hamburg, Germany, 2013.
21. Wolfs, J.; Nickisch, R.; Wanner, L.; Meier, M.A.R. Sustainable one-pot cellulose dissolution and derivatization via a tandem reaction in the DMSO/DBU/CO$_2$ switchable solvent system. *J. Am. Chem. Soc.* **2021**, *143*, 18693–18702. [CrossRef]
22. Jain, S.; Sandhu, P.S.; Malvi, R.; Gupta, B. Cellulose derivatives as thermoresponsive polymer: An overview. *J. Appl. Pharm. Sci.* **2013**, *3*, 139–144.
23. Gelinski, S.; Winter, M.; Wichmann, H.; Bock, R.; Herrmann, C.; Bahadir, M. Development and testing of a novel glycerol/chitosan based biocide-free hydraulic fluid. *J. Clean. Prod.* **2016**, *112*, 3589–3596. [CrossRef]
24. Rahman, M.S.; Hasan, M.S.; Nitai, A.S.; Nam, S.; Karmakar, A.K.; Ahsan, M.S.; Shiddiky, M.J.A.; Ahmed, M.B. Recent developments of carboxymethyl cellulose. *Polymers* **2021**, *13*, 1345. [CrossRef]
25. Heinze, T.; Pfeiffer, K. Studies on the synthesis and characterization of carboxymethylcellulose. *Die Angew. Makromol. Chem.* **1999**, *266*, 37–45. [CrossRef]
26. Casaburi, A.; Montoya Rojo, U.; Cerrutti, P.; Vázquez, A.; Foresti, M.L. Carboxymethyl cellulose with tailored degree of substitution obtained from bacterial cellulose. *Food Hydrocoll.* **2018**, *75*, 147–156. [CrossRef]
27. Elliot, J.H.; Ganz, A.J. Some rheological properties of sodium carboxymethylcellulose solutions and gels. *Rheol. Acta* **1974**, *13*, 670–674. [CrossRef]
28. Lopez, C.G.; Rogers, S.E.; Colby, R.H.; Graham, P.; Cabral, J.T. Structure of sodium carboxymethyl cellulose aqueous solutions: A SANS and rheology study. *J. Polym. Sci. Part B Polym. Phys.* **2015**, *53*, 492–501. [CrossRef]
29. Xiquan, L.; Tingzhu, Q.; Shaoqui, Q. Kinetics of the carboxymethylation of cellulose in the isopropyl alcohol system. *Acta Polym.* **1990**, *41*, 220–222. [CrossRef]
30. Kulicke, W.M.; Kull, A.H.; Kull, W.; Thielking, H.; Engelhardt, J.; Pannek, J.B. Characterization of aqueous carboxymethylcellulose solutions in terms of their molecular structure and its influence on rheological behaviour. *Polymer* **1996**, *37*, 2723–2731. [CrossRef]
31. Kono, H.; Oshima, K.; Hashimoto, H.; Shimizu, Y.; Tajima, K. NMR characterization of sodium carboxymethyl cellulose: Substituent distribution and mole fraction of monomers in the polymer chains. *Carbohydr. Polym.* **2016**, *146*, 1–9. [CrossRef]
32. Holtzapple, M.T.; Caballero, B. Cellulose. In *Encyclopedia of Food Sciences and Nutrition*, 2nd ed.; Academic Press: Oxford, UK, 2003; pp. 998–1007.
33. Mondal, M.I.H.; Yeasmin, M.S.; Rahman, M.S. Preparation of food grade carboxymethyl cellulose from corn husk agrowaste. *Int. J. Biol. Macromol.* **2015**, *79*, 144–150. [CrossRef]
34. Abdelrahim, K.A.; Ramaswamy, H.S.; Doyon, G.; Toupin, C. Effects of concentration and temperature on carboxymethylcellulose rheology. *Int. J. Food Sci. Technol.* **1994**, *29*, 243–253. [CrossRef]
35. Abdelrahim, K.A.; Ramaswamy, H.S. High temperature/pressure rheology of carboxymethyl cellulose (CMC). *Food Res. Int.* **1995**, *28*, 285–290. [CrossRef]
36. Guan, W.; Ke, G.; Tang, C.; Liu, Y. *Study on Lubrication Properties of Carboxymethyl Cellulose as a Novel Additive in Water-Based Stock*; Technical Report; American Society of Mechanical Engineers: New York, NY, USA, 2005; Volume 42010.
37. Kulicke, W.M.; Clasen, C.; Lohman, C. Characterization of water-soluble cellulose derivatives in terms of the molar mass and particle size as well as their distribution. *Macromol. Symp.* **2005**, *223*, 151–174. [CrossRef]
38. Morris, E.R. Assembly and rheology of non-starch polysaccharides. In *Advanced Dietary Fibre Technology*; Wiley-Blackwell: Hoboken, NJ, USA, 2001; pp. 30–41.
39. Herschel, W.H.; Bulkley, R. Konsistenzmessungen von Gummi-Benzollösungen. *Kolloid-Zeitschrift* **1926**, *39*, 291–300. [CrossRef]
40. Cross, M.M. Rheology of non-Newtonian fluids: A new flow equation for pseudoplastic systems. *J. Colloid Sci.* **1965**, *20*, 417–437. [CrossRef]
41. Ostwald, W. Ueber die rechnerische Darstellung des Strukturgebietes der Viskosität. *Kolloid-Zeitschrift* **1929**, *47*, 176–187. [CrossRef]
42. Barus, C. Isothermals, isopiestics and isometrics relative to viscosity. *Am. J. Sci.* **1893**, *s3-s45*, 87–96. [CrossRef]
43. Arrhenius, S. Über die Reaktionsgeschwindigkeit bei der Inversion von Rohrzucker durch Säuren. *Z. FüR Phys. Chem.* **1889**, *4U*, 226–248. [CrossRef]
44. Habig, K.H.; Knoll, G. Schmierung. In *Tribologie-Handbuch: Tribometrie, Tribomaterialien, Tribotechnik*; Czichos, H., Habig, K.H., Eds.; Springer Fachmedien Wiesbaden: Wiesbaden, Germany, 2020; pp. 209–229. [CrossRef]
45. Hamrock, B.J.; Dowson, D. Isothermal elastohydrodynamic lubrication of point contacts: Part III–Fully flooded results. *J. Lubr. Tech.* **1977**, *99*, 264–275. [CrossRef]
46. Stambaugh, R.L. Viscosity index improvers and thickeners. In *Chemistry and Technology of Lubricants*; Mortier, R.M., Orszulik, S.T., Eds.; Springer: Boston, MA, USA, 1992; pp. 124–159.
47. Glass, J.E.; Schulz, D.N.; Zukoski, C.F. Polymers as Rheology Modifiers. In *Polymers as Rheology Modifiers*; ACS Symposium Series; Schulz, D.N., Glass, J.E., Eds.; American Chemical Society: Washington, DC, USA, 1991; Volume 462, pp. 2–17.
48. Miehle, E.; Bader-Mittermaier, S.; Schweiggert-Weisz, U.; Hauner, H.; Eisner, P. Effect of physicochemical properties of carboxymethyl cellulose on diffusion of glucose. *Nutrients* **2021**, *13*, 1398. [CrossRef]
49. Morris, E.R.; Cutler, A.N.; Ross-Murphy, S.B.; Rees, D.A.; Price, J. Concentration and shear rate dependence of viscosity in random coil polysaccharide solutions. *Carbohydr. Polym.* **1981**, *1*, 5–21. [CrossRef]
50. Clasen, C.; Kulicke, W.M. Determination of viscoelastic and rheo-optical material functions of water-soluble cellulose derivatives. *Prog. Polym. Sci.* **2001**, *26*, 1839–1919. [CrossRef]

51. Wagoner, T.B.; Çakır Fuller, E.; Drake, M.; Foegeding, E.A. Sweetness perception in protein-polysaccharide beverages is not explained by viscosity or critical overlap concentration. *Food Hydrocoll.* **2019**, *94*, 229–237. [CrossRef]

52. Charpentier, D.; Mocanu, G.; Carpov, A.; Chapelle, S.; Merle, L.; Müller, G. New hydrophobically modified carboxymethylcellulose derivatives. *Carbohydr. Polym.* **1997**, *33*, 177–186. [CrossRef]

53. Arinaitwe, E.; Pawlik, M. Dilute solution properties of carboxymethyl celluloses of various molecular weights and degrees of substitution. *Carbohydr. Polym.* **2014**, *99*, 423–431. [CrossRef] [PubMed]

54. Yang, X.H.; Zhu, W.L. Viscosity properties of sodium carboxymethylcellulose solutions. *Cellulose* **2007**, *14*, 409–417. [CrossRef]

55. Lopez, C.G.; Colby, R.H.; Graham, P.; Cabral, J.T. Viscosity and scaling of semiflexible polyelectrolyte NaCMC in aqueous salt solutions. *Macromolecules* **2017**, *50*, 332–338. [CrossRef]

56. Barba, C.; Montané, D.; Farriol, X.; Desbrières, J.; Rinaudo, M. Synthesis and characterization of carboxymethylcelluloses from non-wood pulps II. Rheological behavior of CMC in aqueous solution. *Cellulose* **2002**, *9*, 327–335. [CrossRef]

57. Benchabane, A.; Bekkour, K. Rheological properties of carboxymethyl cellulose (CMC) solutions. *Colloid Polym. Sci.* **2008**, *286*, 1173. [CrossRef]

58. Cook, R.L.; King, H.E.J.; Peiffer, D.G. High-pressure viscosity of dilute polymer solutions in good solvents. *Macromolecules* **1992**, *25*, 2928–2934. [CrossRef]

59. Shi, Y.; Minami, I.; Grahn, M.; Björling, M.; Larsson, R. Boundary and elastohydrodynamic lubrication studies of glycerol aqueous solutions as green lubricants. *Tribol. Int.* **2014**, *69*, 39–45. [CrossRef]

60. Sequeira, M.C.; Pereira, M.F.; Avelino, H.M.; Caetano, F.J.; Fareleira, J.M. Viscosity measurements of poly(ethyleneglycol) 400 (PEG 400) at temperatures from 293 K to 348 K and at pressures up to 50 MPa using the vibrating wire technique. *Fluid Phase Equilibria* **2019**, *496*, 7–16. [CrossRef]

61. Joseph, G.G.; Hunt, M.L. Oblique particle-wall collisions in a liquid. *J. Fluid Mech.* **2004**, *510*, 71–93. [CrossRef]

62. Wang, H.; Liu, Y.; Li, J.; Luo, J. Investigation of superlubricity achieved by polyalkylene glycol aqueous solutions. *Adv. Mater. Interfaces* **2016**, *3*, 1600531. [CrossRef]

63. Wan, G.T.Y.; Spikes, H.A. The elastohydrodynamic lubricating properties of water-polyglycol fire-resistant fluids. *ASLE Trans.* **1984**, *27*, 366–372. [CrossRef]

64. Wan, G.T.Y.; Kenny, P.; Spikes, H.A. Elastohydrodynamic properties of water-based fire-resistant hydraulic fluids. *Tribol. Int.* **1984**, *17*, 309–315. [CrossRef]

65. Dowson, D. Elastohydrodynamic and micro-elastohydrodynamic lubrication. *Wear* **1995**, *190*, 125–138. [CrossRef]

66. Dowson, D. Paper 10: Elastohydrodynamics. *Proc. Inst. Mech. Eng. Conf. Proc.* **1967**, *182*, 151–167. [CrossRef]

67. Arnšek, A.; Vižintin, J. Lubricating properties of rapeseed-based oils. *J. Synth. Lubr.* **2000**, *16*, 281–296. [CrossRef]

68. Mobarak, H.M.; Niza Mohamad, E.; Masjuki, H.H.; Kalam, M.A.; Al Mahmud, K.A.H.; Habibullah, M.; Ashraful, A.M. The prospects of biolubricants as alternatives in automotive applications. *Renew. Sustain. Energy Rev.* **2014**, *33*, 34–43. [CrossRef]

69. Attia, N.K.; El-Mekkawi, S.A.; Elardy, O.A.; Abdelkader, E.A. Chemical and rheological assessment of produced biolubricants from different vegetable oils. *Fuel* **2020**, *271*, 117578. [CrossRef]

70. Hutter, C.; Wang, Y.; Chubarenko, I. Phenomenological coefficients of water. In *Physics of Lakes*; Springer: Berlin/Heidelberg, Germany, 2011; pp. 389–418.

71. Lyytikäinen, J.; Laukala, T.; Backfolk, K. Temperature-dependent interactions between hydrophobically modified ethyl(hydroxyethyl)cellulose and methyl nanocellulose. *Cellulose* **2019**, *26*, 7079–7087. [CrossRef]

72. Wilson, B. Lubricants and functional fluids from renewable sources. *Ind. Lubr. Tribol.* **1998**, *50*, 6–15. [CrossRef]

73. Barton, W.R.S.; Payne, J.; Baker, M.; O'Connor, B.; Qureshi, F.; Huston, M.; Knapton, D. Impact of viscosity modifiers on gear oil efficiency and durability. *SAE Int. J. Fuels Lubr.* **2012**, *5*, 470–479. [CrossRef]

74. Veltkamp, B.; Jagielka, J.; Velikov, K.; Bonn, D. Lubrication with non-newtonian fluids. *Phys. Rev. Appl.* **2023**, *19*, 014056. [CrossRef]

75. Campen, S.; Green, J.; Lamb, G.; Atkinson, D.; Spikes, H. On the increase in boundary friction with sliding speed. *Tribol. Lett.* **2012**, *48*, 237–248. [CrossRef]

76. Liu, S.; Guo, D.; Xie, G. Nanoscale lubricating film formation by linear polymer in aqueous solution. *J. Appl. Phys.* **2012**, *112*, 104309. [CrossRef]

77. Chatterjee, A.; Das, B. Radii of gyration of sodium carboxymethylcellulose in aqueous and mixed solvent media from viscosity measurement. *Carbohydr. Polym.* **2013**, *98*, 1297–1303. [CrossRef]

78. Vengudusamy, B.; Grafl, A.; Novotny-Farkas, F.; Schöfmann, W. Comparison of frictional properties of gear oils in boundary and mixed lubricated rolling-sliding and pure sliding contacts. *Tribol. Int.* **2013**, *62*, 100–109. [CrossRef]

79. Antti, G.; Pentti, P.; Hanna, K. Ultrasonic degradation of aqueous carboxymethylcellulose: Effect of viscosity, molecular mass, and concentration. *Ultrason. Sonochem.* **2008**, *15*, 644–648. [CrossRef] [PubMed]

80. Almeida, N.; Rakesh, L.; Zhao, J. Phase behavior of concentrated hydroxypropyl methylcellulose solution in the presence of mono and divalent salt. *Carbohydr. Polym.* **2014**, *99*, 630–637. [CrossRef] [PubMed]

81. Schweiger, R.G. New cellulose sulfate derivatives and applications. *Carbohydr. Res.* **1979**, *70*, 185–198. [CrossRef]

 *lubricants*

Article

Tribological Behavior and Cold-Rolling Lubrication Performance of Water-Based Nanolubricants with Varying Concentrations of Nano-TiO$_2$ Additives

Linan Ma [1], Luhu Ma [2,3], Junjie Lian [2,3], Chen Wang [2,3], Xiaoguang Ma [2,3] and Jingwei Zhao [2,3,*]

[1] School of Mechanical Engineering, Taiyuan University of Science and Technology, Taiyuan 030024, China
[2] College of Mechanical and Vehicle Engineering, Taiyuan University of Technology, Taiyuan 030024, China; wangchen0023@link.tyut.edu.cn (C.W.); maxiaoguang@tyut.edu.cn (X.M.)
[3] Engineering Research Center of Advanced Metal Composites Forming Technology and Equipment, Ministry of Education, Taiyuan 030024, China
* Correspondence: jzhao@tyut.edu.cn

Abstract: This study aimed to investigate the effect of water-based nanolubricants containing varying concentrations (1.0–9.0 wt.%) of TiO$_2$ nanoparticles on the friction and wear of titanium foil surfaces. Water-based nanolubricants containing TiO$_2$ nanoparticles of varying concentrations were prepared and applied in friction and wear experiments and micro-rolling experiments to evaluate their performance regarding friction and wear properties. The findings indicated that the best results were achieved with a 3.0 wt.% TiO$_2$ nano-additive lubricant that significantly improved the tribological properties, with reductions in the COF and wear of 82.9% and 42.7%, respectively, compared to the dry conditions without any lubricant. In addition, nanolubricants contribute to a reduction in rolling forces and an improvement in the surface quality of titanium foils after rolling. In conclusion, nanolubricants exhibit superior lubricating properties compared to conventional O/W lubricants, which is attributed to the combined effect of the rolling effect, polishing effect, mending effect and tribo-film effect of the nanoparticles.

Keywords: nanolubricants; titanium foil; tribological behavior; micro rolling

Citation: Ma, L.; Ma, L.; Lian, J.; Wang, C.; Ma, X.; Zhao, J. Tribological Behavior and Cold-Rolling Lubrication Performance of Water-Based Nanolubricants with Varying Concentrations of Nano-TiO$_2$ Additives. *Lubricants* **2024**, *12*, 361. https://doi.org/10.3390/lubricants12110361

Received: 15 July 2024
Revised: 17 October 2024
Accepted: 19 October 2024
Published: 22 October 2024

1. Introduction

Owing to their high specific strength, toughness and corrosion resistance, titanium foils (TFs) are used in many fields such as medical science and aerospace engineering [1–3]. Numerous advanced rolling methods have been developed to produce high-quality TF, but friction and wear are unavoidable during rolling, impacting the quality of the formed product and resulting in energy loss [4,5]. To minimize the effects of friction and wear, lubricants are commonly used to form a protective layer between contact pairs during the forming process [6]. However, traditional oil-based lubricants cause considerable environmental damage during the rolling process, which limits their application in industry [7]. In order to minimize the environmental pollution of lubricants, environmentally friendly lubricants have been developed to replace traditional oil-based lubricants. With the application of nanomaterials in lubrication technology and the growing insights into the special characteristics of nano-functional materials, nanolubricants exhibit excellent tribological behavior during the forming process, which leads to a decrease in energy loss caused by friction and an improvement in the surface finish of the formed products [8].

As a result of high stability and excellent tribological properties, nanoscale metallic oxide is frequently added to water-based lubricants to improve their overall tribological performance [9]. Thapliyal et al. [10] examined the service life of AISI E-52100 steel balls using copper nanofluid lubrication via rolling contact fatigue tests. The results indicated a significant enhancement in the fatigue life of the bearing balls with the use of nanofluids.

Bao et al. [11] conducted a series of hot-rolling experiments based on a water-based lubricant containing SiO_2 nanoparticles and found that more refined grains were formed within the superficial layer of the rolling strips. Kong et al. [12] studied the frictional characteristics of a graphene-based hybrid lubricant using a ball-and-disk tester, demonstrating its effectiveness in reducing the coefficient of friction (COF) between the contact surfaces. Du et al. [13] examined the tribological performances of GO-TiO_2 nanofluids using a four-ball tribometer. The enhanced lubricating characteristics of GO-TiO_2 nanocomposites are attributed to the formation of absorption films, carbonaceous protective films and transfer films. Srivyas et al. [14] investigated the tribological performance of graphite, graphene nanoplatelets and h-BN. The results showed that the hybrid graphene nanoplatelets plus h-BN-based lubricating oil exhibited superior friction and wear properties with minimized COF. Nassef et al. [15] investigated the tribological and chemical–physical behavior of palm oil grease, which was improved by reduced graphene oxide (rGO) and zinc oxide (ZnO) nano-additives at different concentrations, and found that the addition of ZnO and rGO to palm oil increased the load-bearing capacity by 30% and 60%, respectively. In addition, the coefficient of friction was lowered by as much as 60%, which can be attributed to an absorbing layer formed by the 2D graphene nanoparticles that increased the load capacity, and ZnO chemically reacted with the metal surface layer to form zinc compounds that formed a lubrication film that served as a protective boundary.

Of the different types of nanoparticles used in nanolubricants, TiO_2 nanoparticles have shown great potential owing to their unique properties such as high hardness, excellent antioxidant capabilities and thermal stability, making them ideal candidates for enhancing the lubricating properties of water-based lubricants [16,17]. By adding TiO_2 nanoparticles to water-based lubricants, the lubricating efficiency and reductions in friction and wear resistance can be improved in mechanical systems [18]. Beel et al. [19] investigated the tribological effects of TiO_2 nanostructured particle dispersions in automotive engine base oils using a disk–disk tribometer. Their findings demonstrated a notable reduction in frictional force between the bearing steel and carbon steel surfaces. Wu et al. [20] studied the tribological behavior of innovative nano-TiO_2 additive lubricants on ferritic stainless steel 445 using a ball-on-disk tribometer. They found that the proposed lubricants exhibited excellent frictional characteristics due to the formation of a tribofilm of the TiO_2 nanoparticles. Sharma et al. [21] explored the TiO_2 surface modifications and found that the introduction of boron atoms promotes the formation of protective anti-wear films at the surface. Wang et al. [22] investigated the lubrication effect of a mixture of sodium polyphosphate and nano-TiO_2 as a water-based lubricant in hot rolling. It was found that the rotational and protective action of TiO_2 particles provided effective lubrication. Xia et al. [23] detected the contact angle to investigate the role of a nano-TiO_2 additive in O/W lubricant. The results showed that the minimum contact angle was obtained with the addition of 4.0 wt.% of the nano-TiO_2 additive to the O/W lubricant.

In this study, tribological tests were conducted to explore the performance of nano-TiO_2 additives as nanolubricants in terms of friction and wear characteristics. Additionally, the tribological performance of nano-TiO_2 additive lubricants was studied in the micro rolling of TFs. This study aims to clarify the friction reduction and anti-wear capabilities of water-based nanolubricants on TF surfaces compared to conventional oil-in-water (O/W) lubricants. It also aims to illustrate the lubrication mechanisms of water-based lubricants that incorporate TiO_2 nanoparticles.

2. Experimental Procedures

2.1. Materials

In this study, GCr15 steel balls and disks fabricated from pure TF were employed, and the detailed chemical compositions are shown in Tables 1 and 2, respectively. The disks were machined to the dimensions of $100 \times 25 \times 0.1$ mm^3, and Figure 1 illustrates the surface morphologies and 3D profiles of the contact pair. Before the experiment, both balls and TF disks were cleaned with alcohol to ensure consistent surface conditions.

Table 1. Chemical compositions of GCr15 steel (wt.%).

Element	C	Mn	Si	P	Cr	S	Ni
Content	0.96	0.31	0.22	0.014	1.52	0.002	0.004

Table 2. Chemical compositions of TF disks (wt.%).

Element	Ti	Fe	N	H	O	C
Content	≥99.6	≤0.014	≤0.008	≤0.0013	≤0.046	≤0.010

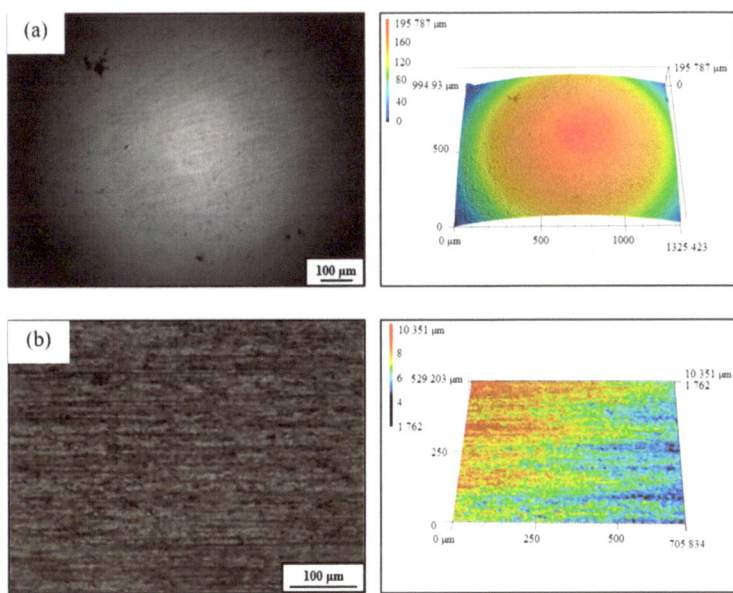

Figure 1. Surface morphologies and 3D profiles of (**a**) GCr15 steel ball and (**b**) TF disk.

In addition, pure TFs were machined to dimensions of $150 \times 5 \times 0.1 \ mm^3$ for micro-rolling tests. Before conducting rolling experiments, the specimens were cleaned with alcohol to avoid residues left from the machining processes.

The detailed procedure for the preparation of the lubricant is presented as follows. First, for the nano-TiO_2 additive lubricants, sodium dodecylbenzene sulfonate (SDBS) was uniformly dissolved in water and stirred in a disperser at 8000 rpm for 10 min. Subsequently, polyacrylic acid sodium salt (PAAS) was added and dissolved in the suspension using a disperser under the same conditions. Then, the nano-TiO_2 particles were added to the suspension and stirred with the disperser at 8000 rpm for 10 min to make a homogeneous dispersion. Finally, the suspension was ultrasonically stirred for 10 min to disperse any remaining agglomerates. SDBS is a potent anionic surfactant that electrostatically prevents nanoparticle aggregation [24,25]. PAAS is a new functional polymeric material that facilitates the enhancement of lubricant viscosity [26]. The TiO_2 nanoparticles are rutile and are approximately 30 nm in diameter. For the O/W lubricants, 1.0 wt.% of oil was dispersed in water and stirred with the disperser at 8000 rpm for 10 min, followed by 10 min of ultrasonic vibration. The lubrication conditions for the tribological and micro-rolling tests are listed in Table 3. The concentrations of the nano-TiO_2 additive lubricants varied from 1.0 to 9.0 wt.%. To assess the lubrication performance of the nano-TiO_2 additive lubricants, dry (DR) and O/W conditions were used as benchmarks.

Table 3. Varying compositions of lubricants.

Lubricant Conditions	Description
DR	Dry
O/W	1.0 wt.% oil + balance water
L1	1.0 wt.% TiO_2 + 0.2 wt.% SDBS + 0.3 wt.% PAAS + balance water
L2	3.0 wt.% TiO_2 + 0.6 wt.% SDBS + 0.3 wt.% PAAS + balance water
L3	5.0 wt.% TiO_2 + 1.0 wt.% SDBS + 0.3 wt.% PAAS + balance water
L4	7.0 wt.% TiO_2 + 1.4 wt.% SDBS + 0.3 wt.% PAAS + balance water
L5	9.0 wt.% TiO_2 + 1.8 wt.% SDBS + 0.3 wt.% PAAS + balance water

2.2. Tribological and Micro-Rolling Tests

The tribological properties of the prepared nanolubricants were assessed with the reciprocating motion module using the MFT-5000 Multi-Environment Lubrication Test System (as shown in Figure 2). Before the tribological tests, the balls and TF disks were wiped with alcohol to standardize the surface condition. The tribological tests were conducted at room temperature, and the experimental conditions that were employed are listed in Table 4. Each lubrication condition was tested using three specimens to ensure accurate results through averaging.

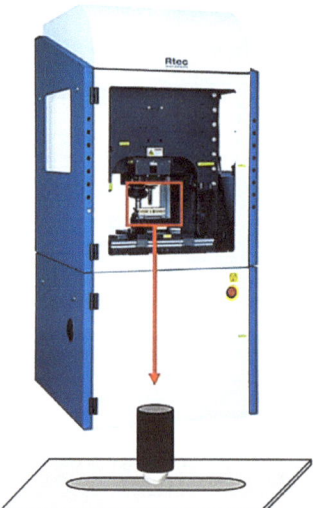

Reciprocating motion module

Figure 2. MFT-5000 Multi-Environment Lubrication Test System.

Table 4. Tribological test conditions.

Sliding Distance/mm	Load/N	Frequency/Hz	Time/s
10	10	1	600

Micro-rolling tests were conducted using a precise four-high micro-rolling mill with work rolls measuring 22 mm in diameter and 44 mm in barrel length. The micro mill is installed with high-precision piezoelectric force sensors, and the roll gap adjustment is accurately controlled by a self-designed PLC program. The tests were performed at a rate of 1 m/min with a 5% reduction under the various lubrication conditions detailed in Table 2. Before each test, the work roll surface was cleaned and uniformly sprayed with lubricant to ensure a consistent liquid film. Each micro-rolling test was repeated three times to minimize fluctuations in rolling force data and to derive average values for subsequent analysis.

2.3. Analytical Techniques

The surface morphologies of steel balls and rolled TFs were examined using a KEYENCE VK-X1000 3D laser scanning microscope (KEYENCE, Osaka, Japan). In addition, elemental distributions on the surfaces of the TF disks and rolled TFs were obtained using a JEOL-IT500 (Oxford, British) scanning electron microscope (SEM) that includes an energy-dispersive spectroscopy (EDS) detector. Nanoparticle distribution and micromorphology of TFs were also observed using SEM.

3. Results and Discussion

3.1. Tribological Behavior of Nanolubricants

Figure 3 shows the average COF values of rolled TFs under varying lubrication conditions. The highest COF value (0.473) was obtained under the DR condition. The average COF values of rolled TFs were significantly lower when using O/W or nano-TiO_2 additive lubricants compared to DR conditions. The minimum COF value of 0.081 was achieved when using nanolubricants containing 3.0 wt.% nanoparticles. Nevertheless, a remarkable increase in COF values were observed when the nanoparticle concentration increased from 3.0 wt.% to 9.0 wt.%. Compared to the DR and O/W conditions, a maximum reduction of 82.9% and 22.4% can be achieved by nanolubricants, indicating an outstanding reduction in the frictional behavior of nanolubricants.

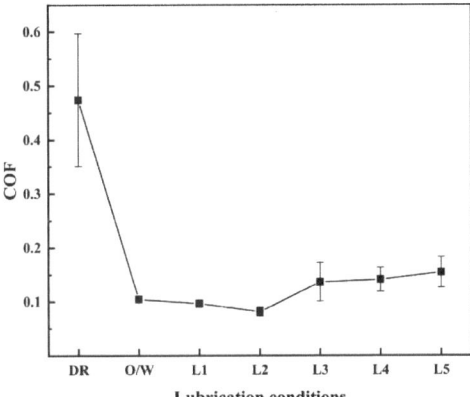

Figure 3. Variations in averaged COF values under varying lubrication conditions.

Figure 4 presents the surface profiles of the worn balls after the tribological tests at varying lubrication conditions. Under DR conditions, obvious scratches were induced on the surface of the worn ball. Compared to the DR condition, the wear scar under the action of the nanolubricants was smaller and smoother with clean boundaries, indicating enhanced wear resistance. The smallest wear scar diameter (361.318 μm) was observed on the balls when using nanolubricants containing 3.0 wt.% nanoparticles, which was reduced by approximately 42.7% compared to the wear diameter of the worn ball under DR conditions (630.514 μm). However, wear became more severe as the concentration of TiO_2 nanoparticles increased to 9.0 wt.%, indicating a poor tribological performance of the nanolubricants at higher nanoparticle concentrations.

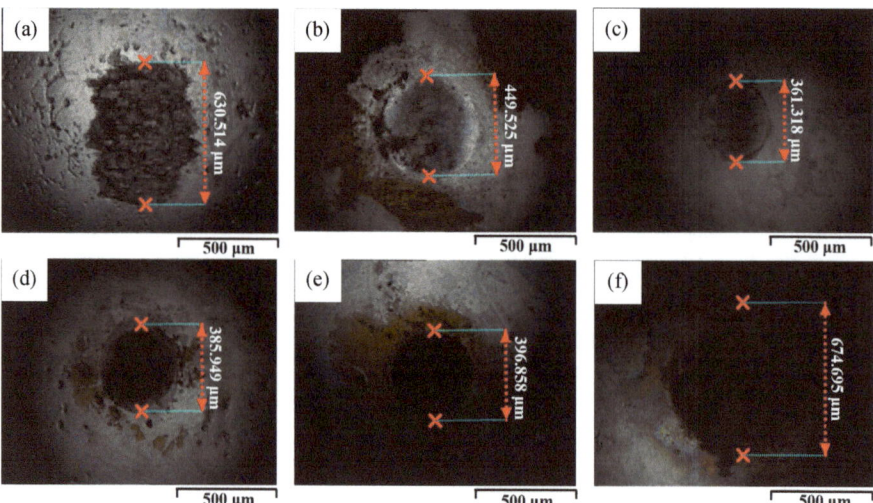

Figure 4. Surface morphologies of worn balls under the following lubrication conditions: (**a**) DR, (**b**) L1, (**c**) L2, (**d**) L3, (**e**) L4 and (**f**) L5.

The SEM images and EDS mappings of the center zone of the wear tracks generated after tribological testing under varying lubrication conditions are illustrated in Figure 5. From the distribution of O elements within the EDS images, the distribution of nanoparticles can be analyzed. During the rolling processes, the nanoparticles with a spherical shape rolled at the contact area, reducing friction and polishing the surface, thereby exhibiting excellent anti-friction and anti-wear performance [27,28]. In Figure 5a, nanolubricants with 1.0 wt.% TiO$_2$ particles reveal that only a small number of particles entered the contact zone, which limited their ability to decrease the COF and ball wear. With an increase in the concentration of nanoparticles to 3.0 wt.%, more TiO$_2$ nanoparticles remained within the contact area, leading to an improved lubricating performance of the nanolubricants. In addition, the formation of a thin TiO$_2$ film on the worn surfaces of the disk was also clearly observed, as shown in Figure 5b. As shown in Figure 5c–e, as the concentration of TiO$_2$ nanoparticles further increased from 3.0 wt.% to 9.0 wt.%, a film formed by the deposition of TiO$_2$ nanoparticles was observed over almost the entire worn surfaces of the disk. The nanoparticles aggregated around the contact zone, forming a barrier that limited the supply of external nanoparticles [29]. Furthermore, nanoparticles that were in excess tended to aggregate and formed a dense film that resulted in an increased COF and more pronounced wear on the surface of the ball [30].

3.2. Tribological Performance of Nanolubricants in Micro Rolling

Owing to the superior tribological performance of nanolubricants containing 3.0 wt.% nanoparticles, their behavior in micro rolling was evaluated in comparison to their behavior in DR and O/W conditions. Figure 6 illustrates the rolling forces during TF micro rolling under varying lubrication conditions. In contrast to the DR condition, both O/W lubricants and nano-TiO$_2$ additive lubricants were effective in reducing the rolling force during micro rolling. A minimum rolling force (3.914 kN) was obtained when nanolubricants containing 3.0 wt.% nanoparticles were used. However, as the concentration of nano-TiO$_2$ additives increased from 3.0 wt.% to 9.0 wt.%, the rolling force increased from 3.914 kN to 5.382 kN. This phenomenon can be attributed to nanoparticle aggregation [5]. The surface roughness of the rolled TFs under varying lubrication conditions is depicted in Figure 7. The application of O/W lubricant reduced the surface roughness of the rolled TF, and a further reduction in the surface roughness of the rolled TF was observed when the same concentration of TiO$_2$ nanolubricant was applied. With a further increase in the concentration of TiO$_2$ additives to 3.0 wt.%, the surface roughness of the TF reached a minimum of 0.107 μm at this point. Therefore, a further increase in TiO$_2$ concentration led to an increase in TF surface roughness, which may have been caused by nanoparticle aggregation [23]. Overall, the optimal nanoparticle concentration that led to an enhanced lubricating behavior of the nanolubricants, i.e., the lowest rolling force and surface roughness, was 3.0 wt.%.

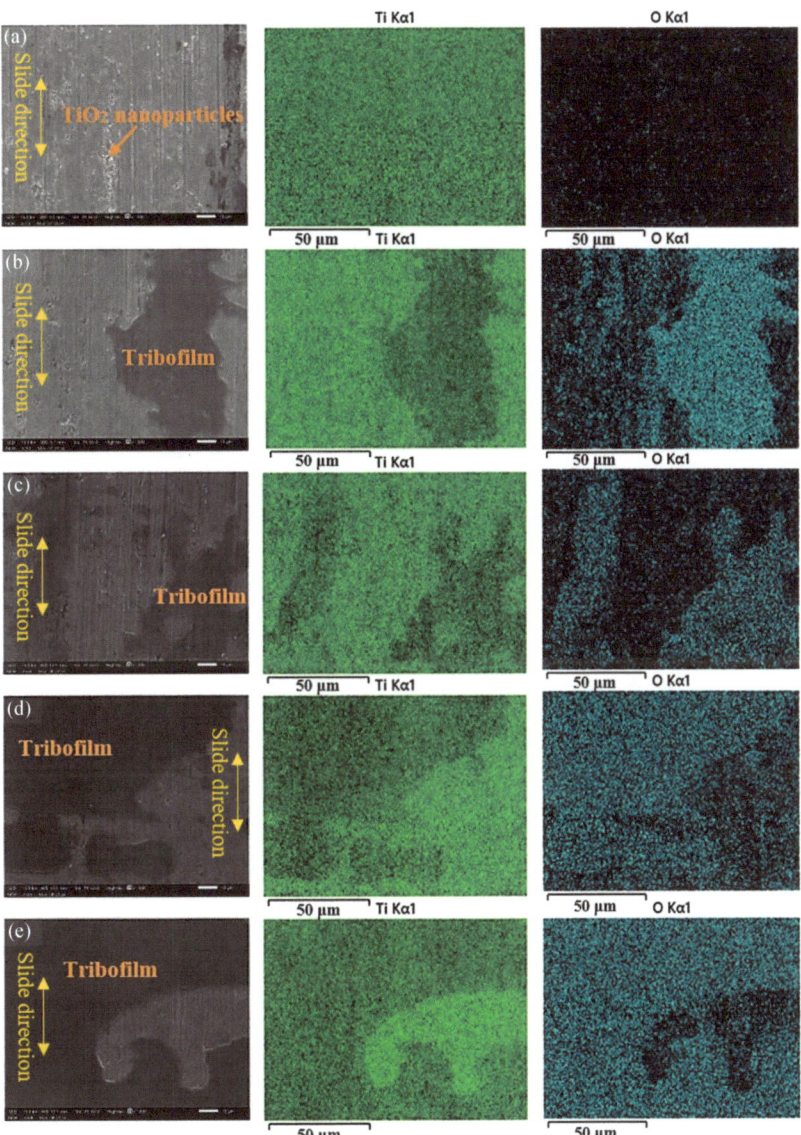

Figure 5. SEM images and EDS mappings of the worn surfaces of disks under the following lubrication conditions: (**a**) L1, (**b**) L2, (**c**) L3, (**d**) L4 and (**e**) L5.

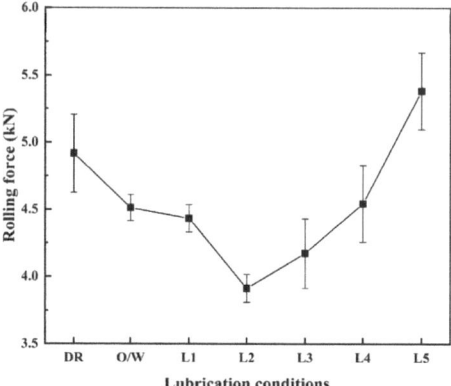

Figure 6. The rolling forces of rolled TFs under varying lubrication conditions.

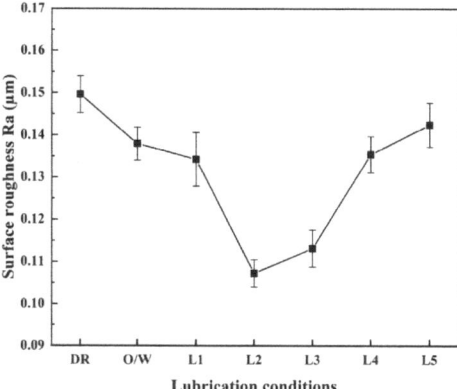

Figure 7. Surface roughness of TFs rolled under varying lubrication conditions.

3.3. SEM–EDS Analysis

Figure 8 shows the surface profile and 3D morphologies of the rolled TFs. Characteristic undulations formed at the surface of rolled specimens under the DR condition (as shown in Figure 8a). When lubricated with nanolubricant containing 3.0 wt.% TiO_2 nanoparticles, the surface quality was improved (as shown in Figure 8d), indicating that the use of these nanoparticles can positively affect the surface quality of TFs. However, increasing the nanoparticle concentration to 9.0 wt.% resulted in a significant decrease in surface quality, which exactly matches the results presented in Figure 7.

Figure 9 presents SEM images and EDS mappings of the rolled TFs under various lubrication scenarios. In the case of samples rolled with lubricants that included 1.0 wt.% TiO_2 nanoparticles, Figure 9a illustrates a surface with more deformation and fewer scattered nanoparticles. As for the rolled specimens with lubricant containing 3.0 wt.% TiO_2 nanoparticles, more O elements are uniformly distributed on the rolled foil surface (as shown in Figure 9b), indicating an improved lubricating effect and leading to less rolling energy consumption and reduced surface roughness of foils. Nevertheless, more aggregated nanoparticles can be found on the surface as TiO_2 nanoparticles concentration increases to 9.0 wt.% (as shown in Figure 9e), resulting in significant contamination of the rolled foil surface [14].

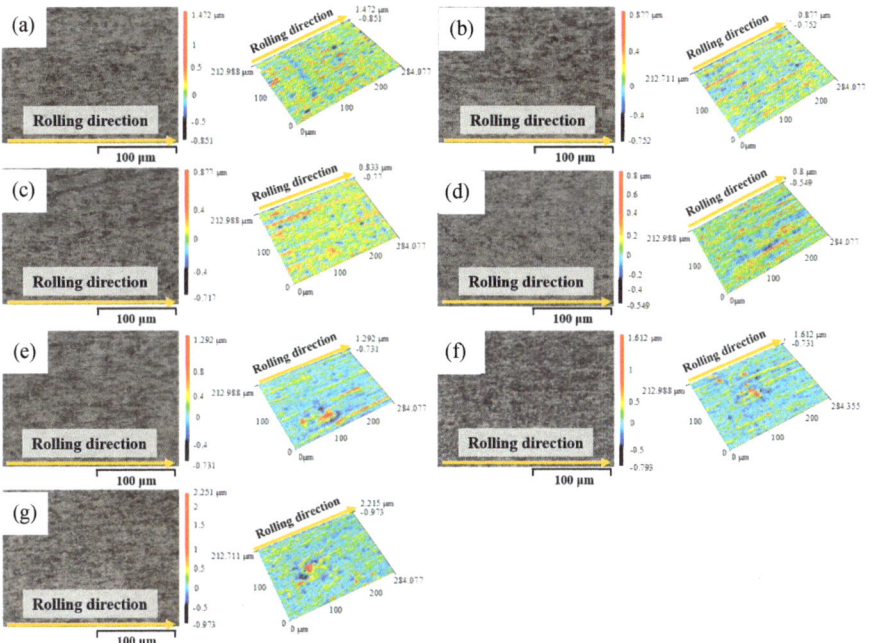

Figure 8. Surface morphologies of the rolled TFs under the following varying lubrication conditions: (**a**) DR, (**b**) O/W, (**c**) L1, (**d**) L2, (**e**) L3, (**f**) L4 and (**g**) L5.

3.4. Lubrication Mechanisms

To reveal the effects of the nanoparticles on the surface of TFs, a comparison of the lubricating mechanism in the DR condition and the nanolubricant condition is illustrated in Figure 10. Under the DR condition, severe abrasive wear occurred, and metal debris worn from the ball and disk surfaces remained in the contact area, as shown in Figure 10a. When using nanolubricants with 1.0 wt.% TiO_2 nanoparticles, few nanoparticles entered the frictional contact zone. The spherical TiO_2 nanoparticles suspended in the lubricant functioned akin to ball bearings within this zone. Additionally, the rolling and polishing effects of the nanoparticles contributed to a decrease in the COF between the frictional surfaces, resulting in a reduction in wear of the steel balls (as shown in Figure 10b). As the TiO_2 nanoparticle concentration rose from 1.0 wt.% to 3.0 wt.%, a protective frictional film was created by the nanoparticles between the contacting surfaces (as shown in Figure 10c), which prevented direct contact between the balls and the disk and thus enhanced the tribological behavior of the lubricant. Additionally, the scratches formed due to intense wear on the surface of the TF disk, allowing TiO_2 nanoparticles to fill the existing grooves. These grooves, together with the nanoparticles supplemented in the external area, filled the surface valleys through the so-called "mending" effect and kept the COF at a relatively low level, thus effectively reducing the wear of the disk after the running-in stage [31]. However, further increasing the concentration of TiO_2 nanoparticles to 9.0 wt.% accelerated the agglomeration of TiO_2 nanoparticles, as shown in Figure 10d. When the steel ball slid on the relatively rough TF disk, a number of nanoparticles agglomerated around the friction region to form a barrier, preventing a continuous supply of nanoparticles to the friction region and leading to an elevation in the COF and wear of the steel ball [32]. In addition, the agglomerated nanoparticles rolled between the roll and TF during micro rolling, which deteriorated the surface quality of the rolled TFs [33]. Overall, the optimal concentration of nano-TiO_2 additives in the lubricant, which was 3.0 wt.%, exhibited improved friction reduction and anti-wear effects on the surface of the TFs, which can be attributed to the combination of the rolling effect, repairing effect, polishing effect and tribo-film effect.

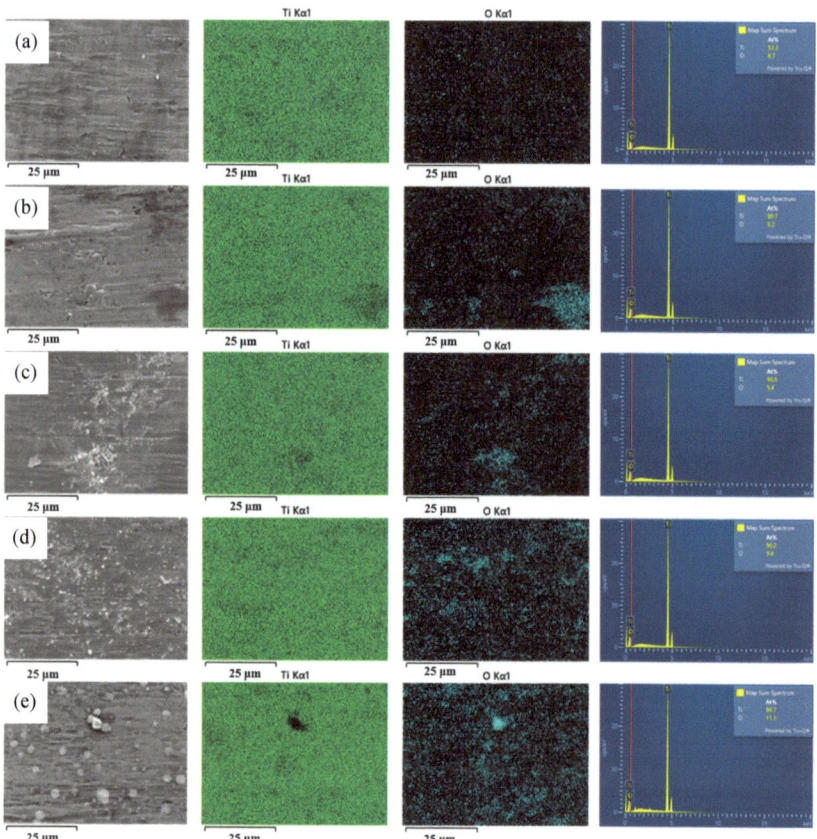

Figure 9. The SEM images and EDS mappings of the rolled TFs under the following varying lubrication conditions: (**a**) L1, (**b**) L2, (**c**) L3, (**d**) L4 and (**e**) L5.

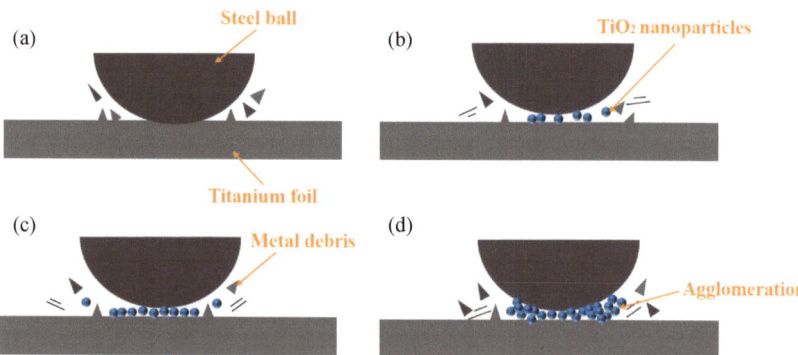

Figure 10. The lubricating mechanisms under the following lubrication conditions: (**a**) DR, (**b**) L1, (**c**) L2 and (**d**) L5.

4. Conclusions

In this manuscript, the tribological performance of water-based nanolubricants was investigated. The main conclusions are as follows:

1. Nanolubricants with an optimal concentration of TiO_2 nanolubricant additives, which was 3.0 wt.%, exhibited excellent tribological performance, reducing the COF and ball wear by 82.9% and 42.7%, respectively, compared to the DR conditions. This phenomenon can be attributed to the formation of a nano-TiO_2 friction film between the contact pairs.

2. Compared with conventional O/W lubricants, the nanolubricants exhibited an improved lubrication performance during the micro-rolling processes, which confirms the superior lubricating effect of the nanolubricants.

3. Applying nano-TiO_2 nanolubricants at a concentration of 3.0 wt.% led to a decrease in the rolling force and an enhancement in the surface quality of the TFs, which is primarily ascribed to the combined effects of the rolling effect, repairing effect, polishing effect and tribo-film effect.

Author Contributions: Conceptualization, J.Z.; formal analysis, J.L.; investigation, C.W.; methodology, L.M. (Luhu Ma); writing—original draft, L.M. (Linan Ma); writing—review and editing, X.M. All authors have read and agreed to the published version of the manuscript.

Funding: This research was supported by the National Natural Science Foundation of China (No. 52275359), the Central Government Guided Local Science and Technology Development Fund Project (No. YDZJSX2021A006), the Major Science and Technology Project of Shanxi Province of China (202101120401008) and the Natural Science Foundation of Shanxi Province (No. 202103021223286).

Data Availability Statement: All data gathered regarding this publication are presented in the article.

Conflicts of Interest: The authors declare no competing interests.

References

1. Hiroyasu, K.; Yoshimasa, T.; Hideyuki, I.; Tatsushi, K.; Takayuki, Y. Application of titanium and titanium alloys to fixed dental prostheses. *J. Prosthodont. Res.* **2019**, *63*, 266–270.
2. Zhao, J.; Lv, L.; Liu, G.; Wang, K. Analysis of deformation inhomogeneity and slip mode of TA15 titanium alloy sheets during the hot tensile process based on crystal plasticity model. *Mater. Sci. Eng. A* **2017**, *707*, 30–39. [CrossRef]
3. Ghosh, A.; Singh, A.; Gurao, N.P. Effect of rolling mode and annealing temperature on microstructure and texture of commercially pure-titanium. *Mater. Charact.* **2017**, *125*, 83–93. [CrossRef]
4. Nagendramma, P.; Kaul, S. Development of ecofriendly/biodegradable lubricants: An overview. *Renew. Sustain. Energy Rev.* **2012**, *16*, 764–774. [CrossRef]
5. Huo, M.; Wu, H.; Xie, H.; Zhao, J.; Su, G.; Jia, F.; Li, Z.; Lin, F.; Li, S.; Zhang, H.; et al. Understanding the role of water-based nanolubricants in micro flexible rolling of aluminum. *Tribol. Int.* **2020**, *151*, 106378. [CrossRef]
6. Zhao, J.; Huo, M.; Ma, X.; Jia, F.; Jiang, Z. Study on edge cracking of copper foils in micro rolling. *Mater. Sci. Eng. A* **2019**, *747*, 53–62. [CrossRef]
7. Ma, X.; Zhao, J.; Du, W.; Zhang, X.; Jiang, Z. Effects of rolling processes on ridging generation of ferritic stainless steel. *Mater. Charact.* **2018**, *137*, 201–211. [CrossRef]
8. Zhu, Y.; Zhang, H.; Li, N.; Jiang, Z. Friction and Wear Characteristics of Fe_3O_4 Nano-Additive Lubricant in Micro-Rolling. *Lubricants* **2023**, *11*, 434. [CrossRef]
9. Zhao, J.; Huang, Y.; He, Y.; Shi, Y. Nanolubricant additives: A review. *Friction* **2021**, *9*, 891–917. [CrossRef]
10. Thapliyal, P.; Thakre, G.D. Influence of Cu nanofluids on the rolling contact fatigue life of bearing steel. *Eng. Fail. Anal.* **2017**, *78*, 110–121. [CrossRef]
11. Bao, Y.; Sun, J.; Kong, L. Effects of nano-SiO_2 as water-based lubricant additive on surface qualities of strips after hot rolling. *Tribol. Int.* **2017**, *114*, 257–263. [CrossRef]
12. Kong, N.; Zhang, J.; Zhang, J.; Li, H.; Wei, B.; Li, D.; Zhu, H. Chemical- and Mechanical-Induced Lubrication Mechanisms during Hot Rolling of Titanium Alloys Using a Mixed Graphene-Incorporating Lubricant. *Nanomaterials* **2020**, *10*, 665. [CrossRef] [PubMed]
13. Du, S.; Sun, J.; Wu, P. Preparation, characterization and lubrication performances of graphene oxide-TiO_2 nanofluid in rolling strips. *Carbon* **2018**, *140*, 338–351. [CrossRef]
14. Srivyas, P.D.; Charoo, M.S. Nano lubrication behaviour of Graphite, h-BN and Graphene Nano Platelets for reducing friction and wear. *Mater. Today Proc.* **2021**, *44*, 7–11. [CrossRef]

15. Nassef MG, A.; Nassef, B.G.; Hassan, H.S.; Nassef, G.A.; Elkady, M.; Pape, F. Tribological and Chemical–Physical Behavior of a Novel Palm Grease Blended with Zinc Oxide and Reduced Graphene Oxide Nano-Additives. *Lubricants* **2024**, *12*, 191. [CrossRef]

16. Singh, A.; Chauhan, P.; Mamatha, T.G. A review on tribological performance of lubricants with nanoparticles additives. *Mater. Today Proc.* **2020**, *25*, 586–591. [CrossRef]

17. Jason YJ, J.; How, H.G.; Teoh, Y.H.; Chuah, H.G. A Study on the Tribological Performance of Nanolubricants. *Processes* **2020**, *8*, 1372. [CrossRef]

18. del Río, J.M.L.; Mariño, F.; López, E.R.; Gonçalves, D.E.P.; Seabra, J.H.O.; Fernández, J. Tribological enhancement of potential electric vehicle lubricants using coated TiO_2 nanoparticles as additives. *J. Mol. Liq.* **2023**, *371*, 121097. [CrossRef]

19. Beel, A.; Gottschalk, M.; Huetten, A.; Toensing, K.; Anselmetti, D. Tribological Performance of TiO_2-Nanostructured Particles as Oil-Lubricant Additives for Different Iron-Carbon Alloys. *Mater. Today Proc.* **2017**, *4*, S75–S80. [CrossRef]

20. Wu, H.; Zhao, J.; Cheng, X.; Xia, W.; He, A.; Yun, J.; Huang, S.; Wang, L.; Huang, H.; Jiao, S.; et al. Friction and wear characteristics of TiO2 nano-additive water-based lubricant on ferritic stainless steel. *Tribol. Int.* **2018**, *117*, 24–38. [CrossRef]

21. Sharma, V.; Timmons, R.B.; Erdemir, A.; Aswath, P.B. Interaction of plasma functionalized TiO_2 nanoparticles and ZDDP on friction and wear under boundary lubrication. *Appl. Surf. Sci.* **2019**, *489*, 372–383. [CrossRef]

22. Wang, L.; Tieu, A.K.; Zhu, H.; Deng, G.; Cui, S.; Zhu, Q. A study of water-based lubricant with a mixture of polyphosphate and nano-TiO2 as additives for hot rolling process. *Wear* **2021**, *477*, 203895. [CrossRef]

23. Xia, W.; Zhao, J.; Wu, H.; Zhao, X.; Zhang, X.; Xu, J.; Hee, A.C.; Jiang, Z. Effects of Nano-TiO2 Additive in Oil-in-Water Lubricant on Contact Angle and Antiscratch Behavior. *Tribol. Trans.* **2017**, *60*, 362–372. [CrossRef]

24. Wang, X.; Zhu, D.; Yang, S. Investigation of pH and SDBS on enhancement of thermal conductivity in nanofluids. *Chem. Phys. Lett.* **2009**, *470*, 107–111. [CrossRef]

25. Wang, B.X.; Zhao, Y.; Zhao, X.P. The wettability, size effect and electrorheological activity of modified titanium oxide nanoparticles. *Colloids Surf. A Physicochem. Eng. Asp.* **2007**, *295*, 27–33. [CrossRef]

26. Wang, C.; Yang, Y. Tunable volume memory poly(acrylic acid sodium) hydrogel by metal ionsd. *Funct. Mater. Lett.* **2022**, *15*, 2250010. [CrossRef]

27. Wu, Y.Y.; Tsui, W.C.; Liu, T.C. Experimental analysis of tribological properties of lubricating oils with nanoparticle additives. *Lubr. Wear* **2007**, *262*, 819–825. [CrossRef]

28. Rapoport, L.; Leshchinsky, V.; Lvovsky, M.; Lapsker, I.; Volovik, Y.; Feldman, Y.; Popovitz-Biro, R.; Tenne, R. Superior tribological properties of powder materials with solid lubricant nanoparticles. *Wear* **2003**, *255*, 794–800. [CrossRef]

29. Su, Y.; Gong, L.; Chen, D. An investigation on tribological properties and lubrication mechanism of graphite nanoparticles as vegetable based oil additive. *J. Nanomater.* **2015**, *2015*, 276753. [CrossRef]

30. Jiao, D.; Zheng, S.; Wang, Y.; Guan, R.; Cao, B. The tribology properties of alumina/silica composite nanoparticles as lubricant additives. *Appl. Surf. Sci.* **2011**, *257*, 5720–5725. [CrossRef]

31. Li, X.; Cao, Z.; Zhang, Z.; Dang, H. Surface-modification in situ of nano-SiO2 and its structure and tribological properties. *Appl. Surf. Sci.* **2005**, *252*, 7856–7861. [CrossRef]

32. Zhu, Z.; Sun, J.; Niu, T.; Liu, N. Experimental research on tribological performance of water-based rolling liquid containing nano-TiO2. *Proc. Inst. Mech. Eng. Part N J. Nanoeng. Nanosyst.* **2015**, *229*, 104–109. [CrossRef]

33. Kim, H.J.; Shin, D.G.; Kim, D.E. Frictional behavior between silicon and steel coated with graphene oxide in dry sliding and water lubrication conditions. *Int. J. Precis. Eng. Manuf.-Green Technol.* **2016**, *3*, 91–97. [CrossRef]

MDPI AG
Grosspeteranlage 5
4052 Basel
Switzerland
Tel.: +41 61 683 77 34

Lubricants Editorial Office
E-mail: lubricants@mdpi.com
www.mdpi.com/journal/lubricants

www.ingramcontent.com/pod-product-compliance
Lightning Source LLC
LaVergne TN
LVHW072337090526
838202LV00019B/2436